Methods in Molecular Biology

Series Editor
John M. Walker
School of Life and Medical Sciences
University of Hertfordshire
Hatfield, Hertfordshire, AL10 9AB, UK

For further volumes:
http://www.springer.com/series/7651

Phenotypic Screening

Methods and Protocols

Edited by

Bridget Wagner

Chemical Biology and Therapeutics Science Program, The Broad Institute of MIT and Harvard, Cambridge, MA, USA

Humana Press

Editor
Bridget Wagner
Chemical Biology and Therapeutics Science Program
The Broad Institute of MIT and Harvard
Cambridge, MA, USA

ISSN 1064-3745 ISSN 1940-6029 (electronic)
Methods in Molecular Biology
ISBN 978-1-4939-7846-5 ISBN 978-1-4939-7847-2 (eBook)
https://doi.org/10.1007/978-1-4939-7847-2

Library of Congress Control Number: 2018940444

Printed on acid-free paper

This Humana Press imprint is published by the registered company Springer Science+Business Media, LLC part of Springer Nature.
The registered company address is: 233 Spring Street, New York, NY 10013, U.S.A.

Preface

There has been a resurgence over the last several years in the enthusiasm for phenotypic screening as an approach for small-molecule discovery. By focusing on a cellular or organismic outcome, rather than targeting a specific protein or other cellular component, a phenotypic approach to high-throughput screening can uncover novel biology and reveal new validated targets for drug discovery. This volume will cover the methods used in phenotypic screening and is aimed at researchers working in chemical biology, cell and molecular biology, biochemistry, and computational biology.

The goals of this volume are twofold. First, we aim to provide the reader a better understanding of the advancements in phenotypic readouts and improved disease models that generate novel biological insights and recapitulate clinically relevant biology. Second, we hope to democratize the phenotypic screening process, by making these approaches more accessible to a wide array of researchers throughout the academic and biotech communities.

We start by learning about phenotypic approaches to modulating infectious diseases, including targeting nutrient stress in *E. coli*, targeting bacterial riboswitches as a novel approach to antibiotic development, and identifying multistage antimalarial compounds. Downstream of phenotypic screening, the relative ease of generating small-molecule-resistant clones in microbes has enabled the rapid identification of novel cellular targets for these models, targets that were not likely to have been chosen in a biochemical approach.

The volume then dives into mammalian cell-based assays, with chapters focused on phenotypic screening targeting mutant thrombopoietin receptors, analysis of the mitotic spindle orientation, proliferation of muscle satellite cells or pancreatic beta cells, enhancement of neurogenesis, and modulating adipocyte physiology. Important aspects of these chapters include a focus on the use of human cell models and the increasing importance of microscopy in phenotypic screening. Indeed, over half of the chapters presented in this volume rely heavily on microscopic approaches, which provide an increasingly sophisticated view of cellular phenotypes over standard endpoint readouts. Consistent with the use of microscopy in cell models, the use of whole organisms in phenotypic screening has become a more commonly observed approach. We then present chapters on assays to measure fat accumulation in *C. elegans*, the threat response in zebrafish, and protein–protein interactions in plant growth and development.

Of course, phenotypic approaches raise the important issues of how to analyze the data, prioritize compounds, and identify similarities in small-molecule activities. We found it important to include chapters on these computational and inferential approaches and present high-dimensional profiling techniques such as cell painting, the use of the Open PHACTS discovery platform, development of a tool score to quantitatively prioritize compounds, and a quantitative systems pharmacology approach to Huntington's disease. Together, these chapters highlight the increasing use of sophisticated computational approaches that parallel and complement the advancement in phenotypic models in phenotypic screening.

Finally, we focus on the next generation of tools and methods that will enable an expansion of the phenotypes feasible for screening. Chapters focused on novel fluorescent probes for cell-based phenotypic screening, a high-throughput patch clamp system for studying neuronal behavior, three-dimensional cancer models that more faithfully mimic human tumors, and techniques that improve the ability to screen in 1536-well format are aimed at augmenting the reader's understanding of what is possible in phenotypic screening, both now and in the future.

Cambridge, MA, USA *Bridget Wagner*

Contents

Preface .... *v*
Contributors .... *ix*

1 Nutrient Stress Small-Molecule Screening Platform for *Escherichia coli* 1
Sara S. El Zahed, Garima Kumar, Madeline Tong, and Eric D. Brown

2 Validation and Development of an *Escherichia coli* Riboflavin Pathway Phenotypic Screen Hit as a Small-Molecule Ligand of the Flavin Mononucleotide Riboswitch 19
Carl J. Balibar, Artjohn Villafania, Christopher M. Barbieri, Nick Murgolo, Terry Roemer, Hao Wang, and John A. Howe

3 Phenotypic Screening of Small Molecules with Antimalarial Activity for Three Different Parasitic Life Stages 41
Nobutaka Kato, Sandra March, Sangeeta N. Bhatia, and Matthias Marti

4 Phenotypic Screening for Inhibitors of a Mutant Thrombopoietin Receptor ... 53
Anna Ngo, Ann Koay, Christian Pecquet, Carmen C. Diaconu, David A. Jenkins, Andrew K. Shiau, Stefan N. Constantinescu, and Meng Ling Choong

5 A Cell-Based Assay for Mitotic Spindle Orientation 67
Elina Glaubke and Holger Bastians

6 Isolation of Skeletal Muscle Stem Cells for Phenotypic Screens for Modulators of Proliferation 77
Aaron C. Hinken and Andrew N. Billin

7 Cell-Based Methods to Identify Inducers of Human Pancreatic Beta-Cell Proliferation 87
Courtney A. Ackeifi, Ethan A. Swartz, and Peng Wang

8 High-Content Imaging Phenotypic Screen for Neurogenesis Using Primary Neural Progenitor Cells 101
Li Sharon Wu and Jingjun Li

9 Discovery of Modulators of Adipocyte Physiology Using Fully Functionalized Fragments 115
Andrea Galmozzi, Christopher G. Parker, Bernard P. Kok, Benjamin F. Cravatt, and Enrique Saez

10 The Ancient Genetic Networks of Obesity: Whole-Animal Automated Screening for Conserved Fat Regulators 129
Wenfan Ke, Anna Drangowska-Way, Daniel Katz, Karsten Siller, and Eyleen J. O'Rourke

11 Modulation of Threat Response in Larval Zebrafish 147
Andrew J. Rennekamp

12 A PLA-iRoCS Pipeline for the Localization of Protein–Protein Interactions In Situ 161
Taras Pasternak, William Teale, Thorsten Falk, Benedetto Ruperti, and Klaus Palme

13 High-Dimensional Profiling: The Theta Comparative Cell Scoring Method 171
Scott J. Warchal, John C. Dawson, and Neil O. Carragher

14 Accessing the Open PHACTS Discovery Platform with Workflow Tools 183
Daniela Digles, Andrei Caracoti, and Edgar Jacoby

15 Quantitative Prioritization of Tool Compounds for Phenotypic Screening 195
Yuan Wang and Jeremy L. Jenkins

16 A Quantitative Systems Pharmacology Approach to Infer Pathways Involved in Complex Disease Phenotypes 207
Mark E. Schurdak, Fen Pei, Timothy R. Lezon, Diane Carlisle, Robert Friedlander, D. Lansing Taylor, and Andrew M. Stern

17 Phenotype-Based High-Content Screening Using Fluorescent Chemical Bioprobes: Lipid Droplets and Glucose Uptake Quantification in Live Cells 223
Young-Hee Shin and Seung Bum Park

18 Developing High-Throughput Assays to Analyze and Screen Electrophysiological Phenotypes 235
Jen Q. Pan, David Baez-Nieto, Andrew Allen, Hao-Ran Wang, and Jeffrey R. Cottrell

19 Testing Susceptibility of Patient-Derived Organoid Cultures to Therapies: Pharmacotyping 253
Richard A. Burkhart, Lindsey A. Baker, and Hervé Tiriac

20 Techniques to Enable 1536-Well Phenotypic Screening 263
Sinéad Knight, Helen Plant, Lisa McWilliams, and Mark Wigglesworth

Index *279*

Contributors

COURTNEY A. ACKEIFI • *Diabetes, Obesity and Metabolism Institute, Icahn School of Medicine at Mount Sinai, New York, NY, USA*
ANDREW ALLEN • *Stanley Center for Psychiatric Research, Broad Institute of MIT and Harvard, Cambridge, MA, USA*
DAVID BAEZ-NIETO • *Stanley Center for Psychiatric Research, Broad Institute of MIT and Harvard, Cambridge, MA, USA*
LINDSEY A. BAKER • *Lustgarten Foundation Pancreatic Cancer Research Laboratory, Cold Spring Harbor Laboratory, Cold Spring Harbor, New York, NY, USA*
CARL J. BALIBAR • *Merck Research Laboratories, Kenilworth, NJ, USA*
CHRISTOPHER M. BARBIERI • *Merck Research Laboratories, Kenilworth, NJ, USA*
HOLGER BASTIANS • *Goettingen Center for Molecular Biosciences and University Medical Center, Goettingen, Germany*
SANGEETA N. BHATIA • *Broad Institute of MIT and Harvard, Cambridge, MA, USA; Institute for Medical Engineering and Science, Massachusetts Institute of Technology, Cambridge, MA, USA; Howard Hughes Medical Institute, Cambridge, MA, USA; Koch Institute for Integrative Cancer Research, Cambridge, MA, USA*
ANDREW N. BILLIN • *Gilead Sciences, Inc., Foster City, PA, USA*
ERIC D. BROWN • *Department of Biochemistry and Biomedical Sciences, McMaster University, Hamilton, ON, Canada; Michael G. DeGroote Institute for Infectious Disease Research, McMaster University, Hamilton, ON, Canada*
RICHARD A. BURKHART • *Division of Hepatobiliary and Pancreatic Surgery, Department of Surgery, Johns Hopkins Hospital, Baltimore, MD, USA; Lustgarten Foundation Pancreatic Cancer Research Laboratory, Cold Spring Harbor Laboratory, Cold Spring Harbor, New York, NY, USA*
ANDREI CARACOTI • *BIOVIA, A Dassault Systèmes Brand, Cambridge, UK*
DIANE CARLISLE • *Department of Neurological Surgery, University of Pittsburgh, Pittsburgh, PA, USA*
NEIL O. CARRAGHER • *Cancer Research UK Edinburgh Centre, Institute of Genetics and Molecular Medicine, University of Edinburgh, Edinburgh, UK*
MENG LING CHOONG • *Experimental Therapeutics Centre, Agency for Science Technology and Research (A*STAR), Singapore, Singapore*
STEFAN N. CONSTANTINESCU • *Ludwig Institute for Cancer Research, Université catholique de Louvain and de Duve Institute, Brussels, Belgium*
JEFFREY R. COTTRELL • *Stanley Center for Psychiatric Research, Broad Institute of MIT and Harvard, Cambridge, MA, USA*
BENJAMIN F. CRAVATT • *Department of Molecular Medicine, The Skaggs Institute for Chemical Biology, The Scripps Research Institute, La Jolla, CA, USA*
JOHN C. DAWSON • *Cancer Research UK Edinburgh Centre, Institute of Genetics and Molecular Medicine, University of Edinburgh, Edinburgh, UK*
CARMEN C. DIACONU • *Stefan S. Nicolau Institute of Virology, Romanian Academy, Bucharest, Romania*
DANIELA DIGLES • *Department of Pharmaceutical Chemistry, University of Vienna, Vienna, Austria*

ANNA DRANGOWSKA-WAY • *Department of Biology, University of Virginia, Charlottesville, VA, USA*
SARA S. EL ZAHED • *Department of Biochemistry and Biomedical Sciences, McMaster University, Hamilton, ON, Canada; Michael G. DeGroote Institute for Infectious Disease Research, McMaster University, Hamilton, ON, Canada*
THORSTEN FALK • *BIOSS Centre for Biological Signaling Studies, University of Freiburg, Freiburg, Germany; Technical Faculty, Department of Computer Science, University of Freiburg, Freiburg, Germany*
ROBERT FRIEDLANDER • *Department of Neurological Surgery, University of Pittsburgh, Pittsburgh, PA, USA*
ANDREA GALMOZZI • *Department of Molecular Medicine, The Skaggs Institute for Chemical Biology, The Scripps Research Institute, La Jolla, CA, USA*
ELINA GLAUBKE • *Goettingen Center for Molecular Biosciences and University Medical Center, Goettingen, Germany*
AARON C. HINKEN • *GlaxoSmithKline, Muscle Metabolism Discovery Performance Unit, King of Prussia, PA, USA*
JOHN A. HOWE • *Merck Research Laboratories, Kenilworth, NJ, USA*
EDGAR JACOBY • *Janssen Research and Development, Beerse, Belgium*
DAVID A. JENKINS • *Small Molecule Discovery Program, Ludwig Institute for Cancer Research, La Jolla, CA, USA*
JEREMY L. JENKINS • *Novartis Institutes for BioMedical Research Inc., Cambridge, MA, USA*
NOBUTAKA KATO • *Broad Institute of MIT and Harvard, Cambridge, MA, USA*
DANIEL KATZ • *Department of Biomedical Engineering, University of Virginia, Charlottesville, VA, USA*
WENFAN KE • *Department of Biology, University of Virginia, Charlottesville, VA, USA*
SINÉAD KNIGHT • *Discovery Sciences, IMED Biotech Unit, AstraZeneca, Cambridge, UK*
ANN KOAY • *Experimental Therapeutics Centre, Agency for Science Technology and Research (A*STAR), Singapore, Singapore*
BERNARD P. KOK • *Department of Molecular Medicine, The Skaggs Institute for Chemical Biology, The Scripps Research Institute, La Jolla, CA, USA*
GARIMA KUMAR • *Department of Biochemistry and Biomedical Sciences, McMaster University, Hamilton, ON, Canada; Michael G. DeGroote Institute for Infectious Disease Research, McMaster University, Hamilton, ON, Canada*
D. LANSING TAYLOR • *Department of Computational and Systems Biology, University of Pittsburgh Drug Discovery Institute, University of Pittsburgh, Pittsburgh, PA, USA*
TIMOTHY R. LEZON • *Department of Computational and Systems Biology, University of Pittsburgh, Pittsburgh, PA, USA; Drug Discovery Institute, University of Pittsburgh Pittsburgh, PA, USA*
JINGJUN LI • *Lilly China Research and Development Center, Eli Lilly and Company, Shanghai, China*
SANDRA MARCH • *Broad Institute of MIT and Harvard, Cambridge, MA, USA; Institute for Medical Engineering and Science, Massachusetts Institute of Technology, Cambridge, MA, USA; Howard Hughes Medical Institute, Cambridge, MA, USA; Koch Institute for Integrative Cancer Research, Cambridge, MA, USA*
MATTHIAS MARTI • *Wellcome Centre for Molecular Parasitology, University of Glasgow, Glasgow, UK; Department of Immunology and Infectious Diseases, Harvard TH Chan School of Public Health, Boston, MA, USA*
LISA MCWILLIAMS • *Discovery Sciences, IMED Biotech Unit, AstraZeneca, Cambridge, UK*

NICK MURGOLO • *Merck Research Laboratories, Kenilworth, NJ, USA*
ANNA NGO • *Experimental Therapeutics Centre, Agency for Science Technology and Research (A*STAR), Singapore, Singapore*
EYLEEN J. O'ROURKE • *Department of Biology, University of Virginia, Charlottesville, VA, USA*
KLAUS PALME • *Faculty of Biology, Institute of Biology II/Molecular Plant Physiology, University of Freiburg, Freiburg, Germany; BIOSS Centre for Biological Signaling Studies, University of Freiburg, Freiburg, Germany; Center for Biological Systems Analysis, University of Freiburg, Freiburg, Germany*
JEN Q. PAN • *Stanley Center for Psychiatric Research, Broad Institute of MIT and Harvard, Cambridge, MA, USA*
SEUNG BUM PARK • *Department of Chemistry, Seoul National University, Seoul, South Korea; Department of Biophysics and Chemical Biology, Seoul National University, Seoul, South Korea*
CHRISTOPHER G. PARKER • *Department of Molecular Medicine, The Skaggs Institute for Chemical Biology, The Scripps Research Institute, La Jolla, CA, USA*
TARAS PASTERNAK • *Faculty of Biology, Institute of Biology II/Molecular Plant Physiology, University of Freiburg, Freiburg, Germany*
CHRISTIAN PECQUET • *Ludwig Institute for Cancer Research, Université catholique de Louvain and de Duve Institute, Brussels, Belgium*
FEN PEI • *Department of Computational and Systems Biology, University of Pittsburgh, Pittsburgh, PA, USA; Drug Discovery Institute, University of Pittsburgh, Pittsburgh, PA, USA*
HELEN PLANT • *Discovery Sciences, IMED Biotech Unit, AstraZeneca, Macclesfield, Cheshire, UK*
ANDREW J. RENNEKAMP • *Department of Medicine, Massachusetts General Hospital, Boston, MA, USA; Department of Medicine, Harvard Medical School, Boston, MA, USA*
TERRY ROEMER • *Merck Research Laboratories, Kenilworth, NJ, USA*
BENEDETTO RUPERTI • *Department of Agronomy, Food, Natural Resources, Animals and Environment, University of Padova, Agripolis, Legnaro (Padova), Italy*
ENRIQUE SAEZ • *Department of Molecular Medicine, The Skaggs Institute for Chemical Biology, The Scripps Research Institute, La Jolla, CA, USA*
MARK E. SCHURDAK • *Department of Computational and Systems Biology, University of Pittsburgh, Pittsburgh, PA, USA; Drug Discovery Institute, University of Pittsburgh, Pittsburgh, PA, USA*
ANDREW K. SHIAU • *Small Molecule Discovery Program, Ludwig Institute for Cancer Research, La Jolla, CA, USA*
YOUNG-HEE SHIN • *Department of Chemistry, Seoul National University, Seoul, South Korea*
KARSTEN SILLER • *Advanced Computing Services, University of Virginia, Charlottesville, VA, USA*
ANDREW M. STERN • *Department of Computational and Systems Biology, University of Pittsburgh, Pittsburgh, PA, USA; Drug Discovery Institute, University of Pittsburgh, Pittsburgh, PA, USA*
ETHAN A. SWARTZ • *Diabetes, Obesity and Metabolism Institute, Icahn School of Medicine at Mount Sinai, New York, NY, USA*
WILLIAM TEALE • *Faculty of Biology, Institute of Biology II/Molecular Plant Physiology, University of Freiburg, Freiburg, Germany*

HERVÉ TIRIAC • *Lustgarten Foundation Pancreatic Cancer Research Laboratory, Cold Spring Harbor Laboratory, Cold Spring Harbor, New York, NY, USA*
MADELINE TONG • *Department of Biochemistry and Biomedical Sciences, McMaster University, Hamilton, ON, Canada; Michael G. DeGroote Institute for Infectious Disease Research, McMaster University, Hamilton, ON, Canada*
ARTJOHN VILLAFANIA • *Merck Research Laboratories, Kenilworth, NJ, USA*
HAO WANG • *Merck Research Laboratories, Kenilworth, NJ, USA*
HAO-RAN WANG • *Stanley Center for Psychiatric Research, Broad Institute of MIT and Harvard, Cambridge, MA, USA*
PENG WANG • *Diabetes, Obesity and Metabolism Institute, Icahn School of Medicine at Mount Sinai, New York, NY, USA*
YUAN WANG • *Novartis Institutes for BioMedical Research Inc., Cambridge, MA, USA*
SCOTT J. WARCHAL • *Cancer Research UK Edinburgh Centre, Institute of Genetics and Molecular Medicine, University of Edinburgh, Edinburgh, UK*
MARK WIGGLESWORTH • *Discovery Sciences, IMED Biotech Unit, AstraZeneca, Macclesfield, Cheshire, UK*
LI SHARON WU • *School of Life Sciences, Nanchang University, Nanchang, China*

Chapter 1

Nutrient Stress Small-Molecule Screening Platform for *Escherichia coli*

Sara S. El Zahed, Garima Kumar, Madeline Tong, and Eric D. Brown

Abstract

Clinically approved antibiotics target a narrow spectrum of cellular processes, namely cell wall synthesis, DNA replication, and protein synthesis. Numerous screens have been designed to identify inhibitors that target one of these cellular processes. Indeed, this narrow range of drug mechanisms and a reliance on chemical classes discovered many decades ago are thought to be principally responsible for the current crisis of antibiotic drug resistance. Seeking to expand the target base of antibacterial drug discovery, we developed a nutrient stress screening platform that identifies inhibitors of the growth of in *Escherichia coli* under nutrient limitation. Under nutrient stress, bacteria require an expanded biosynthetic capacity that includes the synthesis of amino acids, vitamins, and nucleobases. Growing evidence suggests that these processes may be indispensable to certain pathogens and at particular sites of infection. Indeed, more than 100 biosynthetic enzymes become indispensable to *E. coli* grown under nutrient stress in vitro. The screening platform described here puts a focus on these novel targets for new antibiotics and prioritizes growth inhibitory compounds that can be suppressed by individual nutrients and pools thereof.

Key words Antibacterials, Nutrient stress, Growth inhibition

1 Introduction

Since the 1960s, antibacterial research has mostly focused on incremental structural modifications to existing antibiotics and resistance to the modified drugs has never been far behind. The last 20 years or so has seen a renaissance in antibacterial drug discovery using modern approaches. Bacterial genomics heralded a gene-to-drug approach where new targets would lead to new compounds that kill bacteria in ways that are unsusceptible to existing resistance mechanisms.

Surprisingly, there have been no novel antibacterial medicines discovered in the last ~30 years. Accounts of failure are anecdotal but have been documented by industry pundits [1, 2]. High-throughput

Sara S. El Zahed and Garima Kumar are co-first authors.

Bridget Wagner (ed.), *Phenotypic Screening: Methods and Protocols*, Methods in Molecular Biology, vol. 1787, https://doi.org/10.1007/978-1-4939-7847-2_1,

biochemical screens of large synthetic chemical collections using a variety of protein targets, many incompletely characterized, failed to generate promising leads for new drugs. So too did parallel screens of whole cells for growth inhibition. The latter produced large numbers of actives but was hampered by a lack of tools for prioritization and follow-up. In the approach described herein, we outline the details of a screening platform for the facile prioritization and characterization of compounds that lead to growth inhibition under conditions of nutrient stress [3].

When bacteria are grown in a medium containing only carbon, nitrogen, and essential salts, they shift their metabolic activities to include the synthesis of essential amino acids, vitamins, and other cofactors. Only 303 genes are essential for growth of *E. coli* on rich medium and 119 additional genes are required for growth under nutrient limitation [4, 5]. Thus compounds that target bacteria under nutrient-limited conditions could serve as leads for novel antibacterials. In fact, we believe that nutrient-limited medium provides a better proxy for the host environment. There have been many reports of impaired growth and attenuated virulence in pathogens due to mutations in vitamin, nucleobase, and amino acid biosynthetic genes [6–13]. The approach herein aims to target these functions by enriching for small molecules that inhibit growth only when bacteria are under nutrient stress.

The methodology reported here describes details for high-throughput chemical screening in nutrient-limited microbiological medium (glucose and ammonium chloride as carbon and nitrogen sources, respectively) followed by selecting only those compounds that are suppressed with the addition of a nutrient pool of amino acids, vitamins, and nucleobases (Fig. 1). These priority actives are subject to further analysis by nutrient supplementation profiling where individual nutrients and pools thereof are systematically tested for suppression of growth inhibition. Mechanistic hypotheses generated by this platform enable the selection of hit compounds for follow-up using more focused efforts, typically involving more conventional biochemical, physiological, and genetic tools.

2 Materials

2.1 Preparation of Bacterial Culture

1. Appropriate frozen stock of bacterial strains.
2. Sterile wooden sticks.
3. Sterile culture test tubes.
4. LB agar plates—Dissolve agar in LB medium (1.5% w/v) and sterilize by autoclaving. Cool the medium to ~50 °C before supplementing with any required antibiotics. Pour ~20 mL of

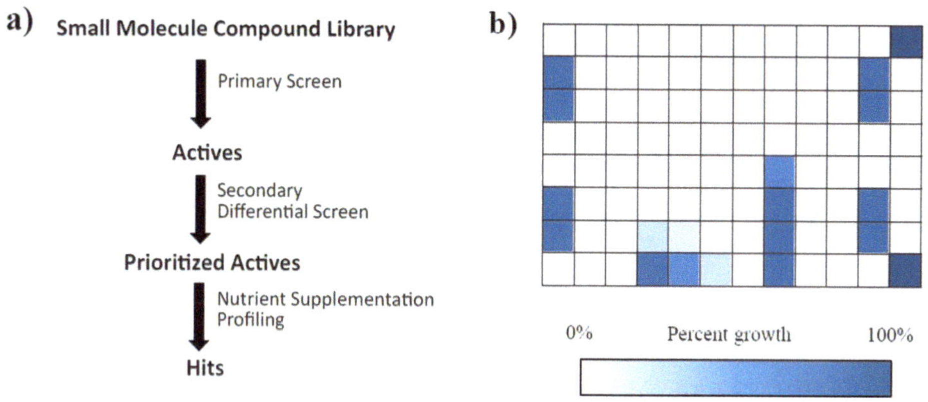

Fig. 1 Screening compound libraries for nutrient biosynthesis inhibitors. (**a**) Schematic of the screening process. (**b**) A nutrient supplementation profile of a prioritized active. Each square represents *Escherichia coli* grown in M9 minimal medium in the presence of a growth inhibitory concentration of the prioritized active and an individual nutrient, or their pools. Blue represents growth; hence the activity of the prioritized active was suppressed in the presence of the respective nutrient(s). On the other hand, white represents no growth; hence the presence of the respective nutrient(s) did not have an effect on the activity of the prioritized active

LB medium per Petri dish and cool it till the agar solidifies. Store the plates at 4 °C.

5. Temperature-controlled incubator.
6. 70% Ethanol.
7. 0.22 μm Filter units.
8. 30 mL Syringes.
9. Sterile screw-top tubes and flasks.
10. 5 × M9 salts stock solution: 64 g $Na_2HPO_4 \cdot 7H_2O$, 15 g KH_2PO_4, 2.5 g NaCl, and 5 g NH_4Cl, to a total of 1 L deionized water (dH_2O). Sterilize by autoclaving.
11. 1 M $MgSO_4$: 24.65 g $MgSO_4 \cdot 7H_2O$ in a total of 100 mL dH_2O, filter sterilize.
12. 1 M $CaCl_2$: 14.7 g $CaCl_2 \cdot 2H_2O$ in a total of 100 mL dH_2O, filter sterilize.
13. 20% Glucose (w/v) in dH_2O, filter sterilize.
14. M9 minimal medium: 200 mL of 5 × M9 salts solution, 2 mL of 1 M $MgSO_4$, 0.1 mL of 1 M $CaCl_2$, and 20 mL of 20% glucose, to 780 mL of sterile dH_2O. This would be the working M9 nutrient-limited (minimal) medium.
15. Temperature-controlled shaking incubator.

2.2 Primary Screen in Nutrient-Limited Medium

1. Working bacterial culture as described in Subheading 3.1, **step** 7.
2. Compound library in 96-well polypropylene plates stored at −20 °C.

3. For *x* compound library plates, set up 2× 96-well sterile lidded clear flat-bottom assay plates (*see* **Note 1**).
4. Reservoirs to hold bacterial cultures.
5. 20 μL and 200 μL 96-racked sterile tips for liquid handler or multichannel pipettors.
6. Plate reader with appropriate detector, measuring absorbance at 600 nm in this case.
7. Temperature-controlled stationary incubator.

2.3 Secondary Screen in Nutrient-Limited and Rich Media

1. 96-Deep well polypropylene plates.
2. 96-Well clear flat-bottom assay plates.
3. M9 minimal medium: Prepare as described in Subheading 2.1, **steps 10–14**.
4. 50× Amino acids stock solution: Weigh out all the amino acids listed in Table 1 and add dH_2O to 80 mL. Add base dropwise until all the compounds go into solution. Add dH_2O to bring up the total volume to 100 mL. Filter sterilize and store at −20 °C in the dark.
5. 100× Nucleobases stock solution: Weigh out all the nucleobases listed in Table 2 and add to 35 mL dH_2O. Add base dropwise until all the compounds go into solution. Add dH_2O to bring up the total volume to 50 mL. Filter sterilize and store at −20 °C in the dark.
6. 1000× Vitamins stock solution: Weigh out all the vitamins listed in Table 3 and add to 8 mL dH_2O. Mix well and add dH_2O to bring up the total volume to 10 mL. Filter sterilize and store at −20 °C in the dark.
7. M9 supplemented medium: 200 mL of 5× M9 salts solution, 2 mL of 1 M $MgSO_4$, 0.1 mL of 1 M $CaCl_2$, 20 mL of 20% glucose, 20 mL of 50× amino acids stock, 10 mL of 100× nucleobases stock, and 1 mL of 1000× vitamins stock, to 750 mL of sterile dH_2O. This would be the working M9 nutrient-rich (supplemented) medium.

2.4 Nutrient Supplementation Profiling of Prioritized Actives

1. Amino acids, metals, nucleobases, nutrient intermediates, and vitamins: Refer to Table 4 for the complete list. Store the solutions at −20 °C.
2. 4 N Hydrogen chloride solutions: Identify the concentration of hydrogen chloride (HCl) on the bottle of commercially available concentrated HCl. Apply $C_1 V_1 = C_2 V_2$, where C_1 is 4 N, V_1 is the 1 L, C_2 is the concentrated HCl concentration, and V_2 is the volume to be aliquoted from the bottle of concentrated HCl. In a 2 L glass beaker, add dH_2O followed by the aliquot of concentrated HCl. Stir the solution, and bring up the volume to 1 L.

Table 1
50× Amino acids stock solution

Amino acids	Working concentration (mg/mL)	50× Stock concentration (mg/mL)	Compound in 100 mL of 50× stock solution (mg)
DL-Alanine	0.1	5	500
L-Arginine	0.022	1.1	110
L-Asparagine	0.1	5	500
L-Aspartic acid	0.1	5	500
L-Cysteine HCl	0.01	0.5	50
Glycine	0.1	5	500
L-Glutamic acid	0.1	5	500
L-Glutamine	0.1	5	500
L-Histidine HCl	0.022	1.1	110
L-Isoleucine	0.02	1	100
L-Leucine	0.02	1	100
L-Lysine HCl	0.088	4.4	440
DL-Methionine	0.02	1	100
L-Phenylalanine	0.02	1	100
L-Proline	0.03	1.5	150
DL-Serine	0.1	5	500
DL-Threonine	0.08	4	400
L-Tryptophan	0.02	1	100
L-Tyrosine	0.02	1	100
L-Valine	0.04	2	200

Table 2
100× Nucleobases stock solution

Nucleobases	Working concentration (mg/mL)	100× Stock concentration (mg/mL)	Compound in 50 mL of 100× stock solution (mg)
Adenine	0.04	4	200
Thymine	0.005	0.5	25
Uracil	0.04	4	200
Guanine	0.04	4	200

Table 3
1000× Vitamins stock solution

Vitamins	Working concentration (mg/mL)	1000× Stock concentration (mg/mL)	Compound in 10 mL of 1000× stock solution (mg)
Biotin	0.0005	0.5	5
Niacin	0.001	1	10
Ca pantothenate	0.001	1	10
Pyridoxine	0.001	1	10
Thiamine	0.001	1	10
Riboflavin	0.2	200	2000
PABA	0.002	2	20
B12	0.01	10	100

3. 5 M Sodium hydroxide solution: Measure 800 mL of dH_2O to a 2 L beaker. Weigh 20 g of sodium hydroxide (NaOH) and transfer to the 2 L beaker to dissolve. Allow the solution to cool down, and then transfer the solution to a graduated cylinder. Bring the volume up to 1 L.
4. Sterile dH_2O.
5. 0.22 μm Filter units.
6. 30 mL Syringes.
7. 96-Deep well polypropylene plates.
8. 96-Well assay plates.

3 Methods

3.1 Preparation of Bacterial Culture

1. Under sterile conditions, streak out the frozen bacterial stock on a LB agar plate with appropriate antibiotics using sterile wood applicators (*see* **Note 2**).
2. Incubate the plate at 37 °C in a stationary incubator for 16–18 h and store it at 4 °C for further use (*see* **Note 3**). This plate can be used for 2 weeks to pick colonies.
3. Using a sterile wood applicator, inoculate 5 mL of M9 minimal medium in a sterile test tube with a single colony from the LB agar plate.
4. Incubate the 5 mL M9 minimal medium culture overnight (16–18 h) at 37 °C in a shaking incubator at 250 rpm.

Table 4
Amino acid, nucleobase, nutrient intermediate, and vitamin solutions

Compound	Stock concentration (mg/mL)	Array concentration (mg/mL)	Working concentration (μg/mL)	Notes
Amino acids				
DL-Alanine	50	2	100	
L-Arginine HCl	11	0.44	22	
L-Asparagine	25	2	100	Dissolve with HCl
L-Aspartic acid	8	2	100	Dissolve in 0.2 M NaOH
L-Cysteine HCl	11	0.2	10	Dissolve with NaOH
Glycine	50	2	100	
L-Glutamic acid	8	2	100	Dissolve with NaOH
L-Glutamine	20	2	100	
L-Histidine HCl	11	0.44	22	
L-Isoleucine	10	0.4	20	
L-Leucine	10	0.4	20	Dissolve with HCl
L-Lysine HCl	44	1.8	88	
DL-Methionine	10	0.4	20	
L-Phenylalanine	10	0.4	20	
L-Proline	15	0.6	30	
DL-Serine	50	2	100	
DL-Threonine	40	1.6	80	
L-Tryptophan	10	0.4	20	Dissolve in 0.5 M HCl
L-Tyrosine	2	0.4	20	Dissolve in 1 M HCl
L-Valine	20	0.8	40	
Nucleobases				
Adenine (A)	6	0.8	40	Dissolve in 0.3 M HCl
Thymine (T)	2.5	0.1	5	
Uracil (U)	6	0.8	40	Dissolve in 0.2 M HCl
Guanine (G)	10	0.8	40	Dissolve in 0.16 M HCl
Nutrient intermediates				
2,6-Diaminopimelate (DAP)	50	2	100	Dissolve with NaOH
L-Homoserine	10	0.4	20	
5-Aminolevulinic acid (5-ALA)	12.5	0.5	25	
3-Dehydroquinate (DHQ)	50	2	100	
Shikimate (SHIK)	50	2	100	

(continued)

Table 4
(continued)

Compound	Stock concentration (mg/mL)	Array concentration (mg/mL)	Working concentration (μg/mL)	Notes
4-Hydroxybenzoate (4-HBA)	4	0.3	15	
2,3-Dihydroxybenzoate (2,3-DHBA)	1	0.04	2	
Citrulline	7.5	0.3	15	
Ornithine	7.5	0.3	15	
Putrescine	10	0.4	20	
Vitamins				
Biotin (Vit. B7)	1	0.01	0.5	Dissolve with NaOH
Niacin (Vit. B3)	10	0.02	1	
Calcium pantothenate (Vit. B5)	50	0.02	1	
Pyridoxine HCl (Vit. B6)	10	0.02	1	
Thiamine HCl (Vit. B1)	10	0.02	1	
Riboflavin (Vit. B2)	10	4	200	Dissolve in 0.1 N NaOH
p-Aminobenzoic acid (PABA)	1	0.04	2	
Cobalamin (Vit. B12)	10	0.2	10	

5. Dilute the overnight liquid culture 1:50 in fresh M9 minimal medium in a sterile culture tube to make a *subculture*. A subculture is generally 5 mL of M9 minimal medium with 100 μL of the overnight culture.
6. Grow the subculture at 37 °C in a shaking incubator at 250 rpm until it reaches mid-log growth phase ($A_{600} = 0.35–0.5$). Typically, this step takes 3–4 h.
7. For the primary screen, dilute the mid-log subculture 1:1,000 in fresh sterile M9 minimal medium to prepare the *working bacterial culture*.
8. For the secondary screen, two different working bacterial cultures are required: Dilute the mid-log subculture 1:1,000 in fresh sterile M9 minimal medium, and dilute the mid-log subculture 1:10,000 in fresh sterile M9 supplemented medium.

3.2 Primary Screen for Small Molecules Targeting Metabolism Under Nutrient Limitation

3.2.1 Labware Preparation

1. Prepare a working bacterial culture as described in Subheading 3.1, **step 7**.
2. Thaw small-molecule compound libraries at room temperature while still sealed in storage plates. Unseal the plates right before use and reseal immediately after use to prevent any uptake of atmospheric water (*see* **Note 4**).
3. Label all the assay plates.
4. Sterilize the reservoir under UV light exposure for 10 min.
5. Transfer the required volume of working bacterial culture to the reservoir.

3.2.2 Liquid Handling

An automated liquid handler or a multichannel pipettor can be used to set up the assay plates in the following manner.

1. Using the 96-racked 20 μL sterile tips, transfer the appropriate volume of compounds (2 μL in this case) (*see* **Note 5**) from each small-molecule storage compound plate to two 96-well assay plates, replicate 1 (R1), and R2.
2. Repeat **step 1** to prepare assay plates for all the compound plates. Change tips after each compound plate.
3. Aliquot the required volume of microbial culture (98 μL per well for our study purposes) from the reservoir into all the wells of the 96-well assay plates. Wash/change tips between handling different assay plates.
4. Reseal all the compound plates and store them at −20 °C.
5. After the completion of liquid transfer for all the assay plates, measure the absorbance of the assay plates at 600 nm for $t = 0$ measurement.
6. Incubate the assay plates at 37 °C for 16 h, and measure the absorbance at 600 nm at the end of incubation for $t = 16$ measurement (*see* **Note 6**).
7. Calculate the growth of bacteria (G_i) in each well as:

$$G_i = A_{600\ (t=16)} - A_{600\ (t=0)}$$

where i represents a well in the assay plate.

8. Normalize the data to identify actives, as described in Chand et al. [14] (*see* **Note 7**).
9. Compounds that result in growth inhibition and are three standard deviations below the mean are identified as active compounds (*see* **Note 8**).

3.3 Secondary Screen to Prioritize the Actives Targeting One of the 119 Minimal Essential Genes

1. Prepare new compound plates for the actives from Subheading 3.2.2, **step 9**, in the following manner: in a 96-well polypropylene plate, add 10 μL of DMSO in all the wells except the first column, A1 through H1. Aliquot 20 μL of 8 compounds into wells A1 through H1. Using a multichannel pipettor, transfer a half volume (10 μL) of compounds from the first column to the second column for all the compounds. Pipet multiple times to mix well. Continue the twofold serial dilution across the plate to column 11 (*see* **Note 9**).
2. Prepare two sets of working bacterial cultures as described in Subheading 3.1, **step 8**.
3. Transfer 2 μL of compound from each plate to two 96-well assay plates, R1 and R2.
4. Add 98 μL of M9 minimal or nutrient-limited working bacterial culture to the entire plate.
5. Repeat **steps 3–4** for the M9 supplemented medium working culture as prepared in **step 2** (*see* **Note 10**).
6. After preparing all the assay plates for all the actives, follow Subheading 3.2.2, **steps 5–7**.
7. Normalize the growth in each well as follows:

$$\text{Normalized growth } (G_{ni}) = \frac{G_i}{G_{\text{DMSO}}}$$

where G_i is the background-corrected bacterial growth of well i, and G_{DMSO} is the background-corrected bacterial growth in the absence of any compound (wells A12 to H12 in this case).

8. For M9 minimal and M9 supplemented media conditions, plot G_{ni} vs. compound concentration. Lowest compound concentration at which no visible growth of bacteria was observed is defined as the minimum inhibitory concentration (MIC). Compounds with at least a fourfold shift in their MIC values in nutrient-rich compared to nutrient-limited conditions are categorized as *priority actives* and followed up for identification of their mechanism of action (*see* **Note 11**).

3.4 Target Determination: Nutrient Supplementation Profiling

Note that the final assay concentrations refer to the working concentrations, or the final concentrations in the assay plate. The array is a 96-deep well polypropylene plate composed of 42 wells of single nutrients (amino acids, metals, nucleobases, nutrient intermediates, and vitamins) and 46 wells of pools thereof. Of the 46 pools, 8 are prepared in duplicates. Unless otherwise stated in Table 4, all solutions are dissolved in sterile dH_2O.

3.4.1 Nutrient Stock Solution Preparation

1. Prepare the following solutions (listed as stock concentration in Tables 4 and 5): 25 mL stock solution of adenine, 10 mL stock solutions of the amino acids, metals, nutrient intermediates,

Table 5
Iron and trace metal pool solutions

Compound	Stock concentration (mg/mL)	In 10 mL M9 minimal medium (μL)
$FeSO_4 \cdot 7H_2O$ (iron)	8	1
Trace metals (pool)		
$FeSO_4 \cdot 7H_2O$	8	1
$AlCl_3 \cdot 6H_2O$	1	1
$ZnSO_4 \cdot 7H_2O$	0.2	1
$CuCl_2 \cdot 2H_2O$	0.1	1
$NaMoO_4 \cdot 2H_2O$	0.2	1
$MnSO_4 \cdot H_2O$	1	1
$CoCl_2$	0.4	1
H_3BO_4	0.05	1

vitamins, thymine, uracil, and guanine, and 50 mL of 20% Casamino acid solution (w/v).

2. Ensure that all solutions are completely dissolved and clear (*see* **Note 12**).
3. Use 0.22 μm filter units and 30 mL syringes to filter sterilize the prepared solutions.

3.4.2 Nutrient Supplementation Array Preparation

1. Using the stock solutions prepared in Subheading 3.4.1, **steps 1–3**, prepare 5 mL of their respective array concentrations. These concentrations are listed as array concentration in Table 4 (*see* **Note 13**).
2. Similarly, prepare 5 mL solutions of the nutrient pools by combining the individual components, as listed in Table 6 (*see* **Note 14**).
3. Ensure that all solutions do not produce any precipitate (*see* **Notes 12 and 15**). Then, filter sterilize the solutions prepared using 0.22 μm filter units and 30 mL syringes.
4. To a 96-deep well polypropylene plate, aliquot 500 μL of the filter-sterilized array solutions into each solutions' respective well [3] (*see* **Note 16**).

3.4.3 Nutrient Suppression Profiling

1. Prepare a bacterial culture as described in Subheading 3.1, **steps 1–6**.
2. Ensure that the array plate is completely thawed at room temperature.
3. The suppression profile for a compound, *n*, is done in duplicates. Accordingly, prepare the following: 20 mg/mL stock solution (in DMSO) of compound *n* (*see* **Note 17**), two assay

Table 6
Nutrient pool solutions: all amino acids are referred to by their three-letter code

Pool	Components at their array concentration					
M9	As described in Subheading 2.1, **steps 10–14**					
M9 all	As described in Subheading 2.3, **steps 4–7**					
Amino acids (AA)	8% Casamino acids	0.4 mg/mL Trp				
Vitamins (VIT)	20 μg/mL Vit. B1, 3, 5, 6	4 mg/mL Vit. B2	10 μg/mL Vit. B7	0.2 mg/mL Vit. B12	40 μg/mL PABA	
Nucleobases (NUC)	0.8 mg/mL Adenine	0.1 mg/mL Thymine	0.8 mg/mL Uracil	0.8 mg/mL Guanine		
AA + VIT	All amino acids and vitamins as listed for AA and VIT, respectively					
AA + NUC	All amino acids and nucleobases as listed for AA and NUC, respectively					
VIT + NUC	All vitamins and nucleobases as listed for VIT and NUC, respectively					
Purines (PUR)	0.8 mg/mL Adenine	0.8 mg/mL Guanine				
Pyrimidines (PYR)	0.1 mg/mL Thymine	0.8 mg/mL Uracil				
PUR HIS	0.8 mg/mL Adenine	0.8 mg/mL Guanine	0.44 mg/mL His			
NUC HIS	0.8 mg/mL Adenine	0.1 mg/mL Thymine	0.8 mg/mL Uracil	0.8 mg/mL Guanine	0.44 mg/mL His	
Tyr-Phe	0.4 mg/mL Tyr	0.4 mg/mL Phe				
Aro AA	0.4 mg/mL Tyr	0.4 mg/mL Phe	0.4 mg/mL Trp			
Aro AA PABA	0.4 mg/mL Tyr	0.4 mg/mL Phe	0.4 mg/mL Trp	40 μg/mL PABA		
Aro	0.4 mg/mL Tyr	0.4 mg/mL Phe	0.4 mg/mL Trp	40 μg/mL PABA	40 μg/mL 2,3-DHBA	0.3 mg/mL 4-HBA
Thr-Met-Lys	1.6 mg/mL Thr	0.4 mg/mL Met	1.8 mg/mL Lys			
Thr-Met-Ile	1.6 mg/mL Thr	0.4 mg/mL Met	0.4 mg/mL Ile			
Thr-Met-DAP-Lys	1.6 mg/mL Thr	0.4 mg/mL Met	2 mg/mL DAP	1.8 mg/mL Lys		
Lys-Met	1.8 mg/mL Lys	0.4 mg/mL Met				

(continued)

Table 6
(continued)

Pool	Components at their array concentration					
Met-Thr	0.4 mg/mL Met	1.6 mg/mL Thr				
Met-Ile	0.4 mg/mL met	0.4 mg/mL Ile				
Ile-Val	0.4 mg/mL Ile	0.8 mg/mL Val				
Thr-Met-Lys-Leu-Ile-Val	1.6 mg/mL Thr	0.4 mg/mL Met	1.8 mg/mL Lys	0.4 mg/mL Leu	0.4 mg/mL Ile	0.8 mg/mL Val
Gly-Ile	2 mg/mL Gly	0.4 mg/mL Ile				
Leu-Val-Ala-Vit.B5-Ile	0.4 mg/mL Leu	0.8 mg/mL Val	2 mg/mL Ala	20 μg/mL Vit. B5	0.4 mg/mL Ile	
Gly-Met-Vit. B5-T-A	2 mg/mL Gly	0.4 mg/mL Met	20 μg/mL Vit. B5	0.1 mg/mL Thymine	0.8 mg/mL Adenine	
Pool 1	0.8 mg/mL Adenine	0.1 mg/mL Thymine	0.44 mg/mL His	0.4 mg/mL Phe	2 mg/mL Gln	
Pool 2	0.8 mg/mL Guanine	0.4 mg/mL Leu	0.4 mg/mL Tyr	2 mg/mL Asn	2 mg/mL Ser	
Pool 3	0.2 mg/mL Cys	0.4 mg/mL Ile	0.4 mg/mL Trp	0.8 mg/mL Uracil	2 mg/mL Glu	
Pool 4	0.4 mg/mL Met	1.8 mg/mL Lys	1.6 mg/mL Thr	2 mg/mL Asp	2 mg/mL DAP	
Pool 5	20 μg/mL Vit. B1	0.8 mg/mL Val	0.6 mg/mL Pro	0.44 mg/mL Arg	2 mg/mL Gly	0.4 mg/mL Ile
Pool 6	0.8 mg/mL Adenine	0.8 mg/mL Guanine	0.2 mg/mL Cys	0.4 mg/mL Met	20 μg/mL Vit. B1	
Pool 7	0.44 mg/mL His	0.4 mg/mL Leu	0.4 mg/mL Ile	1.8 mg/mL Lys	0.8 mg/mL Val	
Pool 8	0.4 mg/mL Phe	0.4 mg/mL Tyr	0.4 mg/mL Trp	1.6 mg/mL Thr	0.6 mg/mL Pro	
Pool 9	2 mg/mL Gln	2 mg/mL Asn	0.8 mg/mL Uracil	2 mg/mL Asp	0.44 mg/mL Arg	
Pool 10	0.1 mg/mL Thymine	2 mg/mL Ser	2 mg/mL Glu	2 mg/mL DAP	2 mg/mL Gly	

plates, labeled *n* R1 and *n* R2 (*see* **Note 18**), and 40 mL M9 minimal medium.

4. In 40 mL M9 minimal medium, prepare a final working concentration of 4× the MIC of compound *n* (*see* **Note 19**). Label this tube M9 with compound *n*.
5. Using a 12-head multichannel pipettor, mix the contents of row A in the 96-deep well array by pipetting up and down.
6. Aliquot 10 μL from row A of the 96-deep well array to row A of assay plate *n* R1.
7. Using the same set of tips, aliquot 10 μL from row A of the 96-deep well array to row A of assay plate *n* R2.
8. Dispose of the tips, and repeat **steps 5–7** for rows B to H of the 96-deep well array to the respective rows B to H in the assay plates. Make sure to change tips when changing rows in the array plate to avoid nutrient cross contamination.
9. Once done, seal the array plate, and store it at −20 °C.
10. Inoculate 40 mL M9 minimal medium, labeled M9 with compound *n*, with bacteria as described in Subheading 3.1, **step 7**. Mix well and add the working bacterial culture, containing compound *n*, to a reservoir.
11. Using a 12-head multichannel pipettor, aliquot 190 μL of the working bacterial culture from **step 10** to assay plates *n* R1 and *n* R2.
12. Measure the absorbance at 600 nm to account for any colored compounds. This measurement corresponds to $A_{600(t=0)}$, the assay plates' absorbance prior to incubation.
13. Incubate the assay plates in a stationary incubator at 37 °C for 16 h, and consequently measure the absorbance at 600 nm to determine growth. This measurement corresponds to $A_{600(t=16)}$, the assay plates' absorbance after incubation.
14. Calculate the background corrected bacterial growth of each well, as described in Subheading 3.2.2, **step 7**.
15. Calculate the percent residual growth for each well, *i*, ($\%G_i$) as follows:

$$\%G_i = \left(\frac{G_i - G_{M9}}{G_{M9\ all} - G_{M9}}\right) \times 100$$

where G_i is the background-corrected bacterial growth of well *i*, G_{M9} is the background-corrected bacterial growth in M9 minimal medium, and $G_{M9\ all}$ is the background-corrected bacterial growth in M9 supplemented minimal medium.

16. Percent residual growth values between 80 and 100% indicate that the corresponding supplemented nutrient suppressed the activity of compound *n*. Therefore, the biosynthetic pathway of the respective nutrient is the target for compound *n*.

4 Notes

1. Clear flat-bottom 96-well plates are used to quantify growth by measuring absorbance at 600 nm. When measuring fluorescence or luminescence, black or white flat-bottom plates are required.
2. Spray working counters with a generous amount of 70% ethanol to ensure sterile conditions. Working in a flow hood or close to a Bunsen burner also helps in preventing contamination.
3. The optimum growth temperature for *E. coli* is 37 °C. Depending on the bacterial strain being investigated, the incubation temperature and time may need to be optimized.
4. Compound libraries are generally dissolved in DMSO. As DMSO is highly hygroscopic in nature, minimize exposure to air to avoid any atmospheric moisture.
5. DMSO can affect the growth of bacteria. Therefore, care should be taken to add as low amount of DMSO as possible. For *E. coli* MG1655, DMSO concentrations of 0.2–5% (v/v) were seen to have negligible effects on bacterial growth. So, addition of 2 μL of compounds dissolved in DMSO to 98 μL of bacterial culture corresponds to 2% DMSO (v/v).
6. To observe the maximal effect of actives in phenotypic screens, the growth of the bacterial strain should be measured at late log or early stationary phase. *E. coli* strain MG1655 requires to grow at 37 °C for 16 h to reach the early stationary growth phase. A growth curve of the strain under investigation would be helpful in determining the optimal incubation time.
7. Analysis of the data can be done in multiple ways. In our case, we normalize the growth of the bacteria to find the mean growth through each assay plate. Alternatively, positive (known antimicrobials) and negative (neat DMSO) controls can be used to prioritize actives.
8. The hit rate should be ~2% of the total number of compounds tested. The stringency of the standard deviation may be adjusted in the case of too many/less hits.
9. These reformatted compound plates would have 11 compound concentrations at twofold intervals. Since MIC values vary for different compounds and bacterial strains, the compound

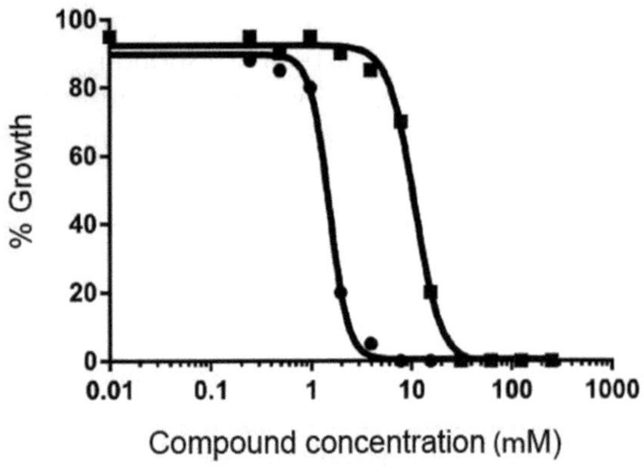

Fig. 2 Representative dose-response curve for a typical nutrient biosynthesis inhibitor. This effect of nutrient inhibitor is observed by fourfold shift in MIC under nutrient-limited (filled circles) vs. nutrient-rich conditions (filled squares)

concentrations in the first wells of these plates need to be fivefold greater in concentration than that of the primary screen concentrations.

10. This results in identical concentrations of prioritized compounds in bacterial cultures grown in M9 minimal and M9 supplemented media. Therefore, we could successfully observe whether nutrient supplementation suppressed the activity of the compound.
11. If the MIC of a compound increases by at least fourfold in nutrient-rich conditions compared to nutrient-limited conditions, this suggests that the compound targets a nutrient biosynthetic pathway which is conditionally essential under nutrient-limited conditions. However, an increase in MIC value that is less than fourfold would mean that the compound is targeting something other than nutrient synthesis (Fig. 2).
12. Many solutions will not have the nutrients completely dissolved after stirring. To dissolve these nutrients, sonicate the solutions (up to 30 min or longer if necessary).
13. The array plate is prepared at 20× the final assay concentrations. This is to facilitate the supplementation of each individual nutrient by dispensing an equal volume of 10 μL from the array plate into the assay plate. For example, aliquot 200 μL of the 50 mg/mL DL-alanine stock solution to 4.8 mL water to obtain an array concentration of 2 mg/mL DL-alanine. Aliquot 500 μL of the 2 mg/mL DL-alanine into the respective well in the array plate [3]. Then, aliquot 10 μL from the array plate to a final volume of 200 μL in the assay plate to obtain an L-alanine final working concentration of 100 μg/mL.

14. The nutrient pools consist of combined nutrients prepared to have the same final nutrient concentration as their individually prepared array concentrations. As an example of how to prepare one of the nutrient pools: The ARO AA pool (Table 6) is composed of 0.4 mg/mL L-phenylalanine, 0.4 mg/mL L-tryptophan, and 0.4 mg/mL L-tyrosine. To prepare the ARO AA pool: add 200 μL L-phenylalanine, 200 μL L-tryptophan, and 1 mL L-tyrosine from the L-phenylalanine, L-tryptophan, and L-tyrosine stock solutions to 3.6 mL water. Mix the solution to ensure homogeneity.
15. In case precipitate forms, add acid (4 N HCl) or base (5 M NaOH) dropwise using a Pasteur pipet, while intermittently mixing, until the precipitate is dissolved.
16. Each 5 mL array concentration solution can be used to prepare ten nutrient supplementation array plates.
17. The compound stock solution can be prepared at higher or lower concentrations as well.
18. *n* R1 represents the compound name and replicate 1, respectively. Similarly, *n* R2 represents the compound name and replicate 2, respectively.
19. An example of how to prepare 4× MIC of a compound *n*: Let compound *n*'s MIC be 2 μg/mL. Keeping in mind that the stock solution was prepared at 20 mg/mL, and applying $C_1V_1 = C_2V_2$, aliquot 16 μL from the 20 mg/mL stock to 40 mL M9 minimal. Mix the solution well.

Acknowledgments

This research was supported by an operating grant from the Canadian Institutes for Health Research (FDN-143215), a Leaders Opportunity Fund grant from the Canada Foundation for Innovation, and a Tier I Canada Research Chair award to E.D.B.

References

1. Payne DJ, Gwynn MN, Holmes DJ, Pompliano DL (2007) Drugs for bad bugs: confronting the challenges of antibacterial discovery. Nat Rev Drug Discov 6:29–40
2. Tommasi R, Brown DG, Walkup GK, Manchester JI, Miller AA (2015) ESKAPEing the labyrinth of antibacterial discovery. Nat Rev Drug Discov 14:529–542
3. Zlitni S, Ferruccio LF, Brown ED (2013) Metabolic suppression identifies new antibacterial inhibitors under nutrient limitation. Nat Chem Biol 9:796–804
4. Baba T, Ara T, Hasegawa M, Takai Y, Okumura Y, Baba M et al (2006) Construction of *Escherichia coli* K-12 in-frame, single-gene knockout mutants: the Keio collection. Mol Syst Biol 2:0008
5. Joyce AR, Reed JL, White A, Edwards R, Osterman A, Baba T et al (2006) Experimental and computational assessment of conditionally essential genes in *Escherichia coli*. J Bacteriol 188:8259–8271
6. Hoiseth SK, Stocker BAD (1981) Aromatic-dependent *Salmonella typhimurium* are

non-virulent and effective as live vaccines. Nature 291:238–239

7. Edwards MF, Stocker BA (1988) Construction of Δ*aroA his* Δ*pur* strains of *Salmonella typhi*. J Bacteriol 170:3991–3995
8. Chamberlain LM, Strugnell R, Dougan G, Hormaeche CE, de Hormaeche RD (1993) *Neisseria gonorrhoeae* strain MS11 harbouring a mutation in gene *aroA* is attenuated and immunogenic. Microb Pathog 15:51–63
9. Bange FC, Brown AM, Jacobs WR (1996) Leucine auxotrophy restricts growth of *Mycobacterium bovis* BCG in macrophages. Infect Immun 64:1794–1799
10. Mei J-M, Nourbakhsh F, Ford CW, Holden DW (1997) Identification of *Staphylococcus aureus* virulence genes in a murine model of bacteraemia using signature-tagged mutagenesis. Mol Microbiol 26:399–407
11. Cersini A, Salvia AM, Bernardini ML (1998) Intracellular multiplication and virulence of *Shigella flexneri* auxotrophic mutants. Infect Immun 66:549–557
12. Polissi A, Pontiggia A, Feger G, Altieri M, Mottl H, Ferrari L et al (1998) Large-scale identification of virulence genes from *Streptococcus pneumoniae*. Infect Immun 66:5620–5629
13. Cuccui J, Easton A, Chu KK, Bancroft GJ, Oyston PCF, Titball RW et al (2007) Development of signature-tagged mutagenesis in *Burkholderia pseudomallei* to identify genes important in survival and pathogenesis. Infect Immun 75:1186–1195
14. Mangat CS, Bharat A, Gehrke SS, Brown ED (2014) Rank ordering plate data facilitates data visualization and normalization in high-throughput screening. J Biomol Screen 19:1314–1320

Chapter 2

Validation and Development of an *Escherichia coli* Riboflavin Pathway Phenotypic Screen Hit as a Small-Molecule Ligand of the Flavin Mononucleotide Riboswitch

Carl J. Balibar, Artjohn Villafania, Christopher M. Barbieri, Nick Murgolo, Terry Roemer, Hao Wang, and John A. Howe

Abstract

A riboflavin biosynthesis pathway-specific phenotypic screen using a library of compounds, all with unspecified antibiotic activity, identified one small molecule later named ribocil, for which intrinsic antibacterial activity against *Escherichia coli* was completely suppressed by addition of exogenous riboflavin to the bacterial growth medium. The ability of riboflavin to suppress the activity of ribocil, and further demonstration that ribocil inhibited riboflavin synthesis ($IC_{50} = 0.3\ \mu M$), supported that a component of the riboflavin synthesis pathway was the molecular target. Remarkably, resistance mutation selection and whole-genome sequencing showed that the target of ribocil was not an enzyme in the riboflavin biosynthesis pathway, but instead the flavin mononucleotide riboswitch, a noncoding structural RNA element in the *ribB* gene that encodes a key riboflavin synthesis enzyme. Although ribocil is structurally distinct from the natural riboswitch regulatory ligand flavin mononucleotide, ribocil binding to the riboswitch results in efficient repression of *ribB* expression and inhibition of riboflavin biosynthesis and bacterial growth. A cell-based riboswitch regulated gene reporter assay as well as an in vitro riboswitch RNA aptamer-binding assay, both of which are described in detail here along with the riboflavin pathway-specific screen, were developed to further validate the mechanism of action of ribocil and to facilitate the discovery of more potent analogues.

Key words Riboflavin, Riboswitch, Antibacterials, Ribocil, RNA aptamer binding, Riboswitch reporter assay

1 Introduction

Pathway-specific phenotypic screens enable simultaneous and unbiased query of all components of essential metabolic biosynthetic pathways for cognate inhibitors. Importantly, phenotypic screens could lead to identification of classes of antibacterials with novel targets that could be used to treat multidrug-resistant bacterial strains. The riboflavin (vitamin B2) biosynthesis pathway represents one essential metabolic pathway that can be investigated for the

Bridget Wagner (ed.), *Phenotypic Screening: Methods and Protocols*, Methods in Molecular Biology, vol. 1787, https://doi.org/10.1007/978-1-4939-7847-2_2,

discovery of novel antibiotics [1]. Riboflavin is the immediate precursor of flavin mononucleotide (FMN) and flavin adenine dinucleotide (FAD), essential cofactors of the ubiquitous flavoenzymes, which play fundamental roles in intermediary metabolism [2]. Riboflavin is synthesized by most pathogenic bacteria de novo, but not by humans who must obtain this essential vitamin from dietary sources. The synthesis of riboflavin in *Escherichia coli*, from guanosine triphosphate and ribulose 5-phosphate, is carried out in a synthetic pathway by five enzymes encoded by the *rib* genes *ribA*, *ribB*, *ribDG*, *ribE*, and *ribH* [3]. FMN and FAD are synthesized sequentially from riboflavin by the bifunctional *ribFC* enzyme that encodes both flavokinase and FAD synthetase activities.

In bacteria, riboflavin levels are regulated by FMN riboswitches that control expression of genes responsible for riboflavin biosynthesis and transport [4–7]. The FMN riboswitch is a well characterized representative of a class of regulatory elements present in mRNAs which bind metabolites, including vitamins, amino acids, and purines, to regulate gene expression in the corresponding metabolic pathway [8, 9]. Riboswitches are composed of two contiguous RNA regions: an RNA aptamer that binds the cognate pathway metabolite and an expression platform that undergoes a conformational change in response to aptamer ligand engagement that results in altered expression of a single gene or operon [10]. In *E. coli*, a Gram-negative bacterium that relies solely on riboflavin synthesis and lacks specific riboflavin transporters, FMN binding to the aptamer of the FMN riboswitch results in suppression of *ribB* expression and inhibition of riboflavin synthesis. The FMN riboswitch therefore provides an elegant negative-feedback circuit to modulate riboflavin levels through control of riboflavin biosynthesis. The FMN riboswitch and the five Rib enzymes mentioned above represent potential targets for inhibition of riboflavin synthesis, all of which are screened simultaneously in a riboflavin pathway-specific screen.

The riboflavin pathway-specific phenotypic screen was carried out using a collection of 57,000 small molecules with antibacterial activity by testing for compounds whose ability to inhibit *E. coli* growth was specifically suppressed by riboflavin supplementation [1]. Among the small molecules tested, only one compound, ribocil, was fully inhibited in the presence of exogenously added riboflavin and demonstrated a dose-dependent reduction in riboflavin levels in *E. coli* ($IC_{50} = 0.3\ \mu M$). To further elucidate the mechanism of action of ribocil the direct target was determined through isolation of ribocil-resistant (ribocil^{R}) colonies and whole-genome sequencing. Remarkably, among all 19 independently isolated ribocil^{R} colonies, mutations were not identified in any of the *rib* genes encoding enzymes responsible for RF biosynthesis, but instead all mutations were found within the FMN riboswitch. Therefore, this genetic data suggested that ribocil targets the FMN riboswitch exclusively to inhibit RF synthesis as an unnatural mimic of the natural FMN ligand of the riboswitch. The discovery of ribocil as a unique regulator of an RNA structural element demonstrates the power of phenotypic screening, coupled with whole-genome sequencing, to reveal new targets and expand our chemical repertoire of antibacterial compounds.

To further validate the FMN riboswitch as the target of ribocil and to facilitate the discovery of more potent ribocil-like molecules, a high-throughput FMN riboswitch-regulated GFP plasmid reporter assay was developed. Finally, detailed characterization of the interaction of ribocil with the FMN riboswitch was carried out utilizing an in vitro biochemical assay to directly measure binding of ribocil to the RNA aptamer through competition with FMN, the natural ligand of the riboswitch. Protocols for the primary riboflavin pathway-specific screen, for the riboswitch gene reporter assays, and RNA aptamer-binding assays, are described in detail in this chapter.

2 Materials

2.1 Primary Phenotypic Screen

1. *E. coli* strain MB5746 [1] overnight culture grown in cation-adjusted Mueller–Hinton broth (CAMHB).
2. CAMHB supplemented with 20 μM riboflavin (RF).
3. 8-Point/twofold titrated screen compounds in DMSO with starting concentration of 10 mM.
4. 1536-well micro-plate (black, clear bottom, low base).
5. Plate reader capable of detecting OD_{600}, such as Safire2.
6. Software for data analysis, such as TIBCO Spotfire.

2.2 Secondary Confirmation

1. *E. coli* strain MB5746 overnight culture grown in CAMHB broth.
2. CAMHB agar incubated in a water bath of 46 °C.
3. 1 mM RF stock solution in 50% ethanol.
4. Screening compounds, 6-point/twofold titrated in DMSO, with starting concentration of 2 mM.
5. 9 × 12 cm Rectangular plate.

2.3 Resistant Mutant Isolation

1. *E. coli* strain MB5746 overnight culture grown in CAMHB.
2. CAMHB soft agar (1.0%) incubated in a water bath of 46 °C.
3. Hit compound(s) in DMSO (5 mg/mL), 24-well plate, petri dish (100 × 15 mm).

2.4 Genomic DNA Preparation and Sequencing Analysis

1. Illumina HiSeq 2500 sequencer for analysis of MB5746 genomic DNA.
2. Velvet algorithm package to assemble fragments [11].

Table 1
PCR primers for constructing pCDF-EcRFN-GFP

Name	Sequence	Template
P1	CTCAAATGCCTGAGGTTTCAGcaggacttgcgtttggacgtc	RFN
P2	GAAAAGTTCTTCTCCTTTACTCATggtaaaaaaacctcactaaaattatg	RFN
P3	GACGTCCAAACGCAAGTCCTGctgaaacctcaggcatttgag	pCDF-1b
P4	CATGGATGAGCTCTACAAATAAgcgcaacgcaattaatgtaag	pCDF-1b
P5	CATAATTTTAGTGAGGTTTTTTTACCatgagtaaaggagaagaacttttc	GFPuv
P6	CTTACATTAATTGCGTTGCGCttatttgtagagctcatccatg	GFPuv

2.5 Plasmid-Based Reporter Assay for High-Throughput Screening of Ribocil Analogues

2.5.1 PCR Reagents

1. Phusion® High-Fidelity Polymerase (2000 U/mL).
2. DMSO.
3. 5× Phusion® GC Buffer.
4. dNTPs (10 mM).
5. Agarose.
6. TAE buffer.
7. QiaQuick® Gel extraction Kit.
8. Ethidium bromide (10 mg/mL).
9. pCDF-1b plasmid (500 ng/μL).
10. pGFPuv plasmid (500 ng/μL).
11. Overnight culture of *E. coli* grown in Mueller–Hinton II broth (CAMHB).
12. PCR primers (Table 1; *see* **Note 1**).

2.5.2 Cloning Reagents

1. DNA loading dye: 50:40:10 Glycerol:TAE buffer:10% Orange G in water (*see* **Note 2**).
2. 5× In-Fusion® HD enzyme premix.
3. One Shot® TOP10 chemically competent *E. coli.*
4. SOC broth.
5. LB broth.
6. LB agar.
7. Spectinomycin.
8. QIAprep Spin Miniprep Kit.

2.5.3 Transformation Reagents

1. Ribocil® *E. coli* strain MB5746-5.
2. ZymoBroth™.
3. MicroPulser™ apparatus.
4. Gene Pulser®/MicroPulser™ electroporation cuvettes.

5. 10% Glycerol.
6. SOB medium.
7. LB agar.
8. Spectinomycin.

2.5.4 Reporter Assay Reagents

1. Mueller–Hinton II broth (CAMHB).
2. Spectinomycin.
3. Ribocil.
4. DMSO.
5. 96-Well deep-well polypropylene plate.
6. 96-Well black polystyrene round-bottom fluorescence plate.
7. 96-Well clear polystyrene round-bottom plate.
8. SpectraMax M4 plate reader.

2.6 Riboswitch RNA Aptamer-Binding Assay

2.6.1 Aptamer DNA Template Synthesis Reagents

1. Phusion® High-Fidelity Polymerase (2000 U/mL) kit.
2. 5× Phusion® GC Buffer.
3. dNTPs (10 mM).
4. MicroAmp® Optical 96-well reaction plates.
5. DMSO.
6. pCDF-EcRFN-GFP plasmid (10 ng/μL).
7. 3 M Sodium acetate (pH 5.2).
8. Isopropanol.
9. Phenol–chloroform.
10. Chloroform.
11. 70% Ethanol.
12. 1.0% Agarose TBE gel.
13. DNA template and primers (Table 2; *see* **Notes 3** and **4**).

Table 2
DNA template and primers for riboswitch RNA aptamer-binding assay

Name	Sequence (*see* Note 3)	Template
RFN aptamer	gcttattctcagggcggggcgaaattccccaccggcggtaa atcaactcagttgaaagcccgcgagcgctttgggtgcga actcaaaggacagcagatccggtgtaattccggggccga cggttagagtccggatgggagagagtaacg	pCDF-EcRFN-GFP
Forward (5′) primer (*see* **Note 4**)	TAATACGACTCACTATAGGgcttattctcagggcg	pCDF-EcRFN-GFP
Reverse (3′) primer	cgttactctctcccatccg	pCDF-EcRFN-GFP

2.6.2 RNA Aptamer Synthesis

1. RNase-free water.
2. RiboMAX® large-scale RNA production T7 system including 5× translation buffer, rATP (100 mM), rCTP (100 mM), rGTP (100 mM), rUTP (100 mM).
3. T7 polymerase mix (RNA Polymerase, Recombinant RNasin® Ribonuclease Inhibitor, and Recombinant Inorganic Pyrophosphatase).
4. RQ1 RNase-free DNase.
5. Citrate-saturated phenol.
6. Chloroform–isoamyl alcohol.
7. GE NAP-5 columns.
8. 3 M Sodium acetate (pH 5.2).
9. Isopropanol.
10. Ethidium bromide (10 mg/mL).
11. Agarose.
12. TAE buffer.

2.6.3 Aptamer Annealing Reagents

1. Aptamer reannealing buffer (5×): 100 mM Potassium phosphate, 320 mM KCl, and 0.5 mM EDTA at pH 7.4 (8 mL of 250 mM KH_2PO_4, 32 mL of 250 mM K_2HPO_4, 16 mL of 2 M KCl, 0.1 mL of 500 mM EDTA, and enough H_2O to make 100 mL).
2. Aptamer: 5 μL of 222 μM aptamer in H_2O.
3. Heat block: Fisherbrand heat block with thermometer.

2.6.4 FMN-Aptamer-Binding Determination Reagents

1. Binding assay buffer (5×): 250 mM Tris–HCl, 500 mM KCl, and 10 mM $MgCl_2$ assay buffer at pH 7.4 (25 mL of 1 M Tris–HCl pH 7.4, 25 mL of 2 M KCl, 1 mL of 1 M $MgCl_2$, and enough H_2O to make 100 mL).
2. FMN stock solution (1000×, 10 mM): Dissolve 10 mg of FMN in 2.2 mL of H_2O. Aliquot and freezer stock solution.
3. Dual monochromator-equipped fluorescence detection plate reader, such as the SpectraMax M5 (Molecular Devices).
4. Fluorescence detection assay plates: Black 384-well flat-bottom polystyrene plates.

2.6.5 FMN Competition Binding Determination Reagents

1. FMN solution (3×): 18 μL of 10 μM FMN stock solution with 200 μL of 5× binding assay buffer diluted in H_2O for 1000 μL total volume.
2. RNA aptamer solution (3×): 22.5 μL of 20 μM reannealed RNA aptamer solution with 200 μL of 5× binding assay buffer diluted in H_2O for 1000 mL total volume.
3. Compound stock solution (3×) 10 mM of putative FMN-binding site competitive compound in 100% DMSO.

3 Methods

3.1 Primary Screen for Compounds Suppressed by Riboflavin

1. Add 2.5 μL of CAMHB or CAMHB with RF to 1536-well plates separately (*see* **Note 5**).
2. Transfer 50 nL compounds and add 2.5 μL of 1:10,000 dilution of overnight bacterial broth.
3. Centrifuge plates for 30 s at 200 × *g*.
4. Incubate for 24 h at 37 °C, 90% humidity.
5. Read OD_{600} with a plate reader and analyze the data to identify hit compounds.

3.2 Secondary Confirmation of Hits

1. Dilute the overnight MB5756 culture 1000× in 30 mL aliquots of 46 °C CAMH with 1.2% agar for a final *E. coli* CFU = ~5 × 10^6.
2. Add 600 μL of the 1 mM RF stock into one aliquot of medium for a final concentration of 20 μM, and 300 μL of 50% ethanol into the other aliquot of medium.
3. Pour into 9 × 12 cm rectangular plates (*see* **Note 6**).
4. After the agar solidifies for about 30 min, spot 5 μL of each test compound onto the above plates, let the compounds dry (~30 min), and incubate for 16 h at 37 °C.
5. Take an image of each plate and measure the diameter of the inhibition zones. Compounds with diminished zones of inhibition on the plates containing RF are putative RF synthesis inhibitors (Fig. 1).

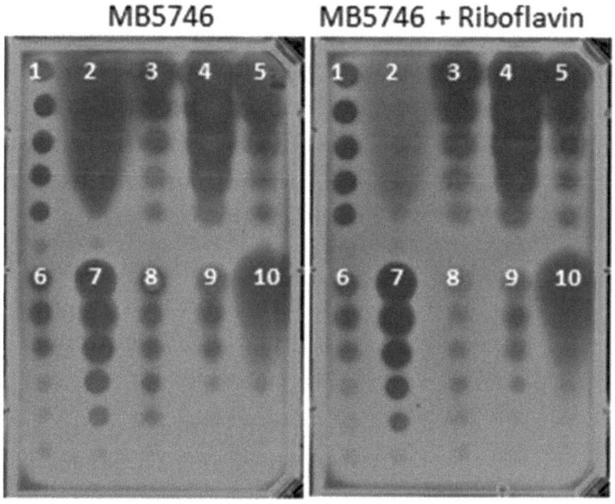

Fig. 1 Whole-cell agar-medium assay in the absence (left) and presence (right) of exogenous riboflavin. The zones of inhibition of compounds 2 and 8 were diminished by supplemental RF. Compound 2 is renamed ribocil

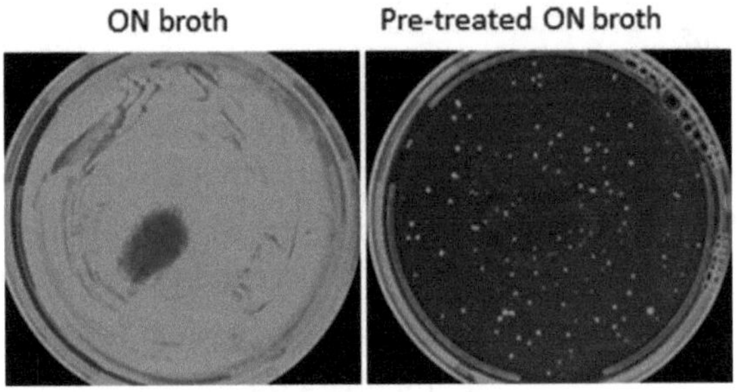

Fig. 2 Ribocil-resistant mutant isolation. Approximately 2.5 × 10^7 cells were spread on each agar plate embedded with 8× MIC of ribocil. Left, untreated overnight culture; right, overnight culture pretreated with ribocil

3.3 Resistant Mutant Isolation

1. Dilute the MB5746 overnight culture 100-fold with fresh CAMH broth and spot 10 μL on each well of a 24-well plate containing 1 mL CAMH soft agar embedded with a twofold titration series of hit compound (starting from 32 μg/mL, and final DMSO concentration 2%).
2. Incubate for 48 h at 37 °C, and record the agar minimum inhibitory concentration (MIC; *see* **Note 7**).
3. Dilute the overnight culture tenfold with fresh CAMH broth containing hit compound (4 × MIC) and incubate for another 6 h (*see* **Note 8**).
4. Spread 100 μL of the above broth (~2.5 × 10^8 cells/mL) onto plates containing 20 mL of CAMH agar embedded with two-fold escalating agar MIC levels of hit compound.
5. Incubate for 48 h at 37 °C.
6. Pick colonies and streak/purify them with CAMH agar plates containing 4× MIC of compound (Fig. 2).

3.4 Genomic DNA Preparation and Sequencing Analysis

1. Prepare genomic DNA from 2 mL of overnight MB5746 cultures of the wild-type and ribocil-resistant mutants using the DNeasy® Blood & Tissue Handbook (Qiagen) and following the pretreatment for Gram-negative bacteria and the DNA purification protocols.
2. Sequence resistant mutant isolates in multiplex format by attaching specific oligonucleotide barcodes to fragmented genomic DNA sequences using the Illumina NEXTflex-96 DNA barcodes kit.
3. Sequence samples using an Illumina HiSeq 2500 sequencer to a depth of at least 50-fold.
4. Separate sequence reads by each barcode tag and align to the wild-type genome assembly using the Burrows Wheeler aligner [12]; mutant alleles of high homogeneity can be called using GATK [13] or VAAL [14].

3.5 Construction of the pCDF-EcRFN-GFP Plasmid

1. Prepare general PCR stock solution for each reaction: 30 μL of water, 10 μL of 5× Phusion® GC Buffer, 1 μL of dNTPs, and 2 μL of DMSO. Keep on ice.
2. Depending on the product being amplified, add 3 μL of each primer and 0.5 μL of template to each PCR stock solution. For PCR of the RFN riboswitch, add 3 μL each of primer P1 and P2 and 0.5 μL of overnight *E. coli* culture.
3. For PCR of the vector backbone, add 3 μL each of primer P3 and P4 and 0.5 μL of pCDF-1b plasmid.
4. For PCR of GFPuv, add 3 μL each of primer P5 and P6 and 0.5 μL of pGFPuv.
5. Add 0.5 μL of Phusion® High-Fidelity Polymerase to each reaction, mix, spin down, and place in the thermal cycler.
6. General PCR cycling conditions are shown in Table 3.
7. For PCR of RFN riboswitch, annealing temperature is 51 °C and extension time is 1 min.
8. For PCR of pCDF-1b vector, annealing temperature is 54 °C and extension time is 3 min.
9. For PCR of GFP, annealing temperature is 51 °C and extension time is 1 min.
10. Add 6 μL of DNA loading buffer to each PCR reaction and load into a 1% agarose gel containing a 1/10,000 dilution of ethidium bromide stock (*see* **Note 9**).
11. Run at 105 V for 30 min.
12. Excise the PCR bands, and purify using the reagents and protocol from the QiaQuick® Gel extraction Kit (*see* **Note 10**).
13. Quantify the DNA concentration using a Nanodrop spectrophotometer (*see* **Note 11**).

Table 3
PCR conditions for construction of pCDF-EcRFN-GFP

Step	Number of cycles	Temperature	Duration
1	1	98 °C	2 min
2		98 °C Amplicon dependent 72 °C	30 s 15 s Amplicon dependent
3	1	72 °C	10 min
4	1	4 °C	∞

3.6 Cloning of the pCDF-EcRFN-GFP Plasmid

1. Mix RFN:GFP:pCDF-1b PCR products in a 2:2:1 mole ratio and add water to a total of 8 μL.
2. Add 2 μL of 5× In-Fusion® HD enzyme premix, incubate at 50 °C for 15 min, and then place on ice.
3. Add 5 μL of In-Fusion® HD Clonase reaction to 50 μL of One Shot® TOP10 chemically competent *E. coli* (*see* **Note 12**).
4. Heat shock at 42 °C for 45 s and immediately place on ice.
5. Add 250 μL of SOC medium, incubate shaking at 37 °C, and then plate 25 μL and 250 μL on LB agar containing 75 μg/mL spectinomycin.
6. Incubate plates at 37 °C overnight.
7. The next day, pick six colonies from the transformation, inoculate into LB medium containing 75 μg/mL spectinomycin, and grow shaking at 37 °C until culture reaches saturation (*see* **Note 13**).
8. Isolate plasmid from the culture using the QIAprep Spin Miniprep Kit (*see* **Note 14**).
9. In order to set up a PCR to check for RFN-GFP fusion insert in the isolated plasmid, mix 30 μL of water, 10 μL of 5× Phusion® GC Buffer, 1 μL of dNTPs (10 mM), 2 μL of DMSO, 3 μL of primer P1, 3 μL of primer P6, 0.5 μL of plasmid, and 0.5 μL of Phusion® High-Fidelity Polymerase, mix, and place in thermal cycler.
10. PCR cycling conditions are shown in Table 4.
11. Add 6 μL of DNA loading buffer to each sample and load on a 1% agarose gel containing a 1/10,000 dilution of the ethidium bromide stock (*see* **Note 9**). Run at 105 V for 30 min.
12. Pick three plasmids for which a band at the proper molecular weight is observed and send for sequencing using primers P1, P2, P5, and P6 (*see* **Note 15**).

Table 4
PCR conditions for cloning of pCDF-EcRFN-GFP

Step	Number of cycles	Temperature (°C)	Duration
1	1	98	2 min
2		98 51 72	30 s 15 s 1.5 min
3	1	72	10 min
4	1	4	∞

13. Use a plasmid with a confirmed proper sequence for construction of the reporter strain. The plasmid is henceforth referred to as pCDF-*Ec*RFN-GFP.

3.7 Construction of Reporter E. coli Strain

1. Grow an overnight culture of a Ribocil^{R} *E. coli* strain MB5746-5 in ZymoBroth™ (*see* **Note 16**).
2. Dilute the overnight culture 1/100 in 20 mL of fresh ZymoBroth™ and grow shaking at 37 °C until OD_{600} is approximately 0.5 (~3 h).
3. Centrifuge the 20 mL culture and wash once with 45 mL of ice-cold water in a 50 mL conical tube (*see* **Note 17**).
4. Transfer to a 1.7 mL microcentrifuge tube and wash three times with 1 mL of ice-cold 10% glycerol.
5. Resuspend the pellet in 200 μL of ice-cold 10% glycerol. These are now competent cells for transformation.
6. Add 1 μL of pCDF-*Ec*RFN-GFP to 50 μL of competent cells, incubate on ice for 3 min, and transfer cells to a 0.2 cm gap MicroPulser™ electroporation cuvette.
7. Electroporate cells on EC2 setting using the MicroPulser™ apparatus, and immediately recover cells in 250 μL of SOC medium and shake at 37 °C for 1 h.
8. After recovery, plate 25 μL and 250 μL on LB agar containing 30 μg/mL spectinomycin and incubate plates overnight at 37 °C (*see* **Note 18**).
9. Pick four colonies from the transformation and confirm that they contain pCDF-*Ec*RFN-GFP as in Subheading 3.6 (*see* **Note 19**).
10. Save the reporter strain by mixing equal volumes of the overnight culture with 1:1 LB:glycerol and freezing at −80 °C (*see* **Note 20**).

3.8 RFN-GFP Reporter Assay

1. Grow an overnight culture of MB5746-5/pCDF-*Ec*RFN-GFP in CAMHB containing 30 μg/mL spectinomycin.
2. Make a 1/5000 dilution of the overnight culture in CAMHB supplemented with 30 μg/mL spectinomycin and dispense 0.5 mL into each well of a 96-well deep-well plate. Include a row of medium only (no cells) as a control.
3. Make fourfold dilutions of the compound to be tested, along with a vehicle control, and add the various dilutions to the 96-well deep-well plate (*see* **Note 21**).
4. Incubate the plate while shaking at 37 °C between 8 and 20 h.
5. Transfer 200 μL of culture from the deep-well plate to a 96-well black polystyrene round-bottom fluorescence plate

Table 5
Typical data obtained from the RFN-GFP reporter assay

	Fluorescence (Ex 395/Em 509)		Absorbance (600 nm)		Corrected fluorescence	Corrected absorbance	FI/Ab	% inhibition
Ribocil (μg/mL)	MB5746-5	Control (background)	MB5746-5	Control (background)	MB5746-5	MB5746-5	MB5746-5	MB5746-5
64.00	1024.0	990.9	0.272	0.091	33.1	0.181	182.4	97
16.00	1071.4	990.6	0.252	0.053	80.8	0.199	405.8	94
4.00	1133.9	1013.6	0.258	0.055	120.7	0.203	594.4	91
1.00	1165.5	1012.1	0.227	0.055	153.4	0.172	890.5	87
0.25	1409.7	1030.3	0.227	0.050	379.3	0.178	2135.9	69
0.06	1764.6	1013.2	0.208	0.050	751.4	0.158	4764.4	31
0.02	1791.7	1021.0	0.193	0.048	770.7	0.146	5282.5	24
0.00	2073.2	1016.5	0.199	0.046	1056.8	0.153	6911.4	0

Corrected fluorescence and corrected absorbance are calculated by subtraction of background which contains only CAMHB and compound, no cells. Percent inhibition is calculated using the no-ribocil control as the maximum fluorescence per absorbance achievable

and measure fluorescence on a SpectraMax M4 using 395 nm excitation and 509 nm emission.

6. Transfer 100 μL of culture from the deep-well plate to a 96-well clear polystyrene round-bottom plate and measure absorbance on a plate reader at 600 nm.
7. Calculate the ratio of fluorescence to absorbance after subtraction of background due to medium and/or compound and plot as a function of compound concentration (see Table 5). The concentration of compound that leads to a 50% reduction in relative fluorescence compared to the untreated control is the IC_{50} (Figs. 3 and 4).

3.9 Aptamer DNA Template Synthesis

1. Prepare PCR reactions as shown in Table 6 (*see* **Note 22**).
2. Make 50 μL aliquots in each well of a 96-well reaction plate. Perform PCR using conditions shown in Table 7.
3. Combine PCR reactions, phenol–chloroform extract, and precipitate DNA with NaOAc (1/10 vol.) and isopropanol (1 volume).
4. Leave at −20 °C for at least 1 h, and microcentrifuge for 20 min at 4 °C.
5. Rinse pellet with cold 70% ethanol and air-dry.
6. Resuspend DNA pellet in water and quantify PCR products using a Nanodrop spectrophotometer. Further check the purity of PCR products on a 1.0% agarose TBE gel.

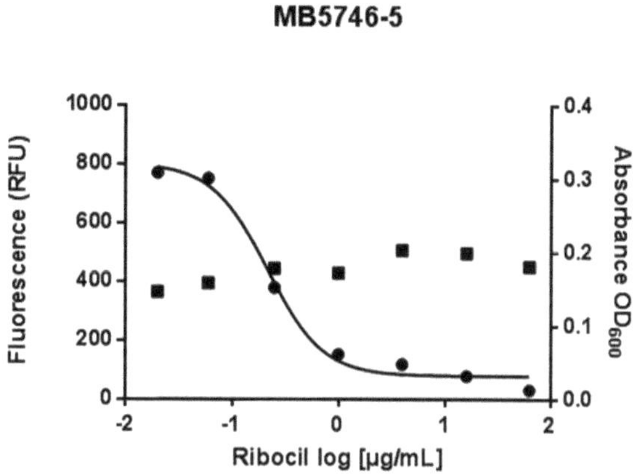

Fig. 3 Dose-dependent response of ribocil on fluorescence (circles) and absorbance (squares). Ribocil has little effect on cell growth of MB5746-5 as indicated by approximately equivalent OD_{600} readings across various concentrations. However, ribocil inhibits fluorescence in a dose-dependent manner

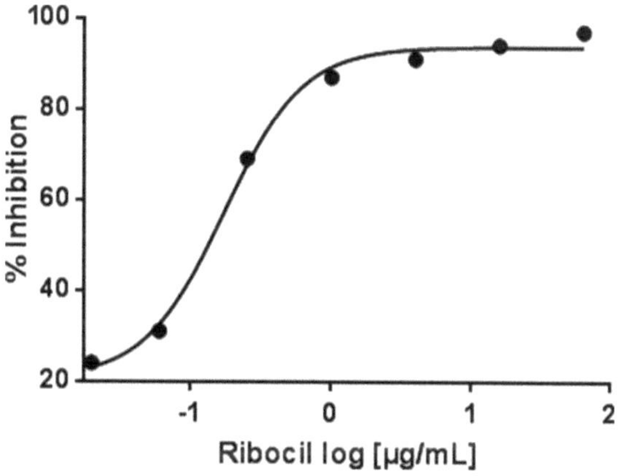

Fig. 4 Dose-dependent response of ribocil on percent inhibition of fluorescence per cell. Normalizing fluorescence to cell density allows for comparison of different compounds on inhibition of RFN-controlled GFP production such that even compounds with small effects on cell growth can be evaluated. The IC_{50} of ribocil is approximately 0.12 µg/mL in this assay

3.10 RNA Aptamer Synthesis

1. Use the RiboMAX (Promega) Large Scale RNA Production T7 System for RNA synthesis as follows: 200 µL of 5× T7 translation buffer, 200 µL of rNTPs (25 mM), 100 µL of enzyme mix, linear template (≈100 µg/1000 µL Rx), water to a total volume of 1000 µL.
2. Gently triturate to mix and incubate for 4 h at 37 °C.
3. Add RQ1 RNase-Free DNase to a concentration of 1 unit/µg of template DNA. Incubate for 15 min at 37 °C.

Table 6
PCR reagents for aptamer DNA template synthesis

Component	50 μL Reaction	100 Reactions	Final concentration
Phusion DNA polymerase	0.5 μL	5 μL	1.0 units
5× Phusion HF or GC buffer	10 μL	1000 μL	1×
10 μM Forward primer	2.5 μL	250 μL	0.5 μM
10 μM Reverse primer	2.5 μL	250 μL	0.5 μM
10 mM dNTPs	1 μL	100 μL	200 μM
Template DNA (*see* **Note 22**)	≈10 ng/reaction	≈1 ug	0.2 μg/mL
DMSO (optional)	1.5 μL	150 μL	3%
Nuclease-free water	To 50 μL	To 5000 μL	

Table 7
PCR conditions for aptamer DNA template synthesis

Step	Number of cycles	Temperature (°C)	Duration
1	1	98	2 min
2		98 50 72	30 s 30 s 30 s
3	1	72	10 min
4	1	4	∞

4. Extract with 1 volume of citrate-saturated phenol/chloroform, vortex for 1 min, and centrifuge at top speed in microfuge.
5. Extract with 1 volume of chloroform:isoamyl alcohol, vortex for 1 min, and centrifuge at top speed in microfuge. Back-extract with 100 μL water if required.
6. Remove unincorporated nucleotides with GE NAP-5 columns. Take caps off the bottom and top of the columns (use 1 column/500 μL sample), let buffer run out of the column, and add 10 mL of RNase-free water. Run water out of the column.
7. Add the 500 μL samples to each column and let run into column and when samples are fully into the column elute with 1 mL RNase-free water.
8. Each 500 μL sample yields 2× 750 μL of eluate. To each of these samples, add 75 μL of 3 M sodium acetate and 1 volume of isopropanol and store at −20 °C overnight.

9. Centrifuge for 15 min at full speed in a microfuge at 4 °C, wash pellet with cold 70% ethanol, and let air-dry.
10. Resuspend pellets in 100 μL of water, combine, and bring up to 600 μL total volume. Measure OD and run on agarose gel (1.2%).
11. Store RNA at −80 °C after checking an RNA agarose gel with ethidium bromide (heat sample in RNA loading buffer for 10 min at 70 °C).

3.11 Aptamer Annealing

1. Prepare a 20 μM RNA aptamer solution by diluting from stock solution into aptamer reannealing buffer. Add appropriate amount of H_2O to adjust the reannealing buffer to 1× concentration.
2. Incubate in preequilibrated heat block set at 95 °C for 5 min. If tubes are not fit tightly into holder, water can be used to aid in thermal conduction.
3. Promptly remove tube and incubate at room temperature for at least 15 min.
4. Maintain reannealed aptamer at room temperature until ready for assay. Do not refreeze or chill on ice (*see* **Note 23**).

3.12 FMN-Aptamer Binding

1. Prepare 100 μL of 3× solutions of 15-point 1.25-fold serial dilution of reannealed aptamer in binding assay buffer, for a final concentration range from 450 to 19.8 nM. Also prepare a no-aptamer control solution.
2. Prepare 1500 μL 3× solutions of FMN at concentrations of 360, 180, and 90 nM in binding assay buffer. Add appropriate amount of H_2O to adjust the binding assay buffer to 1× concentration.
3. Add 10 μL of each 0.6% DMSO solution to 6 columns (96 wells) of 384-well assay plate. The final DMSO concentration in the 30 μL reaction is 0.2%.
4. Add 10 μL of each 3× FMN solution to two columns (32 wells) of a 384-well assay plate. The final FMN concentrations in the 30 μL reaction are 120, 60, and 30 nM.
5. Add 10 μL of each 3× aptamer and control solution to the plated FMN solutions and centrifuge plates in a tabletop centrifuge at ~1000 × *g* for 1 min.
6. Incubate at room temperature for 120 min to allow FMN-aptamer interactions to reach equilibrium.
7. Transfer plate to plate reader, with excitation monochromator set to 455 nm with 10 nm bandwidth, emission monochromator set to 525 nm with 10 nm bandwidth, and 515 nm cutoff filter.

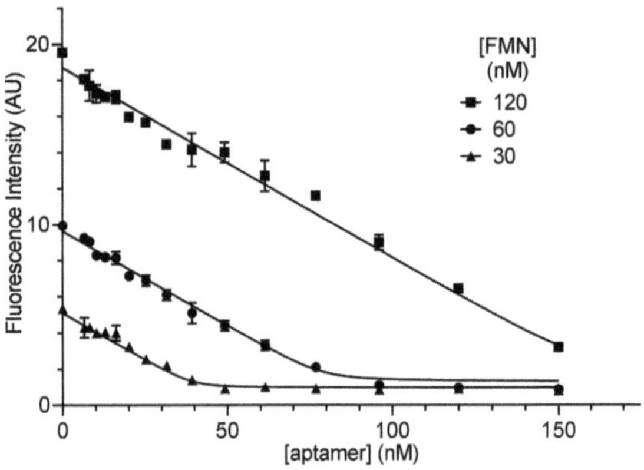

Fig. 5 Fluorescence intensity of FMN is diminished upon binding to the aptamer. Points represent fluorescence intensities of three indicated concentrations of FMN following the addition of various concentrations of aptamer. Lines represent the nonlinear least squares analysis of the data describing the change in FMN fluorescence upon binding. This best-fit curve was derived using Eq. 1 (*see* **Note 24**) to determine the binding constant, $K_d^{FMN} = 1.2 \pm 0.3$ nM, for the FMN-aptamer interaction

8. Record fluorescence emission data in kinetic mode for 2 min with 20 s between reads to ensure that interaction has reached equilibrium and take average value.
9. Export data and analyze using software with curve-fitting capabilities, such as GraphPad Prism. For this binding interaction, where K_d is less than the concentrations of FMN used, use a quadratic equation, solving for the concentration of unbound FMN, to fit the data (*see* **Note 24** and Fig. 5).

3.13 FMN-Aptamer Competition Binding

1. Prepare 50 μL of 3× solutions of 14-point twofold serial dilutions in binding assay buffer of the compounds to be tested. A 0.6% DMSO concentration should be maintained by adding 0.6% DMSO to the dilution buffer. A vehicle control should also be prepared.
2. Prepare a 3× solution of 1000 μL of 180 nM FMN in binding assay buffer.
3. Prepare a 3× solution of 1000 μL of 450 nM reannealed RNA aptamer in binding assay buffer. A no-RNA control should also be prepared.
4. Transfer 10 μL of 3× compound serial dilutions and both controls to 384-well assay plate.
5. Add 10 μL of 3× FMN and 10 μL of 3× aptamer solution to each compound well and centrifuge plates at ~1000 × *g* for 1 min. The final FMN and aptamer concentrations in the final 30 μL reaction volume are 60 nM and 150 nM, respectively.

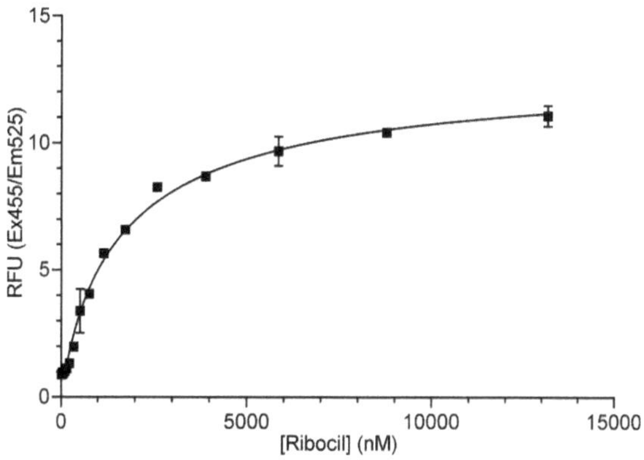

Fig. 6 Ribocil competes with FMN for binding to aptamer. Points represent fluorescence intensities of solutions containing 30 nM FMN, 150 nM aptamer, and various concentrations of ribocil. Solid line represents the nonlinear least squares analysis of the data describing the change in FMN fluorescence upon displacement from aptamer by ribocil. This best-fit curve was derived using Eq. 2 (*see* **Note 25**) to determine the binding constant, $K_d^{comp} = 16 \pm 3$ nM, for the ribocil-aptamer interaction

6. Incubate at room temperature for 120 min to allow aptamer-FMN/aptamer-compound interactions to reach equilibrium.
7. Transfer plate to SpectraMax M5 with excitation monochromator set to 455 nm with 10 nm bandwidth, emission monochromator set to 525 nm with 10 nm bandwidth, and 515 nm cutoff filter.
8. Record fluorescence emission data in kinetic mode for 2 min with 20 s between reads to ensure that interaction has reached equilibrium and take average value.
9. Export data and analyze using software with curve-fitting capabilities, such as GraphPad Prism. For this binding interaction, where K_d of the aptamer and the K_d of the FMN-competitive binding compound are less than the concentrations of FMN used, use a cubic equation, solving for the concentration of unbound FMN, to fit the data (*see* **Note 25** and Fig. 6).

4 Notes

1. The transition between capital and lowercase lettering in the primer indicates the fusion point between the plasmid/RFN element, RFN element/GFP, or GFP/plasmid. The In-Fusion® HD enzyme mix requires 15 bp of overlapping sequence in order to facilitate cloning. Although homologous overlaps longer than 21 bp are not recommended, these primers still

result in successful cloning and allow for alternative cloning methods, such as overlap extension PCR cloning, to be used if In-Fusion cloning is unsuccessful.

2. Orange G is used as the only dye in the loading buffer because it runs at the bottom of the gel, below where a 50 bp fragment would run, preventing obscuring of any desired PCR product bands.
3. Lowercase lettering in Table 2 indicates RFN element sequence in this case from *E. coli*. Uppercase lettering indicates T7 promoter-specific sequences which are added onto the RFN coding element.
4. The final RNA product will have two additional G residues at the 5′ end that were added to the template during the PCR amplification step (T7 transcription is more efficient when starting with GGG). If this is undesired remove the two additional G residues following the T7 promoter sequence with a 33-mer forward primer: Alt 5' TAATACGACTCACTATAgcttattctcagggcg.
5. This is a two-panel, whole-cell phenotypic screen selectively targeting compounds with inhibitory activity against riboflavin biosynthesis. Positive hits are compounds whose activity is suppressed by supplemental RF.
6. The agar plate assay is more robust and reproducible than liquid microplate assay for antibacterial study. Although its throughput is lower, it is perfect to be used for secondary and confirmation assays.
7. Based on our experiences, the agar MIC for majority of antibacterials is higher than liquid MIC. To increase efficiency of resistant mutant isolation, the agar MIC is measured and used for calculating compound concentrations in agar plates. A 24-well plate can accommodate two compounds with 11-point/twofold titration for each.
8. Even if the bacteria are treated with potent compound inhibiting the biosynthesis of riboflavin, the preexisting riboflavin and flavin enzymes can still support the growth for several generations, rendering high background which will interfere with the growth of drug-resistant mutants. This pretreatment is to exhaust the preexisting riboflavin and flavin enzymes (Fig. 2).
9. Be sure to use TAE buffer in making the agarose gel itself, as using water will lead to smearing of bands during electrophoresis.
10. Using the primers listed in Table 1, the sizes of the PCR products for the RFN element, GFP, and the pCDF-1b vector should be 550 bp, 720 bp, and 2050 bp, respectively. When

using a transilluminating UV light to visualize the bands for excision from the gel, work as quickly as possible to avoid introducing mutations into the DNA during cloning. Use water when eluting DNA from the QiaQuick® spin columns because use of elution buffers such as EB can affect downstream applications such as In-Fusion cloning.

11. Typical yields after PCR purification are 40–60 ng/μL.
12. When adding DNA to TOP10 cells, mix by flicking not by pipetting, as the cells are very sensitive to shear forces.
13. A handheld UV light (long wavelength) can be used to visualize which colonies are expressing GFP so that it is more likely that the transformants picked will have the desired plasmid construct.
14. As in **Note 10**, use water when eluting DNA from the QiaQuick® spin columns so that elution buffers do not affect downstream PCR and sequencing reactions.
15. Using primers P1 and P6, the size of the PCR product should be 1270 bp.
16. Strain MB5746-5 is used as the reporter background because it is more permeable (*lpxC101*), efflux deficient (*tolC*), and ribocil resistant (RFN_{G37U}). The permeability and efflux mutations increase the intracellular accumulation of test compounds and mitigate issues with assessing on-target RFN binding due to physicochemical properties that affect the influx of molecules and can cause poor whole-cell activity. The ribocil-resistant mutation allows for assessment of effects on RFN-controlled GFP expression without inhibiting cell growth.
17. All centrifuge steps should be performed at no more than 5000 × *g* to avoid cell lysis.
18. The concentration of spectinomycin used for selection (30 μg/mL) is less than typical concentrations used for *E. coli* selection because the MB5746-5 strain is more susceptible to this antibiotic due to its efflux and permeability defects (*see* **Note 16**).
19. Make sure to use lower spectinomycin concentrations (30 μg/mL) in propagating the MB5746-5/pCDF-*Ec*RFN-GFP strain.
20. When reviving this strain for use in the reporter assay, it is best to streak it out on agar medium and select a colony that shows robust fluorescence under a handheld UV lamp.
21. The solvent typically used to dissolve compounds is DMSO. It is important to keep the DMSO concentration below 2% in the assay medium to avoid affecting growth of the bacteria.

22. RFN elements should be cloned into a small plasmid (e.g., pTOPO series, Invitrogen) for use as templates. Alternatively, in vitro-synthesized DNA copies of the RFN element can be obtained from a commercial vendor.

23. It is preferable to reanneal any oligomeric RNA immediately prior to use. Freezing and thawing procedures along with concentration and dilution of oligomeric RNA stock solutions might enable subpopulations of the RNA molecules to become kinetically trapped in nonfunctional conformational states. Although divalent cations are often required for proper RNA function, they should be avoided when possible during reannealing since they can facilitate sample degradation.

24. Due to the tight binding nature of the interaction between the riboswitch aptamer and FMN, accurate concentrations of FMN and aptamer are necessary for the calculation of the binding constant, K_d^{FMN}. The concentrations of FMN can be confirmed photometrically using the extinction coefficient FMN at 446 nm (12,200 $M^{-1}\cdot cm^{-1}$). Riboswitch aptamer concentration is first estimated photometrically at 95 °C in a temperature-controlled spectrophotometer using the nearest-neighbor calculated extinction coefficient at 260 nm of 1,997,600 $M^{-1}\cdot cm^{-1}$. Further refinement of this value is derived from curve fitting with Eq. 1 using multiple known concentrations of FMN.

$$I_{obs} = I_0 + \frac{I_\infty - I_0}{2FMN_t}\left(\left(\left(RNA_t \times C_f\right) + FMN_t + K_d^{FMN}\right) - \sqrt{\left(\left(RNA_t \times C_f\right) + FMN_t + K_d^{FMN}\right)^2 - 4\left(RNA_t \times C_f\right)FMN_t}\right) \quad (1)$$

In this equation, I_{obs}, I_0, and I_∞ are the observed fluorescence signal, the fluorescence signal of unbound FMN, and the fluorescence signal of fully riboswitch-bound FMN, respectively. FMN_t and RNA_t are the total concentrations of FMN and riboswitch, respectively, at each data point. C_f is a correction factor applied to the RNA_t value that is used to accurately assess the amount of aptamer in solution that productively binds FMN.

25. By taking advantage of the difference in fluorescence intensity of the free and bound states of the natural ligand of this riboswitch the binding constant, K_d^{comp}, of FMN-competitive compounds is determined using the tight binding ligand and competitor set of relationships below:

$$A = K_d^{FMN} + K_d^{comp} + FMN_t + comp_t - RNA_t \times C_f$$
$$B = K_d^{comp}(FMN_t - RNA_t \times C_f) + K_d^{FMN}(comp_t - RNA_t \times C_f) + K_d^{FMN} K_d^{comp}$$
$$C = -K_d^{FMN} K_d^{comp} RNA_t \times C_f$$

$$RNA_f = \frac{-A + 2\sqrt{(A^2 - 3B)}}{3} \cos\left(\frac{\arccos\left(\frac{(-2A^3 + 9AB - 27C)}{2\sqrt{(A^2 - 3B)}^3}\right)}{3}\right) \quad (2a)$$

$$I_{obs} = FMN_t\left(I_0 + \frac{(I_\infty - I_0)RNA_f}{K_d^{FMN} + RNA_f}\right) \quad (2b)$$

26. In these equations, I_{obs}, I_0, I_∞, FMN_t, C_f, and RNA_t are as defined above. $comp_t$ is the total concentration of the competitor compound. *A*, *B*, and *C* are the coefficients of the cubic solution to the explicit function defining RNA_f, which is the concentration of riboswitch aptamer not bound to FMN. To facilitate data analysis any parameters defined from the FMN-aptamer interaction experiments should be fixed in this equation. For further characterization of the variability and interdependence of the fitted values, data from both experiments can be globally fit to Eq. 2.

References

1. Howe JA, Wang H, Fischmann TO, Balibar CJ, Xiao L, Galgocci A et al (2015) Selective small-molecule inhibition of an RNA structural element. Nature 526:672–677
2. Bacher A (1991) Riboflavin kinase and FAD synthetase. In: Müller F (ed) Chemistry and biochemistry of flavoenzymes. CRC Press, Boca Raton, FL, pp 349–370
3. Fischer M, Bacher A (2005) Biosynthesis of flavocoenzymes. Nat Prod Rep 22:324–350
4. Gelfand MS, Mironov AA, Jomantas J, Kozlov YI, Perumov DAA (1999) Conserved RNA structure element involved in the regulation of bacterial RF synthesis genes. Trends Genet 15:439–442
5. Vitreschak AG, Rodionov DA, Mironov AA, Gelfand MS (2002) Regulation of RF biosynthesis and transport genes in bacteria by transcriptional and translational attenuation. Nucleic Acids Res 30:3141–3151
6. Winkler WC, Cohen-Chalamish S, Breaker RR (2002) An mRNA structure that controls gene expression by binding FMN. Proc Natl Acad Sci USA 99:15908–15913
7. Lee ER, Blount KF, Breaker RR (2009) Roseoflavin is a natural antibacterial compound that binds to FMN riboswitches and regulates gene expression. RNA Biol 6:187–194
8. Serganov A, Nudler A (2013) A decade of riboswitches. Cell 152:17–24
9. Lünse CE, Schüller A, Mayer G (2014) The promise of riboswitches as potential antibacterial drug targets. Int J Med Microbiol 304:79–92
10. Mandal M, Breaker RR (2004) Gene regulation by riboswitches. Nat Rev Mol Cell Biol 5:451–463

11. Zerbino DR, Birney E (2008) Velvet: algorithms for de novo short read assembly using de Bruijn graphs. Genome Res 18:821–829
12. Li H, Durbin R (2009) Fast and accurate short read alignment with Burrows–Wheeler transform. Bioinformatics 25:1754–1760
13. McKenna A, Hanna M, Banks E, Sivachenko A, Cibulskis K, Kernytsky A et al (2010) The Genome Analysis Toolkit: a MapReduce framework for analyzing next-generation DNA sequencing data. Genome Res 20:1297–1303
14. Nusbaum C, Ohsumi TK, Gomez J, Aquadro J, Victor TC, Warren RM et al (2009) Sensitive, specific polymorphism discovery in bacteria using massively parallel sequencing. Nat Methods 6:67–69

Chapter 3

Phenotypic Screening of Small Molecules with Antimalarial Activity for Three Different Parasitic Life Stages

Nobutaka Kato, Sandra March, Sangeeta N. Bhatia, and Matthias Marti

Abstract

Malaria remains one of the deadliest infectious diseases globally. Available therapeutic agents are already limited in their efficacy, and drug resistance threatens to diminish further our ability to prevent and treat the disease. Despite a renewed effort to identify compounds with antimalarial activity, the drug discovery and development pipeline lacks target diversity and availability of compounds that target liver- and gametocyte-stage parasites. Phenotypic screens are a powerful and valuable tool for identifying new chemical compounds with antimalarial activity. This chapter highlights recent phenotypic screening methodologies for all three parasitic life stages.

Key words Malaria, *Plasmodium falciparum*, Asexual blood-stage, Liver-stage, Gametocyte-stage, Phenotypic screening

1 Introduction

Malaria remains one of the deadliest infectious diseases [1]. Despite the life cycle of malaria parasites having been described as early as 1897, an eradication strategy having been drafted by the World Health Organization in 1955, and the *P. falciparum* genome having been entirely sequenced in 2002, malaria remains a major public health problem. In addition to multiple-stage activity, an effective armamentarium of new therapeutics should include drugs that can address slowing response rates to artemisinin-based antimalarial endoperoxides that have a low propensity to induce resistance to the parasites, and that are inexpensive to manufacture (US $0.15 per dose).

Available therapeutic agents are already limited in their efficacy, and drug resistance threatens to diminish further our ability to prevent and treat the disease. Despite a renewed effort to identify compounds with antimalarial activity, the drug discovery and development pipeline lacks target diversity and most compounds only target the parasite during the asexual blood stage of infection

Bridget Wagner (ed.), *Phenotypic Screening: Methods and Protocols*, Methods in Molecular Biology, vol. 1787,
https://doi.org/10.1007/978-1-4939-7847-2_3,

[2]. While more than half of the genes in the *Plasmodium* genome have not yet been annotated, and expressing recombinant *Plasmodium* proteins is notoriously difficult, phenotypic screening is validated and the most productive approach to identify compounds with antimalarial activity [3].

Effective treatment of malaria has remained elusive due in part to the complex life cycle of the five species of *Plasmodium* that cause disease in humans [2]. Mosquito-borne sporozoites first migrate to the liver, where they multiply and differentiate before reentering the bloodstream. Drugs that target this liver stage of the life cycle are needed to prevent malaria from developing following initial mosquito-based infection (prophylaxis). The asexual blood-stage parasites also differentiate into sexual forms that can be transmitted to other individuals through the mosquito vector. Drugs that target this sexual (gametocyte) stage of the life cycle are needed to prevent transmission during the course of treatments that target the asexual blood stage. Ideally, therapeutics that are efficacious against liver-, asexual-, and sexual-stage parasites are needed in order for malaria eradication campaigns to be effective [4].

The asexual blood stage is the stage during which patients manifest malaria symptoms. With the exception of primaquine, which is only active against liver- and gametocyte-stage parasites, all antimalarial drugs currently in use have been optimized for blood-stage parasites. Asexual and sexual blood stages of *P. falciparum* are also the only forms that can be cultured in vitro. Thus, the asexual blood-stage screens have the highest throughput and are the most cost effective among phenotypic antimalarial screens.

Since 2008, multiple antimalarial phenotypic high-throughput screens have been conducted, the majority utilizing the SYBR Green growth inhibition assay first described in early 2000s [5–7]. This assay has since replaced the [^{3}H]-hypoxanthine incorporation assay, which requires radioactive materials. The principle behind the SYBR Green growth inhibition assay is the contrast between host red blood cells, which do not contain nucleic acids, and the malaria parasites, which do, and which are thus readily stained with the dyes [7]. The dye is highly fluorescent when it is intercalated into DNA, but is poorly fluorescent when it is not [6], relieving this assay of any requirement for liquid exchange (homogeneous assay).

To achieve malaria eradication, antimalarial drug discovery efforts also need to focus on targeting the liver-stage forms of *Plasmodium* parasites. The liver-stage parasites do not cause malaria pathology, but undergo an asexual replication known as exoerythrocytic schizogony within the hepatocytes. Exoerythrocytic schizogony culminates in the production of merozoites that are released into the bloodstream where they infect red blood cells and cause the symptoms of malaria. The targeting of hypnozoites, the long-lived dormant form of the liver-stage parasites in *P. vivax* and *P. ovale,* is also a crucial step for disease eradication. Hypnozoites are capable of reactivating up to years after the initial transmission, causing chronic, relapsing illness [8–10]. It appears now that this dormant hypnozoite reservoir is the primary barrier blocking the charge to achieve global malaria eradication.

Unlike blood-stage parasites, which can continuously replicate within red blood cells, liver-stage parasites undergo only one replication. Therefore, liver-stage parasites cannot be continuously propagated. Consequently, the establishment of any liver-stage malaria model is ultimately dependent on the constant accessibility of *Plasmodium* sporozoites, which includes a substantial effort to maintain mosquito colonies and to regularly source or produce infectious *Plasmodium* gametocytes. An additional biological complication arises due to the considerable variability in sporozoite numbers and sporozoite infectivity between batches [11]. Recent efforts in adapting cryopreservation techniques for *Plasmodium* sporozoites may facilitate the development of a constant sporozoite production schedule [12, 13].

The use of rodent malaria species models (e.g., *P. berghei, P. yoelii*) to infect hepatoma cell lines (e.g., HepG2, Huh7) is the most cost-effective approach to testing liver-stage antimalarial candidates, and requires lower biosafety regulations for phenotypic screens [14, 15]. However, hepatoma cell lines exhibit different physiologies compared to primary hepatocytes, and these variations can negatively affect the outcome of a phenotypic screen.

Models using human malaria species (*P. falciparum* and *P. vivax*) have also been developed. Existing immortalized cell lines, such as HC04, support infection by the human tropic pathogens, but infection efficiency is low, limiting their miniaturization capacity and thus their potential for use in high-throughput screening [16]. Furthermore, in situ observation of liver-stage development in HC04 cultures is typically obscured after 6 days, due to continued proliferation and overgrowth of the host cells. Moreover, as mentioned above, cell lines exhibit aberrant physiological features. By contrast, primary hepatocytes have been shown to support the development of the liver forms of *P. falciparum* and *P. vivax* in a more physiologic host context [17, 18]. Despite this finding, primary hepatocyte systems are rarely employed and difficult to translate to higher throughput platforms, due to limited cell sourcing and challenges in maintaining their functional phenotype over extended periods of time in vitro [19].

Recent advances have been made toward the generation of more robust models overcoming some critical deficiencies of existing in vitro models of human liver cells. For example, with cocultures of human hepatocytes that are organized into 2D islands and surrounded by nonparenchymal cells using microfabrication methods (MPCC, micropatterned coculture system), it is now possible to maintain functional hepatocyte phenotypes for several weeks

[20]. These MPCCs have been used to establish liver-stage infections for both *P. falciparum* and *P. vivax*, including the observation of small forms (hypnozoite-like forms) up to day 21 post-infection. The platform was validated as a tool for medium-throughput screening [11, 21], and is in development for use in a 384-well format. Another coculturing strategy that combines the sandwich culture method with the use of Matrigel and HepaRG cells has been shown to support the development of *P. falciparum*, and can also maintain infection of *P. cynomolgi* for up to 40 days. Visualization and reactivation of small forms have been achieved in this model [22].

Strategies for preventing transmission of parasites from humans to mosquitos are also essential for efforts aiming at malaria elimination and eradication. During each asexual parasite cycle in red blood cells, a small proportion of parasites stop replicating and initiate development into sexual precursor cells, or gametocytes. Gametocytes also develop in red blood cells, and, at maturity, are the only parasite form capable of establishing an infection in the mosquito vector. Sexual commitment, gametocyte development, and maturation are thus essential steps for transmission of malaria.

Considering the high mortality and morbidity caused by malaria, there is no question that new drugs are needed. It is an exciting time for malaria drug discovery; the combination of new and innovative screens to identify compounds with broad-range activity is hoped to yield new insights into proteins. Several phenotypic assays have been developed that assess the effect of small molecules against transmission stages at different endpoints, and most of them focus on the final maturation process. Assays are either based on reporter bioluminescence, lactate dehydrogenase activity, or exflagellation as endpoints. Here, we describe an assay that allows parallel assessment of biological activity of compounds on asexual parasites, sexual commitment, and sexual development. The assay uses a fluorescent reporter for gametocyte quantification. Use of a reporter renders the assay independent from gametocyte purification steps, and it enables screening the effects of perturbations on the earliest phases of gametocyte development including sexual commitment. Gametocyte development may be perturbed by introducing the compound of interest at any time during early maturation or even before sexual commitment.

2 Materials

2.1 Asexual Blood-Stage Screening

1. Parasites: Any *P. falciparum* strains can be used for this assay (field isolates without prior adaptation need to be viable for at least two intraerythrocytic developmental cycles; *see* **Note 1**).

2. *P. falciparum* maintenance: Fresh O-positive human blood cells (5% hematocrit) in an atmosphere of 93% N_2, 4% CO_2, 3% O_2 at 37 °C in RPMI medium with L-glutamine, 4.16 mg/mL Albumax II, 0.013 mg/mL hypoxanthine, 1.73 mg/mL glucose, 0.18% $NaHCO_3$, 0.031 M HEPES, 2.60 mM NaOH, and 0.043 mg/mL gentamicin (*see* **Note 2**).
3. Screening compounds dissolved in dimethyl sulfoxide (DMSO; *see* **Note 3**).
4. Assay plates: Plates should be sterile, black, clear, flat bottomed with lids. Either 96-, 384-, or 1536-well plates can be used (*see* **Note 4**).
5. Detection reagent (*see* **Note 5**): 10× Dilution of SYBR Green I Nucleic Acid Gel Stain (10,000× concentrate in DMSO) in lysis buffer (20 mM Tris–HCl, 5 mM EDTA, 0.16% (w/v) saponin, 1.6% (v/v) Triton X-100).

2.2 Liver-Stage Screening

1. Sporozoites prepared from rodent malaria parasites: GFP-, RFP-, or luciferase-expressing *P. berghei* and/or *P. yoelii* sporozoites should be dissected out from salivary glands of infected female *Anopheles stephensi* mosquitoes [23] (*see* **Note 6**).
2. Sporozoites prepared from human malaria parasites: Sporozoites can be obtained by dissection of the salivary glands of the infected mosquitoes (*see* **Note 7**).
3. Sporozoites prepared from monkey malaria parasites: *Plasmodium cynomolgi* can also be used as a surrogate model of *P. vivax*. Sporozoites can be obtained from infected *Anopheles stephensi* salivary glands collected on days 14–35 after an infective meal of blood from a *Macaca mulatta*, infected with blood-stage *P. cynomolgi* parasites using membrane-based glass feeders [22].
4. Hepatocytes: Depending on the in vitro model chosen, use hepatoma cell lines or primary hepatocytes (*see* **Note 8**).
5. Hepatocyte culture medium: High-glucose Dulbecco's modified Eagle's medium (DMEM) with 10% (v/v) fetal bovine serum (FBS), 1% (v/v) ITSTM, 7 ng/mL glucagon, 40 ng/mL dexamethasone, 15 mM HEPES, and 1% (v/v) penicillin-streptomycin. Some laboratories increase the concentration of penicillin-streptomycin to 3% (v/v) and add Fungizone to the medium to prevent bacterial or fungal contamination post-sporozoite infection.
6. Assay plates: Sterile, black, clear-bottomed plates with lids (*see* **Note 9**).

7. Detection of infection: Antibody staining for the presence of malaria antigens in hepatic cells followed by high-content imaging [14]. An alternative to staining parasites with fluorescent antibodies is to use, when available, transgenic parasites that stably express a reporter protein like GFP [15], or parasites expressing a luciferase reporter. In the latter case, whole-well luminescence can be used [24, 25].

2.3 Gametocyte-Stage Screening

1. Parasites: Various *P. falciparum* strains can be used and obtained from MR4. In the protocol described here, we use *P. falciparum* line Pf2004/164-tdTom [26].
2. Culture medium and parasite maintenance: See Subsection 2.1, **step 2**, except for the use of human serum instead of Albumax II. If transgenic parasites are used, such as Pf2004/164-tdTom, keep cultures at 37 °C in medium containing 4 nM WR 99210.
3. Assay plates: Sterile, black, clear-bottomed plates with lids. Either 96- or 384-well screening plates can be used; the protocol described here is for 96-well plates. Start the experiment with ~10 mL of culture per plate with 2% parasitemia at a ring stage.
4. Detection of parasitemia: by flow cytometry using SYBR Green DNA dye. Gametocyte detection relies on the expression of the red fluorescent reporter tandem Tomato (tdTom), which is driven by the promoter of the gametocyte-specific gene *etramp10.3* (*PF10_0164/PF3D7_1016900*) [10]. Alternatively, antibodies against the gametocyte-specific antigen Pfs16 can be used.
5. On the MACSQuant VYB, tdTomato is detected in the 615/20 channel (Y2) and SYBR Green is detected in the 525/50 channel (B1). For both the parasitemia and gametocytemia readouts, optimized settings on the MACSQuant VYB are as follows: FSC (forward scatter) channel (trigger on this channel), 270 V; Y2 channel, 375 V; B1 channel, 270 V; use low flow rates (*see* **Note 10**).

3 Methods

3.1 Asexual Blood-Stage Screening

Human red blood cells and *P. falciparum* are biosafety level 2 reagents. Use proper personal protective equipment and a biosafety cabinet for the following procedures:

1. Dispense warmed medium (37 °C) to each well (50, 20, and 4 μL for 96-, 384-, and 1536-well plates, respectively).
2. Dispense screening compounds and a positive control dissolved in DMSO (*see* **Note 11**).

3. Determine the parasitemia of the *P. falciparum* culture and adjust it to 2% parasitemia and 2% hematocrit.
4. Dispense medium containing 2% parasites and hematocrit (50, 20, and 4 μL for 96-, 384-, and 1536-well plates, respectively). The final parasitemia and hematocrit percentage should be 1%.
5. Add lid and place plates in incubator for at least 48 h (72 h is recommended). If using an incubator without humidity control, place a pan filled with water, and avoid opening the incubator door during the incubation period to reduce edge effects (*see* **Note 12**).
6. Dispense detection reagent (25, 10, and 2 μL for 96-, 384-, and 1536-well plate, respectively) and foil-seal the plate.
7. Incubate assay plates for at least 5 h (overnight recommended) in the dark at room temperature.
8. Set up plate reader. SYBR Green I stain is maximally excited at 497 nm, but also has secondary excitation peaks at ~290 and ~380 nm. The fluorescence emission of SYBR Green I stain bound to DNA is centered at 520 nm.
9. Read plates from the bottom. No need to remove the top foil seal.

3.2 Liver-Stage Screening

1. Seed hepatoma cells lines or primary hepatocytes, depending on the in vitro model chosen. An assay is validated if a good rate of infection (minimum of 0.5–1% infection rate, ideally 1–5%) is achievable and robustly reproducible (Z' > 0.5; CV < 20%). These parameters will help to determine if a 96-, 384-, or 1536-well platform can be used.
2. Ensure that you have a suspension of sporozoites that corresponds to 1:5 to 1:10 ratio of attached hepatocyte to infectious sporozoites, resuspended in hepatocyte culture medium. In the case of hepatoma cell lines being used in combination with rodent parasites, these ratios will be different (Table 1).
3. Remove the culture medium from the relevant wells of the plate and replace with the suspension of sporozoites.
4. Centrifuge the plate at 600 × *g* for 5 min at room temperature.
5. Incubate the infected plates at 37 °C and 5% CO_2 for 3 h.
6. Wash the wells with hepatocyte culture medium.
7. Some in vitro cultures require additional steps. For example: the MPCC coculture system requires the seeding of supportive cells [11, 21]. The HepaRG coculture system requires the addition of a Matrigel layer [22].
8. Dispense screening compounds and positive or negative controls (*see* **Note 11**).

Table 1
Desired ratios of attached hepatocyte to infectious sporozoites

Format	# of cells	# of sporozoites
96	10K[a]-80K[b] primary hepatocytes	50–100K human parasites
384	18K HepG2 cells	4K rodent parasites
1536	3K HepG2 cells	1K rodent parasites

[a]MPPC format
[b]Monoculture

9. At the desired endpoint, prepare the plates according to the detection method. Detection of parasites can be performed using a standard immunofluorescence assay to detect various *Plasmodium* antigens expressed during the liver stage, such as heat-shock protein 70 (HSP70). If the parasites in use express the GFP reporter, imaging using a high-content microscope can be used directly. A whole-well luminescence readout can be used if the parasites express the luciferase reporter.

3.3 Gametocyte-Stage Screening

General methods for handling red blood cells and *P. falciparum*, as well plate setup and compound dilutions, are described in Subsection 3.1. Cells should be synchronized twice and 16 h apart to obtain a highly synchronous parasite population (±4 h apart in the cell cycle).

1. Prepare a synchronized ring-stage parasite culture at 0.3% parasitemia in medium with serum.
2. At 28 ± 4 h post-invasion (h.p.i), remove complete medium and add a mixture of 90% parasite-conditioned medium (CM; this is the supernatant from a synchronized parasite culture at 44 ± 4 h.p.i. and 5% parasitemia; it can be stored for at least 90 days at 4 °C or −20 °C) and 10% complete parasite medium to induce gametocytogenesis [26].
3. Seed each well of a 96-well plate with 220 μL of this cell suspension at 0.3% parasitemia and 2.5% hematocrit.
4. Add the compounds now or during **step 5** below (8 ± 4 h.p.i. in the next cycle; *see* **Notes 13** and **14**).
5. At 8 ± 4 h.p.i (24 h after adding CM), spin the plate to pellet cells and replace medium with fresh complete medium. At this point add compounds to be tested for effects on early gametocyte development.
6. At 28 ± 4 h.p.i. resuspend the cells and transfer 10–20 μL of cell suspension from each well to 100 μL of SYBR Green staining solution (SYBR Green stock solution diluted to a

1:5000 ratio in PBS) for determination of parasitemia by flow cytometry.

7. At 66 ± 4 h.p.i. (38 h after **step 6**), stain the cells with SYBR Green. At this time point, young gametocytes have acquired enough tdTom reporter for cytometry-based detection.
8. Determine parasitized erythrocyte multiplication rate (PEMR) for each well by dividing the proportion of SYBR Green-positive cells (parasitemia) from **step 6** by the starting parasitemia evaluated in **step 2**.
9. Determine sexual conversion rates or, when testing compounds active against early gametocytes, gametocyte survival rates for each well by dividing the proportion of tdTom/SYBR double-positive cells (gametocytemia) from **step 7** by the proportion of SYBR Green-positive cells (parasitemia) from **step 6**.

4 Notes

1. Various *P. falciparum* in vitro-adapted strains can also be obtained from the Malaria Research and Reference Reagent Resource Center (MR4) (www.beiresources.org).
2. It is recommended not to add human serum, because this could be a potential source of variability between screens [27]. Some laboratories prefer to use RPMI medium without phenol to reduce the background noise during plate reading.
3. Despite significant investments from public and private funding and advances in high-throughput screening technologies over the past decade, the current global antimalarial drug discovery and development pipeline lacks target diversity [2]. Application of modern methods of asymmetric organic synthesis to create unique chemical matter and use of different screening procedures and assay readouts [28], such as lactate dehydrogenase [29] or luciferase assays [30], will help in the discovery of antimalarial compounds with novel modes of action.
4. We have observed that 384- and 1536-well assay plates with evaporation barrier wells from Aurora (Poway, CA) perform better to reduce edge effects.
5. The SYBR Green growth inhibition assay is an inexpensive, robust, reproducible, homogeneous, gain-of-signal assay, which can be miniaturized up to 1536-well format. Typically, 15- to 20-fold of signal-to-noise ratio (Σ signals of positive control/Σ signals of negative control) and Z' factor of greater than 0.7 should be expected for any plate formats.

6. *P. yoelii and P. berghei* are more infectious and develop more rapidly and completely than the human malaria species. Rodent malaria parasites can infect human cells.

   With the development of in vitro cultures of *P. falciparum* blood stages, production of *P. falciparum* sporozoites is no longer limited by the availability of patient blood samples. However, in vitro culture of blood-stage *P. vivax* has yet to be achieved, which limits the production of *P. vivax* sporozoites for use in experimental models of liver-stage *P. vivax*. Currently, *P. vivax* experiments are typically performed using sporozoites that are isolated from mosquitoes that were infected after feeding on human fresh clinical blood isolates [16, 31]. Alternatively, infected mosquitoes can be obtained after infection with *P. vivax Chesson* strain (via primates) [11, 32]. Cryopreserved human primary hepatocytes are available and highly recommended. Selection of this source allows for the use of the same hepatocyte donors in longitudinal studies/screens. However, pre-selection of a specific, infection-permissive donor lot of hepatocytes must be carried out beforehand [21].
7. 384- and 1536-well format plates have the potential for use in an in vitro system that incorporates hepatoma cell lines infected with a rodent species of *Plasmodium* [27, 28]. 96-Well format plates are recommended for in vitro systems that incorporate human primary hepatocytes.
8. If an alternative flow cytometer is used, the specific settings may need to be optimized to ensure good separation between the uninfected red blood cells, asexual parasite, and gametocyte populations.
9. It is recommended to use primary screening compound concentrations of between 4 and 10 μM, and that each compound is tested at least in duplicate. Assay plates should contain a positive control (e.g., mefloquine) in at least 5% of wells, and another 5% should include a negative control (DMSO alone). For example, >20 wells will contain mefloquine and >20 wells contain DMSO for 384-well format plate. The control wells should be scattered evenly across the assay plate, so that screeners can easily examine the quality of screening results of individual plates after plate reading. Avoid using the two outer wells to reduce edge effects. If using assay plates with evaporation barrier, fill the outermost wells with water.
10. Never stack plates. Stacking plates causes serious edge effects. If you have more plates than can be placed side by side in an incubator, use a plate rack.
11. Positive controls can include compounds with known inhibitory effects on either of the measured phenotypes, i.e., asexual

growth, gametocyte commitment, or gametocyte maturation. For asexual growth and gametocyte maturation, dihydroartemisinin could serve as positive control [26]; there is currently no positive control for a compound blocking commitment.

12. Note that administering compounds during the next cycle will probe for effects on early gametocyte development, whereas compounds added during the present cycle will also test for potential effects on sexual conversion. Include negative control wells (DMSO alone) in every plate.

References

1. World Health Organization WHO (2016) World Malaria Report
2. Kato N, Comer E, Sakata-Kato T et al (2016) Diversity-oriented synthesis yields novel multistage antimalarial inhibitors. Nature 538:344–349
3. Flannery EL, Chatterjee AK, Winzeler EA (2013) Antimalarial drug discovery—approaches and progress towards new medicines. Nat Rev Microbiol 11:849–862
4. Eichborn v JF (1986) Medical and commercial aspects of human interferon. Arzneimittelforschung 36:279–282
5. Baniecki ML, Wirth DF, Clardy J (2007) High-throughput Plasmodium falciparum growth assay for malaria drug discovery. Antimicrob Agents Chemother 51:716–723
6. Bennett TN, Paguio M, Gligorijevic B et al (2004) Novel, rapid, and inexpensive cell-based quantification of antimalarial drug efficacy. Antimicrob Agents Chemother 48:1807–1810
7. Smilkstein M, Sriwilaijaroen N, Kelly JX et al (2004) Simple and inexpensive fluorescence-based technique for high-throughput antimalarial drug screening. Antimicrob Agents Chemother 48:1803–1806
8. White NJ (2011) Determinants of relapse periodicity in Plasmodium vivax malaria. Malar J 10:297
9. Mazier D, Rénia L, Snounou G (2009) A pre-emptive strike against malaria's stealthy hepatic forms. Nat Rev Drug Discov 8:854–864
10. Krotoski WA, Collins WE, Bray RS et al (1982) Demonstration of hypnozoites in sporozoite-transmitted Plasmodium vivax infection. Am J Trop Med Hyg 31:1291–1293
11. March S, Ng S, Velmurugan S et al (2013) A microscale human liver platform that supports the hepatic stages of Plasmodium falciparum and vivax. Cell Host Microbe 14: 104–115
12. Chattopadhyay R, Velmurugan S, Chakiath C et al (2010) Establishment of an in vitro assay for assessing the effects of drugs on the liver stages of Plasmodium vivax malaria. PLoS One 5:e14275
13. Epstein JE, Tewari K, Lyke KE et al (2011) Live attenuated malaria vaccine designed to protect through hepatic CD8$^+$ T cell immunity. Science 334:475–480
14. Meister S, Plouffe DM, Kuhen KL et al (2011) Imaging of plasmodium liver stages to drive next-generation antimalarial drug discovery. Science 334:1372–1377
15. da Cruz FP, Martin C, Buchholz K et al (2012) Drug screen targeted at plasmodium liver stages identifies a potent multistage antimalarial drug. J Infect Dis 205:1278–1286
16. Sattabongkot J, Yimamnuaychoke N, Leelaudomlipi S et al (2006) Establishment of a human hepatocyte line that supports in vitro development of the exo-erythrocytic stages of the malaria parasites Plasmodium falciparum and P. vivax. Am J Trop Med Hyg 74:708–715
17. Mazier D, Beaudoin RL, Mellouk S et al (1985) Complete development of hepatic stages of Plasmodium falciparum in vitro. Science 227:440–442
18. Mazier D, Landau I, Druilhe P et al (1984) Cultivation of the liver forms of Plasmodium vivax in human hepatocytes. Nature 307:367–369
19. Bhatia SN, Balis UJ, Yarmush ML, Toner M (1999) Effect of cell-cell interactions in preservation of cellular phenotype: cocultivation of hepatocytes and nonparenchymal cells. FASEB J 13:1883–1900
20. Khetani SR, Bhatia SN (2008) Microscale culture of human liver cells for drug development. Nat Biotechnol 26:120–126

21. March S, Ramanan V, Trehan K et al (2015) Micropatterned coculture of primary human hepatocytes and supportive cells for the study of hepatotropic pathogens. Nat Protoc 10:2027–2053
22. Dembele L, Franetich J-F, Lorthiois A et al (2014) Persistence and activation of malaria hypnozoites in long-term primary hepatocyte cultures. Nat Med 20:307–312
23. Thiberge S, Blazquez S, Baldacci P et al (2007) In vivo imaging of malaria parasites in the murine liver. Nat Protoc 2:1811–1818
24. Derbyshire ER, Prudêncio M, Mota MM, Clardy J (2012) Liver-stage malaria parasites vulnerable to diverse chemical scaffolds. Proc Natl Acad Sci U S A 109:8511–8516
25. Swann J, Corey V, Scherer CA et al (2016) High-throughput luciferase-based assay for the discovery of therapeutics that prevent malaria. ACS Infect Dis 2:281–293
26. Brancucci NMB, Goldowitz I, Buchholz K et al (2015) An assay to probe Plasmodium falciparum growth, transmission stage formation and early gametocyte development. Nat Protoc 10:1131–1142
27. Plouffe D, Brinker A, McNamara C et al (2008) From the Cover: in silico activity profiling reveals the mechanism of action of antimalarials discovered in a high-throughput screen. Proc Natl Acad Sci U S A 105:9059–9064
28. Diagana TT (2015) Supporting malaria elimination with 21st century antimalarial agent drug discovery. Drug Discov Today 20:1265–1270
29. Gamo F-J, Sanz LM, Vidal J et al (2010) Thousands of chemical starting points for antimalarial lead identification. Nature 465:305–310
30. Vaughan AM, Mikolajczak SA, Camargo N et al (2012) A transgenic Plasmodium falciparum NF54 strain that expresses GFP–luciferase throughout the parasite life cycle. Mol Biochem Parasitol 186:143–147
31. Mikolajczak SA, Vaughan AM, Kangwanrangsan N et al (2015) Plasmodium vivax liver stage development and hypnozoite persistence in human liver-chimeric mice. Cell Host Microbe 17:526–535
32. Collins WE, Sullivan JS, Jeffery GM et al (2012) Mosquito infection studies with Aotus monkeys and humans infected with the Chesson strain of Plasmodiun vivax. Am J Trop Med Hyg 86:398–402

Chapter 4

Phenotypic Screening for Inhibitors of a Mutant Thrombopoietin Receptor

Anna Ngo, Ann Koay, Christian Pecquet, Carmen C. Diaconu, David A. Jenkins, Andrew K. Shiau, Stefan N. Constantinescu, and Meng Ling Choong

Abstract

An inhibitor for the thrombopoietin receptor (TpoR) would be more specific for the treatment of myeloproliferative neoplasms (MPNs) due to constitutively active mutant TpoR compared to the current treatment approach of inhibiting Janus kinase 2 (JAK2). We describe a cell-based high-throughput phenotypic screening approach to identify inhibitors for constitutively active mutant TpoR. A stepwise elimination process is used to differentiate generally cytotoxic compounds from compounds that specifically inhibit growth of cells expressing wild-type TpoR and/or mutant TpoR. We have systematically optimized the phenotypic screening assay and documented in this chapter critical parameters for a successful phenotypic screen, such as cell growth and seeding optimization, plate reproducibility and uniformity studies, and an assay robustness analysis with a pilot screen.

Key words Myeloproliferative neoplasms, Thrombopoietin receptor, Phenotypic screening, Cell-based assay, Cell viability, ATP

1 Introduction

Myeloproliferative neoplasms (MPN) represent a category of disease where acquired mutations in the hematopoietic stem cells lead to hyper-proliferation of progenitors of the myeloid lineage resulting in excessive production of red blood cells, platelets, or granulocytes in the bone marrow. The three major diseases in MPN are polycythemia vera (PV), essential thrombocythemia (ET), and primary or idiopathic myelofibrosis (MF). The acquired somatic JAK2 V617F mutation is implicated in 95% cases of PV and more than 60% cases of ET and MF [1]. Mutations in the thrombopoietin receptor (MPL/TpoR) (W515 L/K/A) account for about 3% of ET and 4% of MF cases [2, 3]. Mutations (insertions and deletions) in exon 9 of calreticulin (CALR) account for 60–80% of ET and MF cases that are negative for JAK2 and TpoR mutations, which

Bridget Wagner (ed.), *Phenotypic Screening: Methods and Protocols*, Methods in Molecular Biology, vol. 1787, https://doi.org/10.1007/978-1-4939-7847-2_4,

amount to 15–30% of ET and MF [4, 5]. CALR mutations induce MPNs by a novel mechanism of persistently activating TpoR signaling in the secretory pathway and at the cell surface [6]. Consequently, JAK2 is key for all three types of mutations, as eventually cytokine-independent JAK2 activation drives the disease.

There is currently one specific inhibitor of JAK2 in the market (ruxolitinib) and several compounds are being evaluated in clinical trials. However, none of these small molecules specifically target cells with the JAK2 V617F mutation or the TpoR W515 mutations. Nonspecific inhibition of blood cell production by the ATP-competitive JAK2 inhibitors leads to anemia and thrombocytopenia, which are documented adverse effects for these compounds [7]. TpoR mutant-specific inhibitors would therefore be expected to specifically target TpoR mutant-driven diseases and avoid undesirable side effects.

This protocol provides information on the systematic approach taken to optimize a phenotypic screening assay for cells expressing wild-type and mutant TpoR. This process involves cell growth and seeding optimization, assay reproducibility and uniformity studies, and an assay robustness analysis with a pilot screen. These tests serve to ensure the assay's robustness, reproducibility, and reliability. We have applied this protocol to screen approximately 500,000 small molecules and identified several molecules that could modulate the growth of Ba/F3 cells expressing wild-type and/or mutant TpoR [8].

2 Materials

2.1 Cell Lines

1. Mouse Ba/F3 pro-B cell lines expressing wild-type TpoR (WT) and mutant TpoR (W515 L) were obtained as described [9].
2. RPMI growth medium: RPMI1640 supplemented with 10% fetal bovine serum (FBS), 10 mM HEPES, 2 mM L-glutamine, 1X penicillin/streptomycin, and 1 mM sodium pyruvate. Filter the prepared RPMI growth medium through a disposable 0.2 μm filtration system and store at 4 °C.
3. Maintain the parental Ba/F3 cells in the above RPMI growth medium supplemented with 0.1 μg/mL IL-3.
4. During the phenotypic screen, culture the parental Ba/F3 cells in 384-well assay plates in the above RPMI growth medium supplemented with 2 ng/ml IL-3 (*see* **Note 1**).
5. Maintain the Ba/F3 TpoR WT cells in the above RPMI growth medium supplemented with 0.1 μg/mL IL-3.
6. During the phenotypic screen, culture the Ba/F3 TpoR WT cells in 384-well assay plate in the above RPMI growth medium

supplemented with 0.1 μg/mL IL-3 or 40 ng/mL TPO (*see* **Note 1**).

7. Culture the Ba/F3 cells expressing TpoR W515 L in the above RPMI growth medium without IL-3 or TPO supplement.
8. Maintain all cells in a 37 °C humidified incubator with 5% CO_2 injection.

2.2 Luminescent Cell Viability Assay

1. The CellTiter-Glo Luminescent Cell Viability Assay kit consists of CellTiter-Glo Buffer and CellTiter-Glo Substrate (lyophilized). Store the unopened assay kit at −20 °C.
2. Thaw CellTiter-Glo Buffer and equilibrate to room temperature prior to use.
3. Equilibrate the lyophilized CellTiter-Glo Substrate to room temperature prior to use.
4. CellTiter-Glo Reagent: Add the entire volume of the CellTiter-Glo Buffer to the CellTiter-Glo Substrate. Dissolve by gently vortexing, swirling, or inverting the contents to obtain a homogeneous solution. Prepare CellTiter-Glo Reagent just prior to use.

3 Methods

This section describes the screening strategies employed to identify TpoR-specific inhibitors and the step-by-step protocol for performing the phenotypic screen. A stepwise elimination process (secondary screens) is used to differentiate pan-cytotoxic compounds from compounds that specifically inhibit growth of cells expressing TpoR WT or TpoR W515 L (Fig. 1).

3.1 Compound Preparation

1. Purchase small molecules from various commercial sources such as ChemDiv, ChemBridge, Microsource Discovery Systems, BioFocus, Enamine, Vitas-M Laboratory, Peakdale Molecular, Asinex, InterBioScreen, Prestwick Chemical, UkrOrgSynthesis, Life Chemicals, and Forma Therapeutics.
2. Dissolve compounds in dimethyl sulfoxide (DMSO) to 10 mM (stock concentration). Sonicate in a water bath at room temperature to dissolve the compounds, if necessary.
3. Aliquot 50 μL of the compounds into 384-well microplates using an automated liquid-handling platform.
4. Seal the plates with aluminum foil.
5. Use commercial compound management software to generate a barcode for each plate. Manage the identity of compounds in each well location within the microplate.
6. Store the barcoded plates at 4 °C until needed.

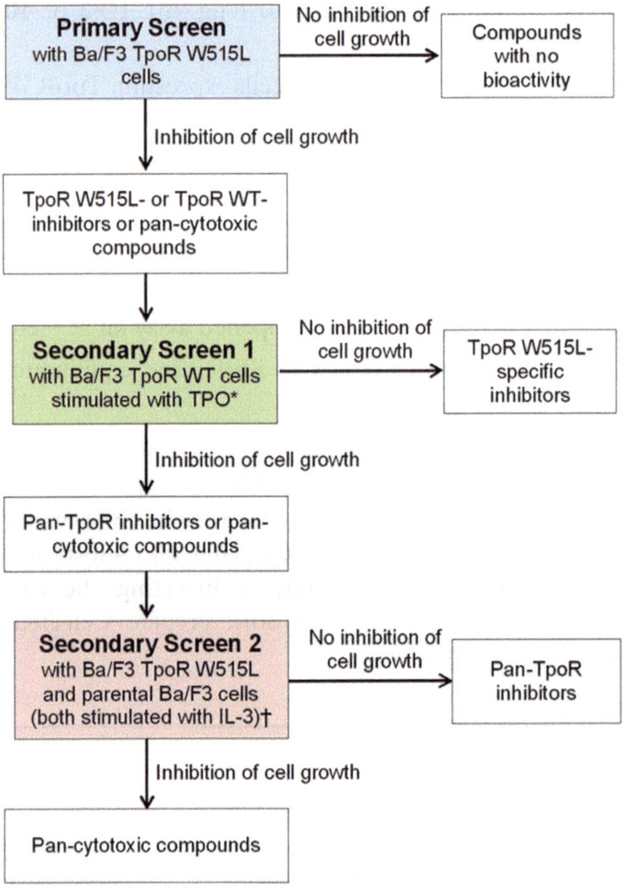

Fig. 1 Stepwise elimination criteria to differentiate general cytotoxic compounds from compounds that specifically inhibit growth of cells expressing TpoR WT or TpoR W515L. *A differential window of twofold or more in 50% growth inhibition (IC_{50}) between TpoR wild-type and TpoR W515L mutant cells is used as a cutoff for selection of TpoR W515L-specific antagonists. †A differential window of twofold or more in IC_{50} between TpoR W515L and parental Ba/F3 cells (both are stimulated to grow via the IL-3 pathway) is used as a cutoff for selection of TpoR W515L-specific antagonists

7. Ruxolitinib (positive control compound): Dissolve to 50 mM in DMSO and store at 4 °C.
8. DMSO is used as the negative control (no treatment) in the phenotypic screen.
9. Perform compound dispensing using an automated liquid-handling system.

3.2 Cell Viability Assay

1. Carry out all procedures at room temperature unless otherwise specified.
2. Grow cells to exponential growth phase. *See* **Note 2** and Fig. 2a, b for cell growth and seeding optimization.

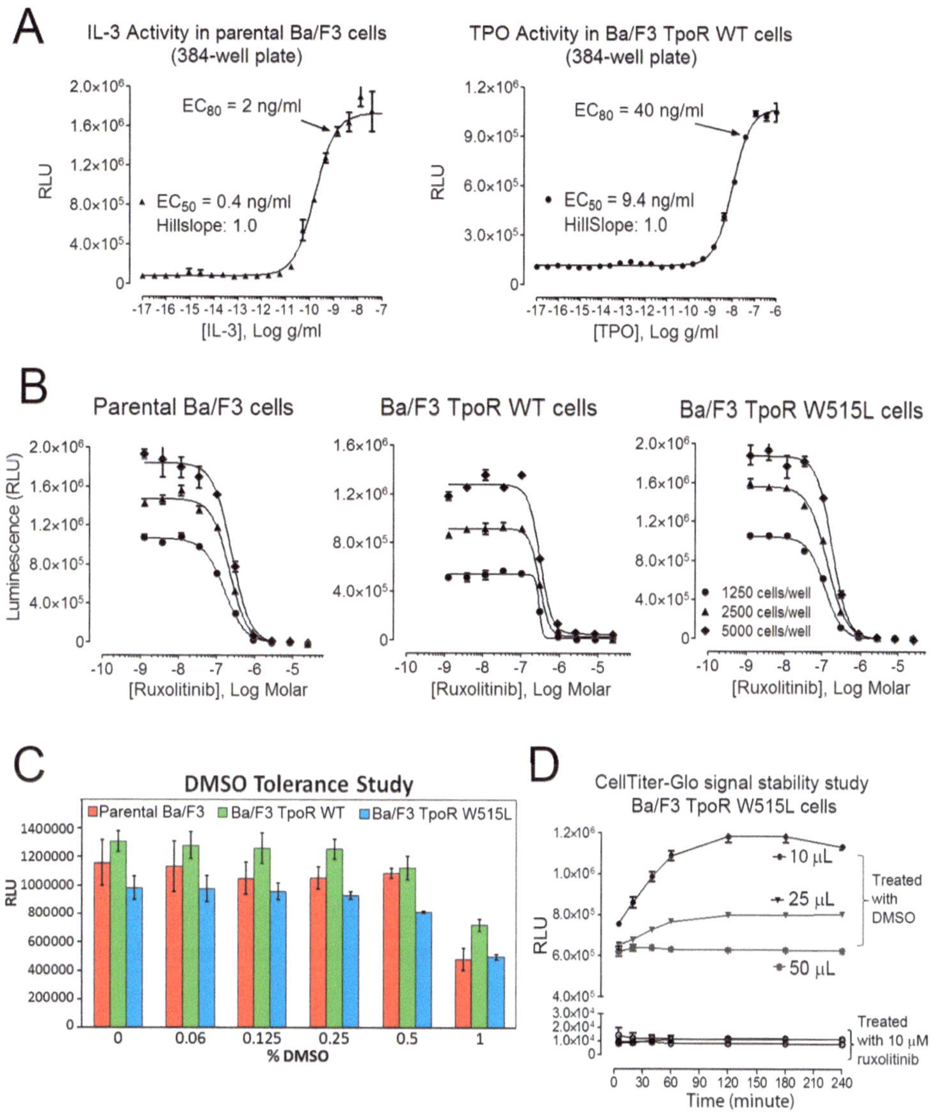

Fig. 2 Cell growth optimization. (**a**) EC_{80} of IL-3 in stimulating growth of parental Ba/F3 cells was 2 ng/mL. EC_{80} of TPO in stimulating Ba/F3 TpoR WT cells was 40 ng/mL. Dose–response studies of ruxolitinib across all three Ba/F3 cell lines showed that comparable assay performances (signal/background ratio, IC_{50} of ruxolitinib) are observed with different cell seeding numbers. (**b**) Cell seeding at 2500 cells/well is chosen for its optimal signal/background ratio. All three Ba/F3 cell lines showed similar tolerance for DMSO with less than 25% loss in cell viability at up to 0.5% DMSO. (**c**) Luminescence signal stability was tested with three different volumes of the CellTiter-Glo assay reagent (10, 25, and 50 μL). Luminescence signals were stable from 1 to 4 h. (**d**) Optimal assay performance based on luminescence signal intensity and signal/background ratio was observed with 10 μL CellTiter-Glo reagent. Mean ± SD of 3 replicates is shown

3. Count and resuspend the cells to 50,000 cells/mL.
4. Add 50 μL cell suspension (2500 cells) to all wells (*see* **Note 3**).
5. Centrifuge plates at 40 × *g* for 30 s.

6. Incubate at 37 °C with 5% CO_2 in humidified incubator for 48 h.
7. Remove all plates from the CO_2 incubator. Allow the plates to equilibrate to room temperature in stacks of five plates on the lab bench (about 1 h).
8. Add 10 μL of CellTiter-Glo Reagent to all wells.
9. Centrifuge plates at 40 × *g* for 1 min.
10. Allow the plate to incubate at room temperature for at least 30 min for an optimal luminescence signal.
11. Measure luminescence signal using a plate reader with signal integration time of 100 ms/well.
12. Luminescence signals are to be read within 4 h (*see* **Note 4** and Fig. 2d for signal stability study).
13. The luminescence signal is reported as relative luminescence units (RLU).

3.3 Assay Validation (Plate Uniformity and Reproducibility)

1. Plate uniformity test is performed with a plate layout as shown in Fig. 3a. Intra-plate, inter-plate, and inter-day plate signals are examined for signal uniformity.
2. Prepare two plates with DMSO and 10 μM ruxolitinib in alternating double columns for each of the three Ba/F3 cell lines.
3. Add 50 μL of cells (2500 cells) to each well and incubate the plates for 48 h.
4. Perform the CellTiter-Glo assay (Subheading 3.2).
5. Plot luminescence signals from cells treated with ruxolitinib and DMSO.
6. We observed that the luminescence signals across the plate are generally uniform without any obvious signal drift or edge effect (Fig. 3b).
7. There is no significant inter-plate or inter-day variation in signals (Fig. 3b).
8. The assay windows for all plates examined are acceptable with Max/Min ratios greater than or equal to 30-fold (Fig. 3c).
9. The Z′ score of all plates has values greater than 0.65 which is indicative of a very robust assay (Fig. 3c).
10. The assay validation steps are critical to identify potential systematic pipetting errors and areas within the CO_2 incubators that have uneven heating or humidity levels.

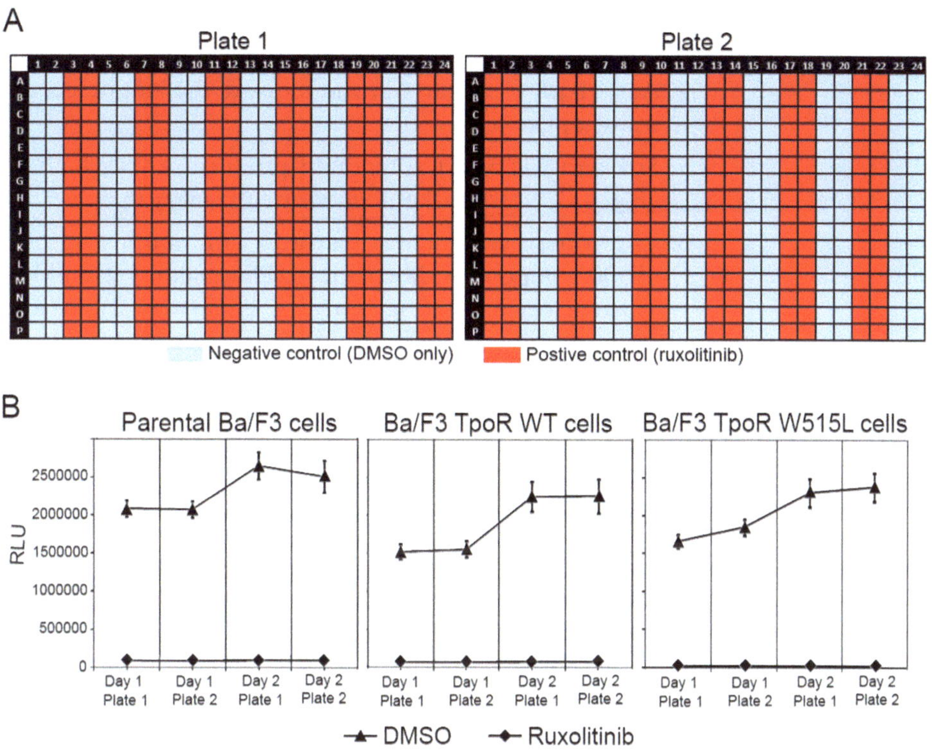

C

		Parental Ba/F3		Ba/F3 TpoR WT		Ba/F3 TpoR W515L	
		Plate 1	Plate 2	Plate 1	Plate 2	Plate 1	Plate 2
Day 1	Max RLU (DMSO) (x1000)	2019 ± 126	2006 ± 131	1518 ± 98	1553 ± 108	1660 ± 84	1848 ± 107
	Min RLU (Ruxolitinib) (x1000)	20 ± 6	20 ± 8	53 ± 7	57 ± 9	16 ± 15	17 ± 15
	Max/Min	104	98	29	27	102	112
	Z'	0.80	0.79	0.78	0.76	0.82	0.80
Day 2	Max RLU (DMSO) (x1000)	2674 ± 209	2514 ± 243	2252 ± 202	2259 ± 229	2300 ± 190	2376 ± 186
	Min RLU (Ruxolitinib) (x1000)	27 ± 22	26 ± 20	69 ± 18	68 ± 13	17 ± 4	16 ± 4
	Max/Min	99	97	33	33	139	146
	Z'	0.74	0.68	0.70	0.67	0.75	0.76

Fig. 3 Assay validation for each of the three Ba/F3 cell lines. (**a**) Cells were seeded at 2500 cells/well and incubated for 48 h. CellTiter-Glo assay was then performed and luminescence signals from cells treated with ruxolitinib and DMSO were plotted. (**b**) No significant inter-plate or inter-day variation was observed. The assay window or Max/Min ratio of all plates examined is acceptable with Max/Min values greater than or equal to 30-fold. *Z'* scores of all plates are within acceptable range with values greater than 0.65 which is indicative of a very robust assay. (**c**) Maximum (max) and minimum (min) RLU values are shown in mean ± SD

3.4 Robustness Testing with Small Library Sets (Pilot Screen)

1. Conduct the robustness test using the Lopac1280 library (Sigma Aldrich) at 2 and 5 μM.
2. To make 2 μM assay plate, add 10 nL of each compound (10 mM stock) to the well using an automated liquid-handling system. To make 5 μM assay plates, add 25 nL of each compound (10 mM stock) to the well using an automated liquid-handling system.

3. Add cells (2500/50 μL) to the plate and incubate at 37 °C in a CO_2 incubator for 48 h.
4. Add 10 μL CellTiter-Glo and measure the luminescence signal using a plate reader.
5. Screening at 5 μM compound produces a very high hit rate (Fig. 4a), which makes follow-up unmanageable.
6. The robustness test was repeated with another small library (1600 compounds, Microsource Pharmakon) at 2 μM on two separate occasions (Fig. 4b, day 1 and day 2). A screening concentration of 2 μM produces a reasonable hit rate and the linear correlation from the two separate tests indicates that the results are reproducible and robust (Fig. 4b).
7. Hence, 2 μM was chosen for the high-throughput phenotypic screening with large compound collections [7].
8. We recommend performing this pilot screen to determine the optimal compound screening concentration and to evaluate the assay reproducibility for your compound collection and particular laboratory conditions.

3.5 Compound Screening Using the Cell Viability Assay

1. Grow cells to exponential growth phase. *See* **Note 2** for cell growth and seeding optimization.
2. Count and resuspend the cells to 50,000 cells/mL.
3. Dispense 10 nL of 10 mM compound stocks to wells A3-P22, 10 nL DMSO to columns 1–2 (negative controls), and 10 nL ruxolitinib (50 mM stock) to columns 23–24 (positive controls; *see* Fig. 5a).
4. Add 50 μL cell suspension (2500 cells) to all wells.
5. The final assay concentration of the compounds is 2 μM in 0.02% DMSO. The final concentration of ruxolitinib is 10 μM.
6. Final DMSO concentration was determined using a DMSO tolerance study (*see* Fig. 2c and **Note 5**).
7. Centrifuge plates at 40 × *g* for 30 s.
8. Incubate at 37 °C with 5% CO_2 in humidified incubator for 48 h.
9. Perform the CellTiter-Glo assay (Subheading 3.2).
10. Compounds which inhibit the growth of Ba/F3 TpoR W515 L cells at two standard deviations (SD) above the median of the overall compound library activity (about 50% cell growth inhibition) are selected as primary hits.

3.6 Dose–Response Study Using the Cell Viability Assay

1. Perform serial dilution of compounds in a 384-well plate (intermediate plate) in duplicate wells using a liquid handler: Add 15 μL of test compounds (10 mM stock) to columns 3–4. Add 10 μL of DMSO to columns 5–22. This is the intermediate plate.

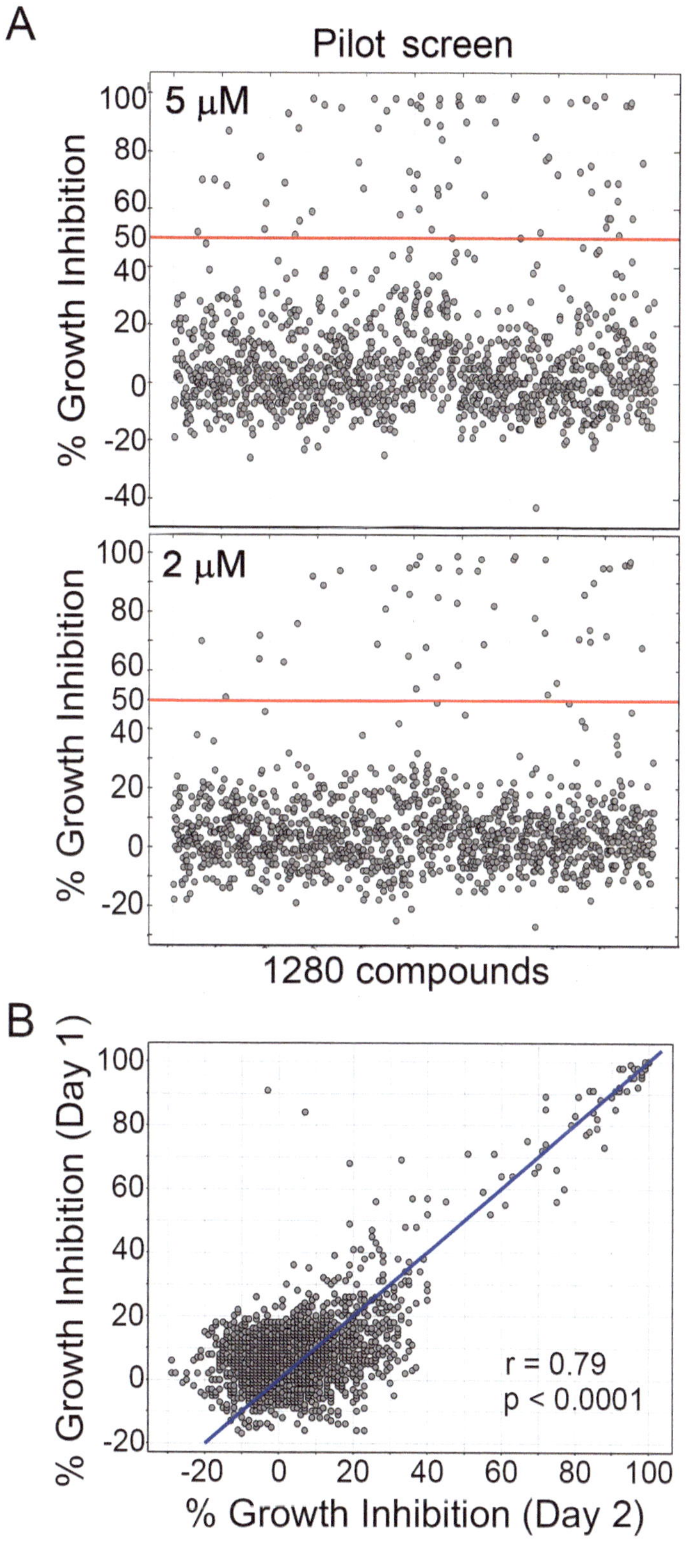

Fig. 4 Assay robustness testing. (**a**) Using a cutoff at 50% growth inhibition (about 2 SD) or more, screening at 2 μM produced 50 hits (3.9% hit rate), while screening at 5 μM produced 83 hits (6.5% hit rate). (**b**) The Pharmakon library (1600 compounds) was subsequently screened at 2 μM on two separate occasions (represented as day 1 and day 2). A linear correlation is observed between the two sets of results (Pearson $r = 0.79$, $p < 0.0001$), indicating that the screening assay is robust and reproducible

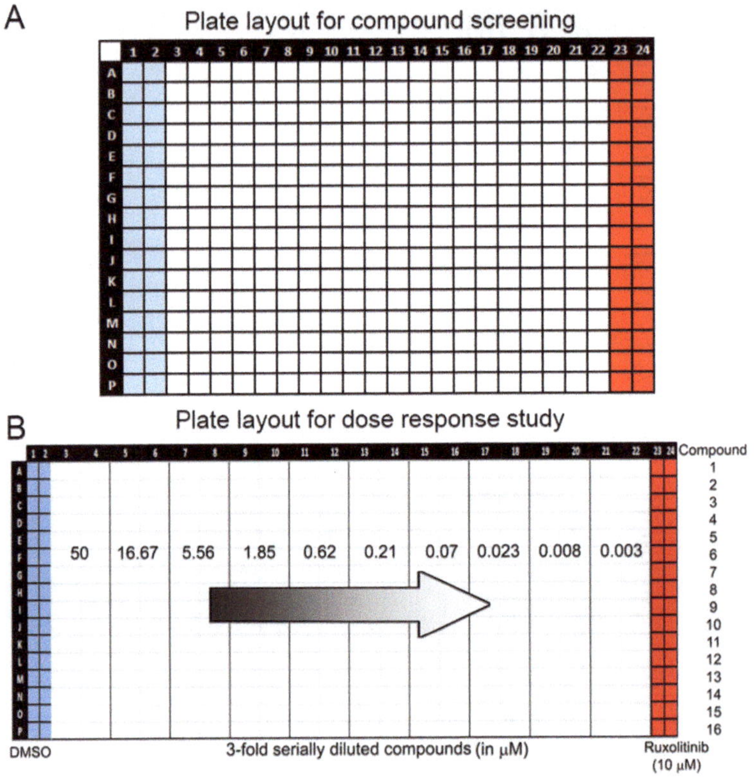

Fig. 5 Cell viability assay layout for 384-well plates. (**a**) High-throughput screening format. Negative control (10 nL DMSO) is dispensed into columns 1 and 2 (light grey). Positive control (10 nL of 50 mM ruxolitinib) is dispensed into columns 23 and 24 (dark grey). 10 nL test compounds (10 mM stock) are dispensed into columns 3–22 (white). After adding 50 μL (2500) cells, the final assay concentration for the compounds is 2 μM with 0.02% DMSO. (**b**) Dose–response format. Compounds are threefold serially diluted in duplicate wells across the plate. Negative control (DMSO) is dispensed into columns 1 and 2 (blue). Positive control (10 μM ruxolitinib) is dispensed into columns 23 and 24 (red)

2. Perform 10-point threefold serial dilutions in duplicate by transferring 5 μL test compound from columns 3 and 4 to columns 5 and 6. Mix well by pipetting up and down.
3. Repeat **step 2** for the remaining 8 concentrations from columns 7–22.
4. Add 10 μL of DMSO (negative control) to columns 1 and 2.
5. Add 10 μL of the 2 mM ruxolitinib (positive control) to columns 23–24. This completes the intermediate plate.
6. Final assay plate: Transfer 0.25 μL of the compound dilution series and controls from the intermediate plate to a new 384-well plate using an automated liquid handler (Fig. 5b).
7. Grow cells to exponential growth phase (*see* **Note 2**).
8. Count and resuspend cells to 50,000 cells/mL.

9. Add 50 μL cell suspension (2500 cells) to all wells in the final assay plate.
10. Final assay plate concentration for each test compounds in the serial dilution series will range between 0.003 and 50 μM in duplicate wells with 0.5% DMSO.
11. The final concentration of ruxolitinib (positive control) is 10 μM in the final assay plate.
12. Centrifuge plates at 40 × g for 30 s.
13. Incubate at 37 °C with 5% CO_2 in humidified incubator for 48 h.
14. Remove all plates from the CO_2 incubator. Allow the plates to equilibrate to room temperature in stacks of five plates on the lab bench (about 1 h).
15. Add 10 μL of CellTiter-Glo Reagent to all wells.
16. Centrifuge plates at 40 × g for 1 min.
17. Allow the plate to incubate at room temperature for at least 30 min to obtain optimal luminescence signal.
18. Measure luminescence signal using a Tecan M1000 plate reader with signal integration time of 100 ms/well.
19. The luminescence signal is reported as relative luminescence units (RLU).

3.7 Selectivity Assay for TpoR W515L-Specific Inhibitors

1. Compounds selected in the primary screen are tested against the Ba/F3 TpoR WT cells (stimulated for cell growth with IL-3) at the same concentration (2 μM) as in the primary screen.
2. Compounds exhibiting comparable growth inhibitory activity on Ba/F3 TpoR W515L cells and Ba/F3 TpoR WT cells are pan-cytotoxic compounds and/or pan-inhibitors of the TpoR signaling pathway.
3. Compounds that do not inhibit growth of Ba/F3 TpoR WT cells but inhibit growth of Ba/F3 TpoR W515L cells are TpoR W515L-specfic inhibitors.
4. Perform a dose–response study to more accurately identify compounds which are active (i.e., $IC_{50} \leq 2$ μM) and showing twofold or more toxicity toward the Ba/F3 TpoR W515L cells compared to the Ba/F3 TpoR WT cells. These are considered the specific inhibitors of TpoR W515L.
5. See Fig. 1 for a summary guide to hit selection.

3.8 Selectivity Assay for Pan-TpoR Inhibitors

1. Grow Ba/F3 TpoR W515 L cells and parental Ba/F3 cells with IL-3 to activate the IL-3 signaling pathway.
2. Compounds selected from the primary screen (see Subheading 3.5) above which also inhibit growth of Ba/F3 TpoR W515 L

cells and parental Ba/F3 cells (both stimulated with IL-3) are pan-cytotoxic compounds.

3. Compounds selected from the primary screen (see Subheading 3.5) above which do not inhibit growth of Ba/F3 TpoR W515L cells and parental Ba/F3 cells (both stimulated with IL-3) are pan-TpoR inhibitors.
4. These pan-TpoR inhibitors are selected for dose–response study to determine their IC_{50} values in both Ba/F3 TpoR WT and Ba/F3 TpoR W515L cells.
5. These compounds are specific inhibitors for active TpoR signaling if they have similar IC_{50} values in both Ba/F3 TpoR W515L and Ba/F3 TpoR WT cells in the dose–response study.
6. See Fig. 1 for a summary guide to hit selection.

3.9 Calculations for Measuring Assay Performance and Compound Activity

The quality and robustness of the single-point high-throughput assay can be assessed by the following three parameters [10]. Negative control in the formulae below refers to treatment with DMSO only and positive control refers to treatment with ruxolitinib.

Signal/background ratio

$$\text{Max/Min ratio} = \frac{(\text{RLU of negative control})}{(\text{RLU of positive control})} \quad (1)$$

Assay robustness

$$Z' = 1 - \left[\frac{3\text{SD}(\text{negative control}) + 3\text{SD}\ (\text{positive control})}{|\text{RLU}(\text{negative control}) - \text{RLU}(\text{positive control})|}\right] \quad (2)$$

$Z' > 0.5$ is indicative of a robust assay

$$\text{Percent cell growth inhibition} = 100 - 100\left[\frac{\text{RLU}(\text{test compound}) - \text{Average RLU}\ (\text{positive control})}{\text{Average RLU}(\text{negative control}) - \text{Average RLU}\ (\text{positive control})}\right] \quad (3)$$

4 Notes

1. The growth of parental Ba/F3 and Ba/F3 TpoR WT cells is IL-3 dependent. Ba/F3 TpoR WT cells can also grow in the presence of TPO. Growth of Ba/F3 TpoR W515L cells is not dependent on IL-3 or TPO.
2. Ba/F3 cell number should not exceed 1.2 million/mL. Once the growth rate accelerates, they should be diluted back to 150,000 cells/mL and cultured in growth medium to achieve a density of 1.2 million cells/mL after 48 h. Based on the

dose–response curves of IL-3 and TPO, the 80% growth-enhancing concentration (EC_{80}) of IL-2 was 2 ng/mL, and the EC_{80} for TPO in stimulating Ba/F3 TpoR WT cells was 40 ng/mL (Fig. 2a). In order to enhance the responsiveness of the cells to compound treatment, it is critical that the cytokine EC_{80} be used instead of the maximum growth-enhancing (Emax) concentration.

3. Comparable ruxolitinib potency and assay performance are observed (i.e., signal/background ratio and IC_{50}) at 1250, 2500, and 5000 cells/well in all three cell lines (Fig. 2b). Thus, 2500 cells/well is chosen for the screening assay. A plate containing a ruxolitinib serial dilution profile should be incorporated into every assay run as an internal quality control check to ensure that the assay performs within an acceptable range of $Z' > 0.5$ and signal/background ratio >10. Plates with any lower values are rejected and re-run.
4. Parental BaF3, Ba/F3 TpoR WT, and Ba/F3 TpoR W515 L cells have similar tolerance for DMSO. These cells can tolerate up to 0.5% DMSO with less than 25% loss in cell viability (Fig. 2c).
5. The luminescence signal from the CellTiter-Glo Assay is stable up to 4 h at room temperature in our experimental setup of culturing 2500 cells/well in 50 μL RPMI assay medium for 48 h in a 384-well plate (Fig. 2d). Based on the results from Fig. 2d, we recommend using 10 μL CellTiter-Glo reagent for the high-throughput screening assay, and that plates be incubated with CellTiter-Glo for at least 30 min prior to reading the luminescence signals.

Acknowledgments

There is no financial conflict of interest among the authors. Support for A.N., A.K., and M.L.C was from an A*STAR core funding to the Experimental Therapeutics Centre. Support for C.C.D. was from UEFISCDI PCCA PN II 133/2012 and OPERATIONAL COMPETITIVITATY PROGRAM 2014-2020 POC-A1-A1.1.4-E-2015/P_37_798/149/2016. S.N.C. and A.K.S. received funding from the Ludwig Institute for Cancer Research. S.N.C. also received supports from FRS-FNRS, Salus Sanguinis Foundation, the Action de Recherche Concertée project ARC10/15-027 of the Université catholique de Louvain, the Fondation contre le Cancer, the PAI Programs BCHM61B5, and Belgian Medical Genetics Initiative.

References

1. Levine RL, Pardanani A, Tefferi A, Gilliland DG (2007) Role of JAK2 in the pathogenesis and therapy of myeloproliferative disorders. Nat Rev Cancer 7:673–683
2. Pardanani AD, Levine RL, Lasho T, Pikman Y, Mesa RA, Wadleigh M et al (2006) MPL515 mutations in myeloproliferative and other myeloid disorders: a study of 1182 patients. Blood 108:3472–3476
3. Pecquet C, Staerk J, Chaligne R, Goss V, Lee KA, Zhang X et al (2010) Induction of myeloproliferative disorder and myelofibrosis by thrombopoietin receptor W515 mutants is mediated by cytosolic tyrosine 112 of the receptor. Blood 115:1037–1048
4. Klampfl T, Gisslinger H, Harutyunyan AS, Nivarthi H, Rumi E, Milosevic JD et al (2013) Somatic mutations of calreticulin in myeloproliferative neoplasms. N Engl J Med 369:2379–2390
5. Nangalia J, Massie CE, Baxter EJ, Nice FL, Gundem G, Wedge DC et al (2013) Somatic CALR mutations in myeloproliferative neoplasms with nonmutated JAK2. N Engl J Med 369:2391–2405
6. Chachoua I, Pecquet C, El-Khoury M, Nivarthi H, Albu RI, Marty C et al (2016) Thrombopoietin receptor activation by myeloproliferative neoplasm associated calreticulin mutants. Blood 127:1325–1335
7. Hobbs GS, Rampal RK (2015) JAK2 mutations and JAK inhibitors in the management of myeloproliferative neoplasms. Contemp Oncol 7:22–28
8. Ngo A, Koay A, Pecquet C, Diaconu CC, Ould-Amer Y, Huang Q et al (2016) A phenotypic screen for small-molecule inhibitors of constitutively active mutant Thrombopoietin receptor implicated in myeloproliferative neoplasms. Comb Chem High Throughput Screen 19:824–833
9. Defour JP, Itaya M, Gryshkova V, Brett IC, Pecquet C, Sato T et al (2013) Tryptophan at the transmembrane-cytosolic junction modulates thrombopoietin receptor dimerization and activation. Proc Natl Acad Sci U S A 110:2540–2545
10. Zhang JH, Chung TD, Oldenburg KR (1999) A simple statistical parameter for use in evaluation and validation of high throughput screening assays. J Biomol Screen 4:67–73

Chapter 5

A Cell-Based Assay for Mitotic Spindle Orientation

Elina Glaubke and Holger Bastians

Abstract

The regulation of mitotic spindle orientation is essential to ensure proper cell division and development (Kiyomitsua and Cheeseman Nat Cell Biol 14:311–317, 2012). For identification of potential spindle orientation regulators, determination of the mitotic spindle angle is a well-known but time-consuming procedure. Here we describe a simple and time-saving phenotypic screening assay for the identification of potential spindle orientation regulators. This screen is based on the analysis of monopolar mitotic spindle structures, which form upon inhibition of the mitotic kinesin Eg5/KSP by the small-molecule inhibitor dimethylenastron (DME) or similar compounds.

Key words Dimethylenastron, Monopolar spindle structure, Spindle orientation, Mitosis, Immunostaining

1 Introduction

Errors in mitotic spindle orientation are associated with altering tissue morphology, tumor development, and chromosomal instability [2–4]. Thus, determining genetic alterations that cause defects in spindle orientation contributes to a better understanding in how this process is regulated. Determination of the mitotic spindle angle is the most common approach in order to identify spindle orientation regulators (Fig. 1). This method requires immunostaining of α-tubulin for visualization of mitotic spindles and γ-tubulin as centrosome marker as well as DNA staining by DAPI [5]. Via fluorescence microscopy samples have to be searched for (pro)metaphase cells from which z-stack images are taken. These images are used to measure the centrosome-to-centrosome distance and the z-stack distance between the centrosomes both required for calculating the spindle angle (Fig. 1). However, analyzing the spindle orientation in individual cells by this method is a time-consuming procedure that makes it unsuitable for large-scale screening of candidates which might be involved in the regulation of spindle orientation.

Bridget Wagner (ed.), *Phenotypic Screening: Methods and Protocols*, Methods in Molecular Biology, vol. 1787, https://doi.org/10.1007/978-1-4939-7847-2_5,

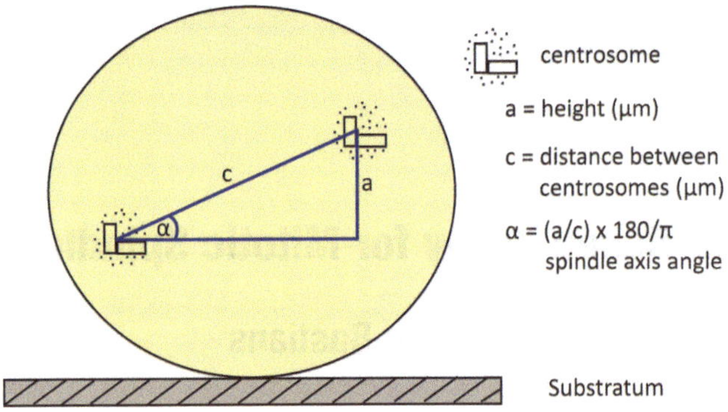

Fig. 1 Determination of the spindle angle. The scheme illustrates how the spindle axis angle of a mitotic cell is measured. The given formula is used for exact calculation of the angle

Searching for alternate ways, we found that there is an intriguing correlation between cells harboring defects in spindle orientation and formation of asymmetric monopolar mitotic spindles upon inhibition of the mitotic kinesin Eg5 [5]. In general, formation of monopolar mitotic spindles can be induced by inhibition of the mitotic kinesin Eg5 (also known as KIFl1, kinesin-5 or KSP), which is involved in centrosome separation starting at the prophase of mitosis [6]. Inhibition of Eg5 can be achieved by treating the cells with small-molecule inhibitors, such as monastrol [7] or dimethylenastron (DME) [8]. Highly symmetric monopolar mitotic spindles that form upon Eg5 inhibition are still able to attach to mitotic chromosomes leading to a rosette-like alignment of chromosomes around the monoaster [6, 7]. In contrast, our studies revealed that chromosomally unstable cancer cells that are characterized by abnormally increased microtubule plus end growth rates and spindle misorientation often show asymmetric monopolar spindles upon Eg5 inhibition [4, 5] (Fig. 2). Studies on candidates for spindle orientation regulation in colorectal cancer cell lines also demonstrated that most candidates that induce spindle misorientation (as analyzed by directly determining the spindle angle) also showed an asymmetric monopolar spindle formation upon DME treatment [5]. This strong correlation between the spindle orientation status and the monopolar spindle structure in mitotic cells opens up the implementation of a simple and time-saving screening assay for potential spindle orientation regulators, which we will describe here.

A further advantage of this screening assay is the ease of automation. Due to the clear phenotypic differences between symmetric and asymmetric spindles and the underlying displacement of the two centrosomes from the center of the cell it is feasible to develop a software that automatically distinguishes between these two spindle structures and/or centrosome positioning relative to the center

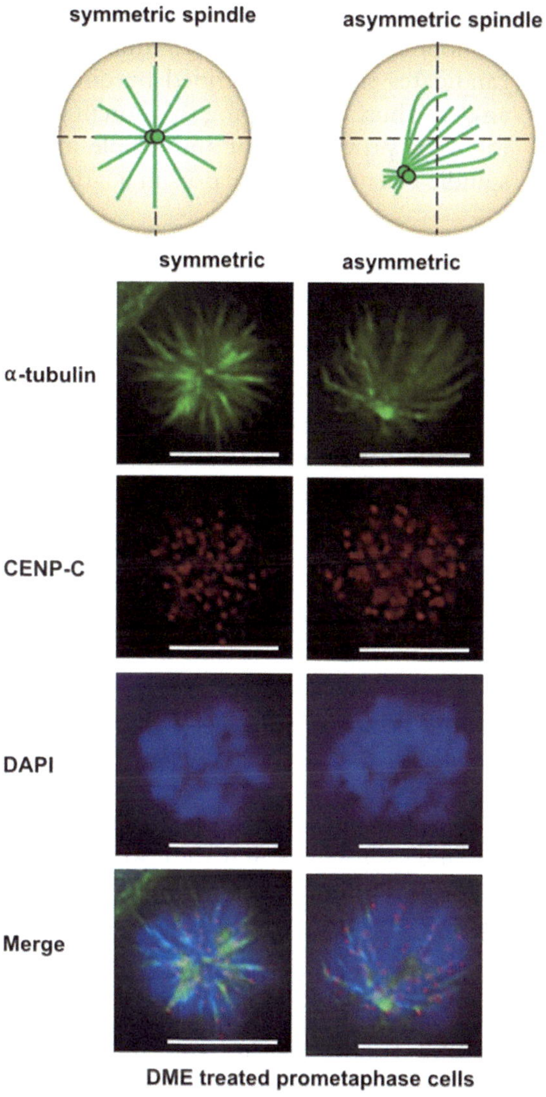

Fig. 2 Scheme and example images of symmetric and asymmetric monopolar mitotic spindles. Treatment of cells with the Eg5 inhibitor DME leads to the formation of symmetric and asymmetric monopolar spindles. Chromosomally unstable cancer cells exhibiting a defect in the regulation of spindle orientation show higher rates of asymmetric spindles than chromosomally stable cancer cells (spindles, anti-α-tubulin: green; kinetochores, anti-CENP-C: red; chromosomes, DAPI: blue; scale bar, 10 μm)

of the cell. If applied, our assay can then be employed for large-scale and even genome-wide screenings to identify spindle orientation regulators in a convenient and simple manner.

However, there are also limitations to this novel screening assay. Since the mechanism that causes the observed differences in monopolar spindle formation are still unclear the screening assay might identify false-positive regulators of spindle orientation. In fact, a few candidates have been identified that show spindle misorientation using classical assays without displaying asymmetric monopolar spindles after Eg5 inhibition [5]. Therefore, identified candidates should be verified by direct measurements of spindle angles. Thus, with keeping this potential pitfall in mind this screening assay provides a fast and simple method to identify potential spindle orientation regulators in a large scale.

2 Materials

2.1 Cell Culture

1. Cell lines (e.g., HCT116, SW48, SW480, SW620).
2. Appropriate cell medium (RPMI 1640 medium complemented with 10% fetal calf serum (FCS), 1% penicillin/streptomycin for HCT116, SW48, SW480, SW620).
3. Trypsin/EDTA.
4. Cover glasses: 10 × 10 mm.
5. 24-Well plates.
6. 10 mM Dimethylenastron (DME) in DMSO.

2.2 Fixation and Immunostaining

1. 2% Paraformaldehyde (PFA) in PBS (*see* **Note 1**): Heat 500 mL of PBS to 70 °C in a water bath and add 10 g of PFA (*see* **Note 2**). Dissolve PFA in PBS by using a magnetic stirrer. Keep the temperature at 70 °C with the water bath. After dissolving PFA let the solution cool down in the water bath. Determine the pH and adjust to 7.4 with 1 M NaOH if necessary. Filter PFA solution through a 0.45 μm filter (*see* **Note 3**).
2. 100% Methanol at −20 °C.
3. 5% FCS in PBS.
4. Anti-α-tubulin: Mouse monoclonal α-tubulin antibody (clone B-5-1-2, 200 μg/mL); 1:700 dilution in 2% FCS in PBS.
5. Anti-CENP-C: Guinea pig polyclonal CENP-C antibody; 1:1000 dilution in 2 % FCS in PBS.
6. Goat anti-mouse IgG H&L, Alexa Fluor 488-conjugated: Goat polyclonal secondary antibody (2 mg/mL); 1:1000 dilution in 2% FCS in PBS.
7. Goat anti-guinea-pig IgG H&L, Alexa Fluor 594-conjugated: Goat polyclonal secondary antibody (2 mg/mL); 1:1000 dilution in 2% FCS in PBS.
8. 20 mg/mL Hoechst33342 in PBS.
9. Mounting medium for fluorescence microscopy.
10. Translucent nail polish.

2.3 Immunofluorescence Microscopy

1. Microscope suitable for immunofluorescence microscopy (e.g., Leica DM6000B microscope).
2. Immersion oil.

3 Methods

3.1 Seeding Cells on Cover Glasses

1. Place uncoated and sterile 10 × 10 mm cover glasses in 24-well plates.
2. Wash cover glasses with 500 μL of PBS (*see* **Note 4**). Afterwards add 500 μL of medium to each well.
3. After detaching cells by using trypsin/EDTA seed an appropriate cell number (*see* **Note 5**) into prepared wells and incubate at 37 °C, 5% CO_2, overnight. Cells should be 50–60% confluent the next day.
4. Next day, remove medium and wash cells with 500 μL of PBS. Add 500 μL of medium containing 2 μM DME (*see* **Note 6**). In order to screen for spindle orientation regulators, specific inhibitors or activators for potential targets can also be added during this step. For example, treatment with low doses of taxol (final concentration 2 nM) increases the amount of symmetric monopolar spindles in chromosomally unstable cancer cells by restoring normal microtubule plus end assembly rates.
5. Always use a sample treated only with 2 μM DME as control. Incubate the cells for 3.5–4 h at 37 °C, 5% CO_2 (*see* **Note 7**).

3.2 Cell Fixation and Immunostaining

1. Remove medium and add 500 μL of 2% PFA in PBS to the wells (*see* **Note 8**). Incubate for 5 min at room temperature.
2. Remove PFA/PBS (*see* **Note 9**) and add 500 μL of 100% methanol (−20 °C) to each well and incubate for 5 min at −20 °C.
3. Remove methanol and wash once with PBS. The cover glasses can be stored in PBS at 4 °C until staining.
4. For blocking add 500 μL of freshly prepared 5% FCS in PBS and incubate for 30 min at room temperature. Remove the solution and wash once with PBS.
5. Centrifuge the anti-α-tubulin and ant-CENP-C antibody solutions (*see* **Note 10**) for 3 min at 4 °C and 21,100 × *g* (*see* **Note 11**).
6. Prepare a moist chamber: Place a suitable piece of Whatman paper in a lightproof box and add water until the paper is completely wet. Then place a suitable piece of parafilm onto the Whatman paper. Pipet 25 μL of the primary antibody solution per slide onto the parafilm. Then place the cover glasses with the cells facing the drop onto the drop. Tweezers

and a needle will help to get the cover glasses out of the well. Close the chamber and incubate for 1.5 h at room temperature.

7. Prepare a 24-well plate with 500 μL of PBS per well and put the cover glasses with the cells head up into the wells. Wash four times with 500 μL of PBS.
8. Pipet 25 μL of the secondary antibody solution (*see* **Note 12**) per cover glass onto the parafilm in the moist chamber. Then place the slides with the cells facing the drop onto the drop. Close the chamber and incubate for 1.5 h at room temperature. Both incubation steps with the primary and the secondary antibody can be done overnight at 4 °C as well.
9. Prepare a 1:20,000 dilution of a 20 mg/mL Hoechst 33342 (or DAPI) solution in PBS. Add 500 μL of this prepared solution into wells of a 24-well plate and put the cover glasses with the cells head up into the wells and incubate for 5 min at room temperature. Afterwards wash the cover glasses four times with PBS and once with water. Place cover glasses with cells head up onto dry Whatman paper and let them air-dry completely.
10. Pipet 1–2 μL of mounting medium onto microscope glass slides and place the cover glasses with the cells facing the drop onto the mounting medium.
11. Use a soft tissue and the cover of the box to squeeze out excess mounting medium.
12. Seal the cover glass with translucent nail polish (*see* **Note 13**).
13. Store the slides at 4 °C (*see* **Note 14**).

3.3 Immunofluorescence Microscopy

1. Apply immersion oil onto the cover glass or the objective depending on the microscope type (upright or inverted microscope).
2. Search for monopolar mitotic spindles by using the green channel (in case the secondary antibody for the anti-α-tubulin antibody is Alexa Fluor 488 conjugated as described above). Example images and a scheme of symmetric and asymmetric monopolar mitotic spindles are shown in Fig. 2 and help to differentiate correctly between symmetric and asymmetric spindles.
3. Hoechst 33342- or DAPI-stained mitotic DNA can be found in the blue channel and visualizes the chromosomes which are aligned to the microtubule plus ends of the mitotic spindle. A closer look at the symmetry of the rosette-like chromosome alignment will further help to differentiate between symmetric and asymmetric spindles (Fig. 2). Additionally, the red channel shows the kinetochore distribution (in case the secondary antibody for the anti-CENP-C antibody is Alexa Fluor

594 conjugated as described above), which also helps to discriminate between symmetric and asymmetric chromosome alignment (*see* **Note 15**).

4. Determine the symmetry state of monopolar mitotic spindles of at least 500 cells per experiment. The experiment should be repeated at least three times.
5. Typically, 50–60% of chromosomally unstable colorectal cancer cell lines (e.g., SW480, SW620) exhibit highly asymmetric monopolar spindles indicative for increased microtubule plus end assembly rates and spindle misorientation. Conversely, only 5–10% of chromosomally stable colorectal cancer cell lines (e.g., HCT116, SW48) show asymmetric monopolar spindles [5].

4 Notes

1. It is important that the 2% PFA solution is adjusted to pH 7.4. Thawing of the PFA solution should be done in the dark and thawing should not be accelerated by heating. PFA degrades when exposed to light or long-term heating. PFA solution that has been thawed can be stored at 4 °C in the dark for 2–3 days.
2. PFA is a carcinogen due to its formaldehyde-releasing potential. Wear a mask when weighing PFA. When transporting the weighed PFA to the fume hood, avoid exposing it to co-workers. Therefore, use another weigh boat as cover for the weigh boat containing the weighed PFA. Transfer the PFA into a 2 L beaker inside the fume hood and then add the PBS heated up to 70 °C with the water bath.
3. 2% PFA solution in PBS can be stored at −20 °C for several months. Make aliquots that fit to the volume that is usually used within 2–3 days.
4. When washing 24-well plates containing cover glasses with PBS make sure that the cover glasses stick to the bottom and do not swim at the surface of the added PBS. Make also sure that there are no bubbles between the cover glasses and the bottom of the well. In case there are bubbles use sterile tweezers or the tip of a sterile needle and push the cover glass closer to the bottom to remove the bubbles.
5. Depending on the cell line used for the experiment and the generation time the cell number can vary. Knowing the usual split ratio for the used cell lines helps when calculating the appropriate cell number that will result in 50–60% confluence the next day.
6. Dimethylenastron (DME) is a small-molecule inhibitor that inhibits Eg5, thereby preventing the establishment of a bipolar

mitotic spindle. This results in a prolonged mitotic arrest of DME-treated cells. Alternatively, other commercially available Eg5 kinesin inhibitors can be used. The concentration to be used has to be determined according to the induction of monopolar spindles during mitosis.

7. Incubation time with DME varies among cell lines. 3.5–4 h is suitable for most cell lines. Incubation with DME to up to 6 h is only recommended for slow-growing cell lines to allow more cells to enter mitosis before fixation. Prolonged treatment with DME results in the formation of asymmetric monopolar spindles even in chromosomally stable cell lines.
8. Fixation with PFA should be done at room temperature; otherwise microtubules will break.
9. During fixation and immunostaining the cover glasses have to be wet at all times. Otherwise antibody binding will be less effective and thus immunostaining will result in low-quality images.
10. When using an automated software that distinguishes between symmetric and asymmetric spindles by determining the centrosome positioning relative to the center of the cell, immunostaining of γ-tubulin instead of CENP-C is required.
11. Centrifuging the primary antibody solution helps to avoid aggregates, which otherwise would become visible as highly fluorescent particles in the sample during microscopy.
12. Depending on the primary antibodies used and the researchers' preference the secondary antibody has be chosen thoughtful: It has to fit to the primary antibody host and the emission spectra of the conjugated fluorescent dyes should not overlap.
13. Some nail polishes show autofluorescence. So, be aware of this when choosing the nail polish.
14. To obtain high-quality images samples should be examined under a fluorescence microscope within a few days. The later the examination the lower the fluorescence signal intensity.
15. The use of different markers (microtubules, kinetochores, chromosomes) ensures correct determination of the spindle structure.

Acknowledgments

We are grateful to the Deutsche Forschungsgemeinschaft and the German Cancer Aid for financial support.

References

1. Kiyomitsu T, Cheeseman IM (2012) Chromosome- and spindle-pole-derived signals generate an intrinsic code for spindle position and orientation. Nat Cell Biol 14:311–317
2. Pease JC, Tirnauer JS (2011) Mitotic spindle misorientation in cancer – out of alignment and into the fire. J Cell Sci 124:1007–1016
3. Bergstralh DT, St Johnston D (2014) Spindle orientation: what if it goes wrong? Semin Cell Dev Biol 34:140–145
4. Ertych N, Stolz A, Stenzinger A, Weichert W, Kaulfuß S, Burfeind P et al (2014) Increased microtubule assembly rates mediate chromosomal instability in colorectal cancer cells. Nat Cell Biol 16:779–791
5. Stolz A, Ertych N, Bastians H (2015) A phenotypic screen identifies microtubule plus end assembly regulators that can function in mitotic spindle orientation. Cell Cycle 14:827–837
6. Kapoor TM, Mayer TU, Coughlin ML, Mitchison TJ (2000) Probing spindle assembly mechanisms with monastrol, a small molecule inhibitor of the mitotic kinesin, Eg5. J Cell Biol 150:975–988
7. Mayer TU, Kapoor TM, Haggarty SJ, King RW, Schreiber SL, Mitchison TJ (1999) Small molecule inhibitor of mitotic spindle bipolarity identified in phenotype-based screen. Science 286:971–974
8. Müller C, Gross D, Sarli V, Gartner M, Giannis A, Bernhardt G et al (2007) Inhibitors of kinesin Eg5: antiproliferative activity of monastrol analogues against human glioblastoma cells. Cancer Chemother Pharmacol 59:157–164

Chapter 6

Isolation of Skeletal Muscle Stem Cells for Phenotypic Screens for Modulators of Proliferation

Aaron C. Hinken and Andrew N. Billin

Abstract

Adult skeletal muscle contains a population of resident stem cells known as muscle stem cells (MuSC) or satellite cells. This population of cells is required for regeneration of functional myofibers after damage. Aging reduces the proliferative response of satellite cells post-injury. This deficient response is thought to contribute to slowed recovery of muscle function after damage in the elderly and may also contribute to age-related loss of muscle function (sarcopenia). Numerous techniques are now available for the isolation of highly purified satellite cells from mice and humans (Sherwood, et al. Cell 119:543–554, 2004; Cerletti, et al. Cell 134:37–47; 2008; Conboy, et al. Methods Mol Biol 621:165–173, 2010; Bareja, et al. PLoS One 9: e90398; 2014; Castiglioni et al. Stem Cell Rep 2:92–106, 2014; Charville, et al. Stem Cell Rep 5:621–632, 2015; Liu et al. Nat Protoc 10:1612–1624, 2015; Sincennes et al. Methods Mol Biol 1556:41–50, 2017), thus opening an opportunity to use satellite cells in phenotypic screens for regulators of satellite cell proliferation and differentiation. In this chapter, we describe a technique for the prospective isolation of mouse satellite cells that we have recently used in a phenotypic screen of a focused set of small molecules.

Key words Skeletal muscle, Satellite cells, Cell sorting, Screen, Proliferation, Differentiation, Stem cells, Aging

1 Introduction

The skeletal muscle satellite cell was identified in 1961 using electron microscopy of striated frog muscle [9]. It was described as a cell with condensed chromatin and little cytoplasm that is sandwiched between the plasma membrane (sarcolemma) of the striated muscle cell and the basement membrane that surrounds the muscle cell. It was suggested that these cells may play a critical role in the process of muscle regeneration and this has indeed proven to be the case. The study of satellite cells was advanced greatly with the advent of methods to culture individual myofibers with attached satellite cells and the discovery of molecular markers of quiescent and mitotically activated satellite cells. The discovery of molecular markers in turn allowed for the development of numerous methods

Bridget Wagner (ed.), *Phenotypic Screening: Methods and Protocols*, Methods in Molecular Biology, vol. 1787,
https://doi.org/10.1007/978-1-4939-7847-2_6,

of prospectively isolating satellite cells from mouse and human skeletal muscle. The original definition of the satellite cell is an anatomic one, referring to the location of the cell in a particular compartment of the skeletal muscle fiber. Thus, myogenic cells isolated from the satellite cell compartment and cultured in vitro are referred to by a variety of names to distinguish them from the cell in its native environment and we will use the term MuSC throughout this chapter when referring to cultures of myogenic stem cells derived from isolated satellite cells.

Skeletal muscle comprises a large fraction of the lean mass of an adult human and it is of obvious importance for carrying out the most basic aspects of locomotion and movement. Additionally, skeletal muscle is an important determinant of whole-organism energy expenditure and is a major site for the disposal of consumed calories in the form of glucose. The homeostatic maintenance of muscle mass can be negatively impacted by numerous diseases and conditions, such as age-related sarcopenia, chronic obstructive pulmonary disease (COPD)-related cachexia, cancer-related cachexia, and HIV-related morbidity.

The role of MuSC dysfunction in the pathobiology of muscle regeneration is of upmost importance in inherited muscular dystrophies and in age-related loss of muscle mass, function, and efficient repair. Thus, therapies that target the MuSC in order to promote productive responses to regenerative signals may be effective in slowing, stopping, or reversing deficient repair or loss of functional muscle.

Given the advances in the isolation and culture of MuSCs, it is now conceivable to perform phenotypic screens in order to identify genes, small molecules, or biologic drug leads that potentiate MuSC responses and thus may open the door to discovering new medicines to treat muscle diseases. We recently reported the results of a small molecule-focused set screen of MuSCs isolated from aged (~24 months old) mice [10]. The goal of the screen was to identify molecules that restored the deficient proliferative response of aged MuSCs to a level similar to that of younger MuSCs. We were able to identify a series of small molecules of interest and establish structure–activity relationship (SAR) using the MuSC isolation and culture methods described below.

MuSCs are a rare cell population and they decrease in numbers as mice and human age. Thus in order to perform a screen and follow-up testing of hits, and to support SAR testing, we developed isolation and culture methods that reduced the time required to isolate the cells and minimized the number of cells per well needed to test compounds. Early reports on techniques for FACS isolation of satellite cells employed time-consuming steps to first obtain single muscle fibers by collagenase digestion and trituration followed by repeated washing and sedimentation of the fibers by gravity. Once the washing procedure is completed the fibers are then digested a second time with collagenase and dispase to release single cells. The released cells were then stained with various fluorescently linked antibodies and FACS was performed to isolate the cells. Typically, two rounds of FACS purification were required to obtain a purified population of MuSCs. Instead of isolating muscle fibers by collagenase digestion and manual mechanical trituration prior to digesting the fibers to release the satellite cells, we developed a semiautomated method to disrupt and digest the muscle tissue to a single-cell suspension in one step. We then utilized magnetic bead depletion of unwanted cells to enrich for the satellite cells prior to isolation of viable satellite cells via fluorescent-activated cell sorting (FACS). MuSCs obtained from a single round of FACS sorting have >95% purity and a 1–2% yield from the depleted cell suspension. This results in 75,000–100,000 MuSCs per young adult mouse. Yields from aged mice (20–-26 months old) are about 30% lower than young adult mouse yields. An additional sort (double-sorted cells) will increase MuSC purity to >98% but will reduce yield by ~30%. Performance (e.g., response to positive and negative controls) of singly and doubly sorted MuSCs in outgrowth experiments is indistinguishable. Therefore, typically only one round of FACS isolation was required to obtain MuSCs of suitable purity for further experimentation. The methods described herein improve MuSC yield and were sufficient to support multiple library screening efforts that required more than 50 million MuSCs.

2 Materials

2.1 Equipment

1. Nutating platform mixer.
2. Incubator.
3. GentleMacs (Miltenyi Biotec).
4. FACS sorter.
5. High-content imager.

2.2 Muscle Tissue Processing and Cell Culture

1. Collagenase type IV, filtered through a 0.45 μm filter prior to use.
2. Dispase.
3. PBS with calcium and magnesium.
4. GentleMacs C tubes (Miltenyi Biotec).
5. AutoMacs separation column, 10× (Miltenyi Biotec).
6. Nylon monofilament mesh 75 μm.
7. Nylon monofilament mesh 35 μm.
8. 70 μm Cell strainer.

9. 40 μm Cell strainer.
10. 5 mL Polystyrene round-bottom tube.
11. 5 mL Polystyrene round-bottom tubes with cell strainer cap.
12. F10 Nutrient Mixture (Ham) medium.
13. Horse serum (heat inactivated before used).
14. Fetal bovine serum (FBS) (heat inactivated).
15. Collagen type I solution, rat tail (5 mg/mL).
16. Mouse laminin, 1 mg.
17. Recombinant human FGF Basic (FGF2), 10 μg.
18. Greiner 384-well plate, polystyrene with micro-clear bottom.

2.3 MuSC Labeling and Detection

1. Rat anti-mouse integrin alpha-7-FITC (Clone: 334908, 25 μg/mL).
2. Rat anti-mouse CD34 PE conjugated Ab (Clone: RAM34, 0.2 mg/mL).
3. Anti-mouse CD31 (PECAM-1) APC-conjugated antibody (Clone: 390, 0.2 mg/mL).
4. Anti-mouse Sca-1 (6A/E) APC-conjugated antibody (Clone: D7, 0.2 mg/mL).
5. Anti-mouse CD45 (Ly-5) PE-Cy7-conjugated antibody (Clone: 30-F11, 0.2 mg/mL).
6. Anti-mouse CD11b PE-Cy7-conjugated antibody (Clone: M1/70, 0.2 mg/mL).
7. Propidium iodide (PI).
8. CellTrace Calcein blue, AM.
9. Hanks' balanced salt solution (HBSS).
10. Anti-APC bead.
11. Anti PE-Cy7 bead.
12. Staining solution: HBSS + 2% heat-inactivated horse serum.
13. DMSO, anhydrous.
14. Paraformaldehyde, 16% solution, EM grade.
15. Hoechst 33342.

3 Methods

Carry out all procedures at room temperature unless otherwise indicated.

3.1 Muscle Harvest and Tissue Processing

1. From a euthanized mouse, deglove the lower and upper extremities. With forceps and scissors, excise all the muscles from the fore limbs and hind limbs including gluteal muscle. Place the

harvested muscle in a 50 mL conical tube containing PBS on ice (*see* **Notes 1** and **2**).

2. Rinse tissues with PBS to remove blood contamination. In the 50 mL tube with the tissue settled, slowly pour off the PBS and floating debris (commonly fat). Add 40 mL of fresh PBS into the tube, permit material to resettle, and then pour off contents again. Repeat for four rinse cycles.
3. Place the tissues into a petri dish resting on ice. Cut muscle into small pieces (~2 mm size) with scissors while trimming away adipose, nervous, and vascular tissues. Cutting muscles into small pieces will facilitate a more complete and homogeneous enzyme digestion.
4. Suction any liquid in the dish and weigh the tissue (*see* **Note 3**). Place 3 g of tissue into a GentleMacs C tube and add 15 mL of collagenase IV and dispase solution (final concentration 600 unit/mL of collagenase type IV and 2.4 unit/mL of dispase) in PBS (with calcium and magnesium) (*see* **Note 4**).
5. Place warmed C-tubes on the GentleMacs tissue block and run the preset homogenizing program B.
6. Place the C-tubes on a nutating mixer in a 37 °C incubator with fixed rotate speed for 45 min. At 15-, 30-, and 45-min time points, run homogenizing program B. After 45-min incubation, transfer muscle slurry from C-tube to a 50 mL conical tube and add 15 mL/tube of cold 20% heat-inactivated FBS in F10 medium to stop the enzyme digestion (*see* **Note 5**).

The following steps are carried out at 4 °C and with media/buffers precooled to 4 °C:

7. Combine the cell suspension into a 500 mL bottle and dilute to 400 mL with F10 medium (*see* **Note 6**).
8. Pass the sample through a 75 μm filter. Gently mix liquid in the filter using a serological pipette, as necessary. Multiple filters may be used to pass entire sample. Then pass the sample through a 35 μm filter.
9. Divide sample evenly into 50 mL tubes and centrifuge for 10 min at 180 × *g*.
10. Slowly pour off, but retain, the supernatant from each tube and resuspend the cell pellet with 1 mL of staining solution. Transfer pellet into a 5 mL polystyrene round-bottom tube. The supernatant may be centrifuged again, with conditions identical to **step 9**, to ensure that sample is not lost. Resuspend cell pellets from additional centrifugations in cold staining solution and combine with the prior cell suspension.
11. Dilute cell suspensions in round-bottom tubes to 5 mL with staining solution and centrifuge for 10 min at 180 × *g*.

12. Slowly pour off the supernatant from each tube and resuspend cell pellet with 1 mL of staining solution. Combine cell suspensions, dilute to 5 mL, and centrifuge for 6 min at 180 × *g*. Repeat once.
13. Slowly pour off the supernatant from the tube and resuspend the cell pellet in staining solution to a final volume of 400 μL. Mix cells well to break any cell aggregates.

3.2 Cell Staining and Sorting

1. Prepare the FACS sorter and run a compensation experiment (refer to the manual for details on compensation) using 20 μL of a diluted aliquot (5 μL cells + 160 μL staining solution) of the cell suspension (from Subheading 3.1, **step 13**), for single-color staining, including calcein and PI. For the remainder of the cells from Subheading 3.1, **step 13**, add the following antibodies, mix well, and incubate at 4 °C for 45 min.
   (a) Anti-mouse integrin alph-7-FITC-conjugated Ab (1:5).
   (b) Anti-mouse CD34 PE-conjugated Ab (1:15).
   (c) Anti-mouse CD31 APC-conjugated Ab (1:300).
   (d) Anti-mouse Sca-1 APC-conjugated Ab (1:300).
   (e) Anti-mouse CD45 PE-Cy7-conjugated Ab (1:300).
   (f) Anti-mouse CD11b PE-Cy7-conjugated Ab (1:300).
2. Resuspend the cells in 4 mL of staining solution and centrifuge for 6 min at 180 × *g*. Pour off the supernatant.
3. Resuspend cells in 250 μL of staining solution and add 30 μL of anti-Cy7 beads and 30 μL of anti-APC beads. Mix well with 1 mL pipette.
4. Incubate sample at 4 °C for 15 min.
5. Resuspend the cells in 4 mL of staining solution and centrifuge for 6 min at 180 × *g*. Pour off the supernatant and resuspend cells in 5 mL of staining solution. Mix well with 1 mL pipette and filter the cells through a 35 μm filter cap with a 1 mL pipette.
6. Run AutoMACS Depletion program and collect the negative fraction or collect the flow-through from a column in a magnetic stand (*see* **Note 7**).
7. Place negative fraction into 5 mL polystyrene tubes and add staining buffer up to 5 mL.
8. Centrifuge for 6 min at 180 × *g*. Slowly discard the supernatant and resuspend the cells with ~1 mL staining buffer.
9. Freshly make 10× calcein blue solution (add 10.48 μL of DMSO to the calcein blue vial for a 10 mM stock solution) and add appropriate amount into the cell suspension.

10. Add PI solution (for 10× working stock solution, mix 500 μL of staining solution with 5 μL of propidium iodide stock) into the cell suspension. Incubate on ice for 2 min and cells are ready to sort (*see* **Note 8**).
11. Run samples on a FACS sorter using positive and negative selection markers, with gating established from unstained and individually stained compensation samples (Fig. 1).

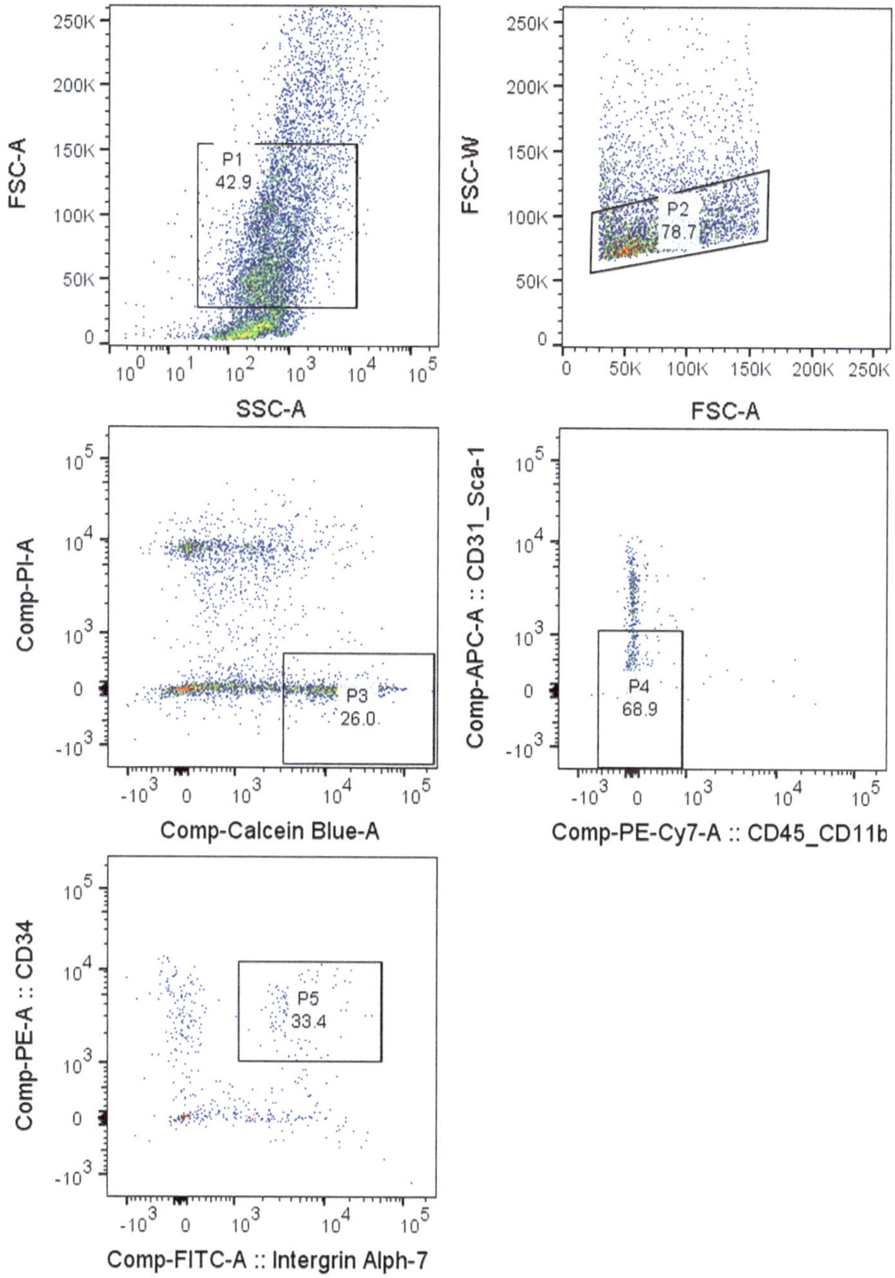

Fig. 1 FACS plot of gating strategy and typical outcomes for MuSC isolation

12. Adjust the flow rate to achieve an event rate of 1000–3000 events per second. Sort cells for ITGA7+, CD34+ Sca1-, CD45-, CD11b-, CD31- muscle precursor cells. First, the cells are sorted based on their sizes. As shown in Fig. 1, MuSCs are relatively small and nongranular in FSC and SSC scatter (P1 and P2 gate), and then cell viability is identified by positive staining with calcein blue and negative staining for propidium iodide (P3 gate). The negative selection gating for APC and PE-CY-7 staining removes hematopoietic cells, monocytes, endothelial cells, and mesenchymal cells. The positive selection gating enriches PE- (P4 gate) and FITC-positive cells (P5 gate) which are ITGA7+, CD34+ positive, markers for muscle satellite cells.
13. A single round of sorting will yield MuSCs with >95% purity and ~2% yield from the initial cell suspension. This results in ~75,000–100,000 MuSCs per young adult mouse (yields from old mice are about 30% lower than young adult mouse). An additional sort (double-sorted cells) will increase MuSC purity to >98% but will reduce yield by ~30% (*see* **Note 9**).
14. Once the pure population of MuSCs is isolated, dilute the cells in 1.5X proliferation medium (F10 medium supplemented with 30% HI horse serum, 7.5 ng/mL FGF2, penicillin/strep) (*see* **Notes 10** and **11**).
15. Plate cells in 384-well microclear plates precoated with collagen and laminin to ensure adherence of cells. Twenty-four hours after plating, add a 2× volume of compound diluted in MuSC proliferation medium to each well (*see* **Note 12**).
16. Treat MuSCs for 4–6 days in proliferation medium (select timing that permits largest window for detection of positive controls).
17. After this period of treatment, fix MuSCs with 1% paraformaldehyde and stain with Hoechst 33342 (*see* **Note 13**).
18. Image MuSCs with a high-content imager (BD Pathway Bioimager or GE InCell 2000). The number of MuSCs per well is determined by counting the number of nuclei (*see* **Note 14**).

4 Notes

1. Muscle-specific origin of MuSC has limited impact on outgrowth potential. Hind limb skeletal muscle MuSC pools and fore/hind/trunk skeletal muscle pools had similar kinetics of activation and outgrowth. Avoid harvesting non-muscle tissues, such as tendon, nerve, major blood vessels, fat, and connective tissue, as much as possible to improve downstream processing.

2. We have used the C57Bl6/J strain of mice and have not tested this protocol on other strains.
3. Typical muscle yield will be about 3 g per young adult male mouse (~25 g). Taring the scale with the petri dish in advance improves fidelity of weight determination.
4. Units/mL of collagenase in the digestion solution need to be titrated when using a new batch of collagenase. Purchasing a number of bottles from a single lot is advised.
5. The progress of the digestion can be monitored by examining on a small sample with a slide or hemocytometer. If muscle fibers are present, then repeat additional cycles of incubation and program B on the GentleMacs.
6. Sample dilution prior to the filtration and centrifugation steps is important. The filters will clog quickly if the sample is not diluted resulting in loss of cells.
7. Depletion prior to sorting significantly improves yield, similar to our previous report for human MuSCs [4].
8. Calcein blue appears to be excluded from MuSCs at a much higher rate than published values for other cells. Addition of calcein blue just prior to the sort improves yield dramatically.
9. Singly and doubly sorted MuSCs performed similarly in outgrowth experiments.
10. MuSCs isolated in the manner described behave similarly at low and high cell densities, potentially due to their highly motile nature. For example, proportional outgrowth kinetics are observed when 250–1000 MuSCs were plated in 384-well plates, resulting in ~11%, ~25%, ~52%, and 95% confluence following 6 days in culture (as determined by IncuCyte imaging). Therefore, screening for both proliferation and differentiation endpoints can be accomplished with low cell densities, though the latter requires larger initial numbers than the former.
11. If cells fail to grow, it may be that bFGF was accidentally omitted from the growth medium or that the bFGF stocks have lost activity. It is recommended that aliquots of bFGF are made up in single-use volumes and frozen at −20 °C. Do not freeze/thaw the bFGF aliquots.
12. MuSCs do not attach well to plates and are easily washed off, even with very gentle wash conditions such as using robotic liquid handling. Best results are obtained when treatments are added to the wells without complete medium removal. Fixing cells prior to washing, when possible, is suggested.
13. Outgrowth kinetics are age dependent, with 250 plated MuSCs growing to ~2000 in number from young, adult mice

(3–5 months of age) with 5 days of incubation, whereas 250 plated MuSCs from aged animals (20–26 months of age) will result in ~250 MuSCs with 5 days of incubation. The latter population demonstrates greater cell death following isolation than MuSCs from younger animals.

14. When imaging labeled nuclei, debris can often interfere with signal detection as MuSC outgrowth commonly occurs with lower cell number. To limit the impact of debris, a size exclusion (e.g., $>30\ \mu m^2$) can be included in the analysis step.

References

1. Sherwood RI, Christensen JL, Conboy IM, Conboy MJ, Rando TA, Weissman IL et al (2004) Isolation of adult mouse myogenic progenitors: functional heterogeneity of cells within and engrafting skeletal muscle. Cell 119:543–554
2. Cerletti M, Jurga S, Witczak CA, Hirshman MF, Shadrach JL, Goodyear LJ et al (2008) Highly efficient, functional engraftment of skeletal muscle stem cells in dystrophic muscles. Cell 134:37–47
3. Conboy MJ, Cerletti M, Wagers AJ, Conboy IM (2010) Immuno-analysis and FACS sorting of adult muscle fiber-associated stem/precursor cells. Methods Mol Biol 621:165–173
4. Bareja A, Holt JA, Luo G, Chang C, Lin J, Hinken AC et al (2014) Human and mouse skeletal muscle stem cells: convergent and divergent mechanisms of myogenesis. PLoS One 9:e90398
5. Castiglioni A, Hettmer S, Lynes MD, Rao TN, Tchessalova D, Sinha I et al (2014) Isolation of progenitors that exhibit myogenic/osteogenic bipotency in vitro by fluorescence-activated cell sorting from human fetal muscle. Stem Cell Rep 2:92–106
6. Charville GW, Cheung TH, Yoo B, Santos PJ, Lee GK, Shrager JB et al (2015) Ex vivo expansion and in vivo self-renewal of human muscle stem cells. Stem Cell Rep 5:621–632
7. Liu L, Cheung TH, Charville GW, Rando TA (2015) Isolation of skeletal muscle stem cells by fluorescence-activated cell sorting. Nat Protoc 10:1612–1624
8. Sincennes MC, Wang YX, Rudnicki MA (2017) Primary mouse myoblast purification using magnetic cell separation. Methods Mol Biol 1556:41–50
9. Mauro A (1961) Satellite cell of skeletal muscle fibers. J Biophys Biochem Cytol 9:493–495
10. Billin AN, Bantscheff M, Drewes G, Ghidelli-Disse S, Holt JA, Kramer HF et al (2016) Discovery of novel small molecules that activate satellite cell proliferation and enhance repair of damaged muscle. ACS Chem Biol 11:518–529

Chapter 7

Cell-Based Methods to Identify Inducers of Human Pancreatic Beta-Cell Proliferation

Courtney A. Ackeifi, Ethan A. Swartz, and Peng Wang

Abstract

Diabetes is the result of the insufficiency or dysfunction of pancreatic beta cells alone or in combination with insulin resistance. The replacement or regeneration of beta cells can effectively reverse diabetes in humans and rodents. Therefore, the identification of novel small molecules that promote pancreatic beta-cell proliferation is an attractive approach for diabetic therapy. While numerous hormones, small molecules, and growth factors are able to drive rodent beta cells to replicate, only a few small molecules have demonstrated the ability to stimulate human beta-cell proliferation. Hence, there is an urgent need for therapeutic agents that induce regeneration and expansion of adult human beta cells. Here, we describe a detailed protocol for coating chamber slides, culturing primary islets, performing islet cell disassociation, seeding cells on chamber slides, treating islet cells with compounds or infecting them with adenovirus, immunostaining of proliferation markers and imaging, and data analysis.

Key words Diabetes, Human pancreatic beta cell, Proliferation, Diabetes therapeutics, Human islets, Drug development

1 Introduction

Type 1 diabetes [T1D] and type 2 diabetes [T2D] result entirely or in part from a reduction in functional pancreatic beta cells, with a resultant loss of adequate insulin secretion [1–5]. In T1D, autoimmune destruction of β-cells leads to failure of insulin secretion [2, 6]. Autopsy studies demonstrate that β-cell mass is reduced by 90% or more in T1D [5]. In the past, it was thought that T2D was the result of insulin resistance; however, recent genome-wide association studies [GWAS] and autopsy studies indicate that beta-cell loss and dysfunction are also central etiologic features of the disease [1, 4, 7]. β-Cell replacement using cadaver islet transplant has been shown to correct diabetes in humans and rodents [8, 9]. Consequently, methods to increase functional beta-cell mass through regenerative and replacement methods are attractive approaches to treating diabetes.

Bridget Wagner (ed.), *Phenotypic Screening: Methods and Protocols*, Methods in Molecular Biology, vol. 1787, https://doi.org/10.1007/978-1-4939-7847-2_7,

Although promising, beta-cell replacement and regeneration face many challenges including low availability of donor islets and resistance of adult human pancreatic beta cells to replicate [3]. Identifying new ways to promote proliferation of adult human beta cells is key for diabetes therapeutic discovery.

Like other mammalian cells, human β-cell replication is controlled by G1/S cell cycle machinery. The cell cycle progresses through four critical phases: G0/G1, DNA synthesis [S], G2, and mitosis. Although human β-cells contain the requisite cyclins, cyclin-dependent kinases [CDKs], and E2F transcription factors that direct cell cycle entry, these cells also express CDK inhibitors [CDKIs] and pocket proteins [pRb, p107, and p130] at high levels that negatively regulate cell cycle progression [10–12]. The identification of proliferating human beta cells relies on immunostaining for key markers including Ki67, bromo-deoxyuridine incorporation [BrdU] or its analog 5-ethynyl-deoxyuridine [EdU], and phospho-histone-H3 [13]. Ki67 is a 360 kDa nuclear protein that is commonly used to detect and quantify replicating cells [14]. Ki67 expression is induced during the G1-S-phase transition and continues throughout G2 and M phases. Ki67 expression is undetectable in cells in G0 phase [15]. BrdU [or EdU] incorporation is also commonly used to detect human beta-cell proliferation. BrdU is an analog of the nucleoside thymidine and DNA incorporation of these reagents marks cells that have passed through the S phase of the cell cycle at any time during the assay. BrdU can accumulate in proliferating cells or recently replicated cells with increased incubation time. Phospho-histone H3 [pHH3] is another cell proliferation marker, limited to detecting cells undergoing mitosis, and therefore detects a very small population of cells [16]. Though, when compared with Ki67, BrdU, or EdU, pHH3 is the least sensitive but the most specific for detecting cell replication. pHH3 staining is a snapshot of the cells that successfully complete the cell cycle and may be a more accurate indication of beta-cell regeneration. In this chapter, we provide a detailed protocol about how we use the four different proliferation markers to screen proliferation in beta cells.

Our group has focused on (1) high- and medium-throughput screening methods to identify small molecules to drive human beta-cell proliferation and (2) development of small-molecule drugs to induce pancreatic beta cells to regenerate in people with T1D and T2D for the purpose of restoring normal insulin secretion. We have successfully identified the compound, harmine, the first small molecule to reproducibly drive significant rates of adult human beta-cell proliferation [17–20]. In this chapter, we describe protocols for coating chamber slides to enable dispersed human islet cell attachment and culture, human islet cell dissociation and seeding, compound treatment and adenoviral infection, as well as detection of beta-cell proliferation. Finally, we discuss image acquisition and analysis of beta-cell replication rates.

To start, we recommend performing initial screens in primary rat islets since they are significantly more homogenous and sensitive. In addition, testing should be completed using at least three doses to maximize the identification of active compounds. Compounds that show an increase in proliferation (hits) from the rat islet screen can subsequently be tested in human islets at multiple doses. There is uncontrollably high variability between human islet donors due to cause of death, donor genetics, and islet isolation and transport. Therefore, we recommend completing multiple replicates with different donors and multiple markers of cell replication.

2 Materials

2.1 Coating the Chamber Slides to Culture Dispersed Human Islet Cells

1. Phosphate-buffered saline (PBS).
2. 15 and 50 mL conical tubes.
3. 1.5 mL Eppendorf tubes.
4. 8-Well chamber slides.
5. Poly-D-lysine (PDL).
6. 10× Hanks' balanced salt solution (HBSS).
7. Laminin Mouse Protein, Natural.
8. Pipettes.
9. 37 °C Sterile incubator.

2.2 Islet Dispersion and Islet Cell Plating

1. Islet culture medium: RPMI 1640 medium containing 10% fetal bovine serum (FBS), 5.5 mM glucose, and 1% penicillin-streptomycin (Pen/Strep).
2. Human cadaveric islets are obtained from the Integrated Islet Distribution Program: (https://iidp.coh.org/secure/isletavail/home.aspx).
3. Vacuum-driven filter (0.22 μm).
4. Accutase.
5. 60 mm Culture dish.
6. PDL-laminin-coated chamber slides or coverslips.

2.3 Treatment of Dispersed Beta Cells

1. Centrifuge (for 50 or 15 mL conical tubes).
2. Compounds and adenovirus of choice ~100–1000× (*see* **Note 1**).
3. Dimethyl sulfoxide (DMSO).
4. 10 mM Harmine in DMSO for positive control.
5. Control viruses.

2.4 Human Beta-Cell Proliferation Assay

1. 16% Paraformaldehyde (PFA) (working concentration: 4% PFA in PBS).
2. Blocking buffer: 5% Normal goat serum (NGS), 1% bovine serum albumin (BSA), 0.5% Triton X-100, in PBS, filtered (*see* **Note 2**).
3. 1 N HCl.
4. Rabbit-anti-Ki67, 1:300 dilution.
5. Guinea pig-anti-insulin, 1:500 dilution.
6. Rat-anti-BrdU, 1:300 dilution.
7. Rabbit-anti-P-histone-H3, 1:500 dilution.
8. Alexa Fluor 594 goat-anti-rat IgG, 1:500 dilution.
9. Alexa Fluor 594 goat-anti-rabbit IgG, 1:500 dilution.
10. Alexa Fluor 488 goat-anti-guinea pig IgG, 1:500 dilution.
11. DAPI (4′,6-diamidino-2-phenylindole, dihydrochloride).
12. Amersham Cell Proliferation Labelling Reagent, 1:100 dilution.
13. Mounting medium.
14. 60 mm Cover slides.

2.5 Microscopy Image Acquisition and Proliferation Data Analysis

1. Fluorescence or confocal microscope.
2. Appropriate data analysis software.

3 Methods

3.1 Coating the Chamber Slides to Culture Dispersed Human Islet Cells

Human islets do not thrive outside of the body; they do not attach well to most tissue culture chamber slides and therefore require additional coating to maintain cell adhesion and promote beta-cell survival, proliferation, and differentiation. In our lab, we have developed a PDL and laminin-coating system to culture, treat, fix, and stain dispersed human islet cells in 8-well chamber slides. In this system, we have achieved increased human beta-cell adhesion, retention, and health throughout the 4–6-day treatments and reduced the number of cells lost during staining protocols.

1. Prepare a 50 μg/mL PDL solution in sterile water; 200 μL is sufficient to cover each well of an 8-well glass chamber slide (*see* **Note 3**).
2. Coat slides with PDL for 2 h at 37 °C.
3. Wash twice with sterile water at room temperature.

4. Laminin must be thawed slowly on ice and freeze–thaws avoided.
5. Prepare a 500 μg/mL laminin solution in 1× HBSS; 120 μL is sufficient to cover each well of an 8-well chamber slide (*see* **Note 4**).
6. Coat slides with laminin overnight 4 °C.
7. Remove laminin at least 1–2 h prior to starting islet dispersion.
8. Wash once with PBS.
9. Use wide-tip aspirator to dry each well thoroughly and be sure not to scrape the surface of the well.
10. Leave the chamber slide lid open in the tissue culture hood for 2–3 h to air-dry. Slides can be used immediately or stored at 4 °C for up to 4 weeks (*see* **Note 5**).

3.2 Islet Dispersion and Islet Cell Plating

Pancreatic islets are spherical clusters of cells distributed throughout the pancreas and are made up of several endocrine cell types, including beta cells, alpha cells, delta cells, and pancreatic polypeptide cells. In order to clearly identify beta-cell proliferation, it is necessary to disperse the islets into single cells and seed them as a monolayer for treatment, staining, imaging, and quantitation.

1. Prepare the islet culture medium and warm to 37 °C; thaw Accutase at room temperature prior to dispersing islets.
2. Unpack the islet package and check the temperature monitor to make sure that the islet package did not experience extreme high or low temperatures. Decontaminate the outside of the islet container with 70% ethanol.
3. Open the container under the tissue culture hood and pour the islets and medium into 50 mL tubes. Allow islets to sink to the bottom of the tube for about 2 min and use 5 mL of medium from the tube to rinse the islet container and return to the 50 mL conical tube.
4. Centrifuge the islets at 180 × *g* for 3 min at room temperature. Aspirate the supernatant top medium into bleach without disturbing the islet pellet.
5. If the islets will not be used immediately, resuspend islets in 5 mL of complete islet culture medium and transfer to a sterile non-treated 60 mm petri dish. With an additional 2 mL of medium, rinse the conical tube and add medium to the dish. Keep islet suspension culture at 37 °C incubator until use. For best results, islets should be allowed to recover at 37 °C for about 24 h prior to dispersion. When ready to disperse islets, transfer islets to conical tube and proceed to **step 7**.
6. For immediate use, keep cells in conical tube and continue with protocol.

7. Add 5 mL of sterile PBS to wash the islet pellet once and centrifuge at 180 × *g* for 3 min at room temperature. Aspirate the supernatant with aspirator. For islet pellets of <5000 islet equivalents (IEQ), add 2 mL of Accutase to the islet pellet. Additional Accutase is required for larger IEQ numbers.
8. Incubate islets in Accutase in incubator at 37 °C for 5 min.
9. Gently disaggregate the islets with a P1000 pipette by pipetting up and down 5–7 times.
10. Return islets to incubator for up to three 5-min incubations. In between each interval, agitate tube and monitor the status of islet cells dispersed into individual single cells under the microscope.
11. Total digestion time should be less than 15 min; over-digestion will decrease the health of the cells and their ability to proliferate; incomplete dispersion also interferes with cell adhesion and accurate quantification of beta-cell number and proliferation.
12. Stop dispersion by adding 3–5× total volume of complete islet culture medium. Mix well and centrifuge at 700 × *g* for 5 min; discard supernatant into bleach.
13. Add 1 mL of complete islet culture medium per estimated 1000–2000 IEQ. Mix the dispersed islet cells by pipetting up and down 5× and count cells. Calculate required cell numbers and medium volumes for plating.
14. For 8-well chamber slides, seed 2.5×10^4 cells/well in 20–25 μL of medium. Cells should be seeded into the center of each chamber well.
15. After seeding cells, place slides in a 37 °C, 5% CO_2, humidified cell culture incubator. Wait for 1–2 h to allow the cells to attach to the slides before adding additional medium or adding treatments. For an 8-well chamber slide, add 500 μL to each well (*see* **Notes 6** and **7**).

3.3 Treatment of Dispersed Beta Cells

In this section, we discuss two common approaches in inducing beta-cell replication: pharmacological treatments and adenoviral transduction. Although we discuss these methods separately, many of the conditions of these assays remain constant. For both treatments, we use ~2.5×10^4 total cells per condition. Since beta-cell proliferation is very slow, we generally treat with the agents for 96 h before we fix and immunostain for one of the proliferation markers. When both adenoviral and pharmacological treatments are required, particularly for mechanistic studies, we recommend transducing first with the adenovirus overnight before adding the compounds. This paradigm helps keep the transduction efficiency consistent and high without interference by other treatments.

We recommend performing initial screens in primary rat islets because of the higher proliferative sensitivity of rat beta cells [21];

testing compounds should be completed ideally at four doses in the range of nM to μM to avoid missing true hits. The protocols for rat and human islet dispersion and treatment are the same; therefore, we will not discuss them separately. Once hit treatments that drive rat beta-cell proliferation are identified, they can be verified in multiple human islet donors. Human islets should also be used to study the mechanism of action of these hits. For pharmacological treatments, DMSO or other vehicles used to reconstitute small molecules should be used as the vehicle control and 10 μM harmine can be used as positive control. For adenovirus transduction treatments, Ad-cre, Ad-lacz, or Ad-scramble shRNA can be used for transduction control and Ad-CDK4/CDK6 can be used for the positive control for each experiment.

3.3.1 Human Islet Treatment: Compound Treatment

1. After dispersing islet cells, calculate how many treatments to test and adjust cell density to ~2.5 × 10^4 cells in 25 μL of medium.
2. Seed 25 μL of cell suspension into the center of each well of a pre-coated chamber slide or onto a coated coverslip on the bottom of 24-well plate. Allow cells to attach for 1–2 h before adding treatment solutions.
3. For best results, wait for at least 1–2 h after plating before treating cells; 2 h post-plating, add 500 μL of complete medium and allow cells to recover from dispersion.
4. Prepare compound solutions: in a 1.5 mL tube, prepare 500 μL–1 mL of each compound solution in complete medium and vortex to mix (*see* **Note 8**).
5. Gently add the solution to the cells.
6. Incubate cell culture at 37 °C for 3–4 days.

3.3.2 Human Islet Treatment: Viral Transduction

To study how the expression of a gene can regulate beta-cell proliferation, adenoviral vectors can be utilized to overexpress or silence specific gene expression. Under optimal conditions, adenoviruses can transduce up to 100% human islet cells. To maximize beta-cell health and adenovirus transduction efficiency, 2.5 × 10^4 cells and 100–300 plaque-forming units (pfu)/cell virus are used for each condition. For initial experiments, it is necessary to complete a dose–response for each virus to identify the most effective pfu/cell that does not decrease the overall health of the cells. All calculations should be done prior to the start of treatment. Virus infection should be in 0–2% serum medium for 2 h in a small volume (~25 μL).

1. Calculate the number of treatments and the particles of virus needed. Dilute and mix the virus in serum-free medium, put on ice (*see* **Note 9**).

2. Separate cells needed for viral infection, and centrifuge at 700 × *g* for 5 min.
3. Remove the complete medium supernatant, resuspend the cells required in serum-free medium, and adjust the cell density to 2.5×10^4 cells in 25 μL of medium. Place 25 μL/condition cells into a 1.5 mL Eppendorf tube, add the appropriate amount of virus to the cells, and mix well with pipet.
4. Seed treated cells as a droplet into the center of each chamber slide well or on a cover slide at the bottom of 24-well plates. Be sure not to disrupt the droplets and gently move the cells to the cell culture incubator (*see* **Note 10**).
5. Incubate for 2 h for attachment and virus infection.
6. After 2 h, add 500 μL of complete islet culture medium to stop the virus infection and return to the incubator for 72–96 h.

3.4 Human Beta-Cell Proliferation Assay

After the 4-day treatment, cells are fixed and co-immunostained with antibodies against a proliferation marker (BrdU incorporation, EdU, Ki67, or pHH3) and insulin. Here, we describe staining for Ki67 and incorporated BrdU. Similar protocols exist for other thymidine analogs such as EdU and for pHH3 immunostaining that can be alternatively used here [22].

3.4.1 Ki67 Immunostaining

In order to visualize and quantify beta-cell proliferation following treatment, cells must be fixed and co-immunostained for a cell division marker, Ki67, and a marker for beta cells, insulin (Fig. 1). This protocol can be applied to other proliferation markers such as pHH3 using the primary and secondary antibodies described in Subheading 2 (Fig. 2).

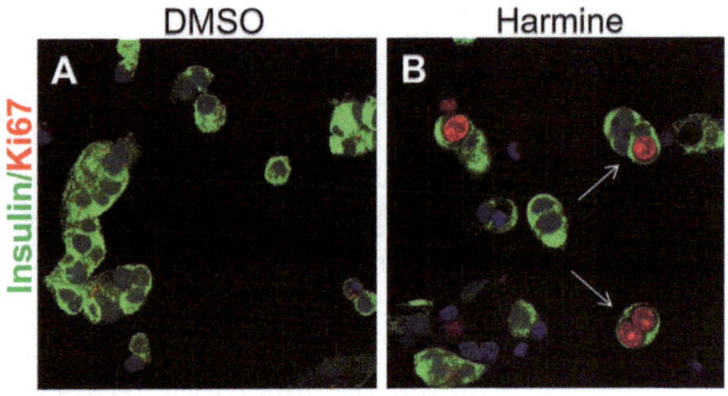

Fig. 1 Insulin and Ki67 immunostained human islets allow for visualization and quantification of beta-cell proliferation. Dispersed human islets immunostained for insulin (green) and Ki67 (red) demonstrate a lack of beta-cell proliferation following the (**a**) vehicle control treatment (DMSO) and (**b**) positive control treatment (10 μM harmine); numerous insulin-positive beta cells with Ki67-positive nuclei can be seen

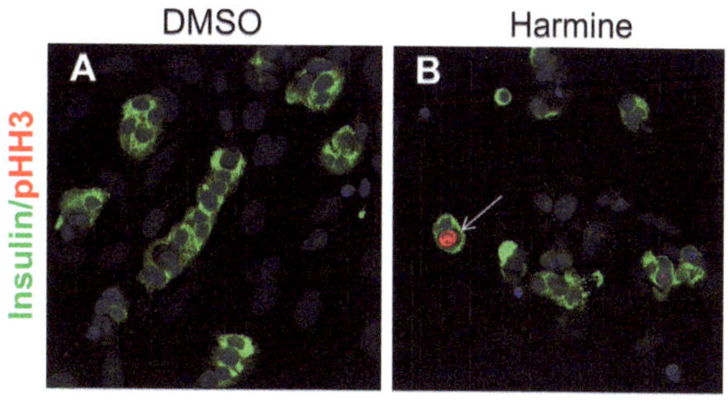

Fig. 2 Immunolabeling for insulin and pHH3 allows for quantitation and visualization of beta-cell replication. Dispersed human islets were treated with the (**a**) vehicle control (DMSO) and (**b**) positive control (10 μM harmine) for 96 h. Islets immunostained for insulin (green) and pHH3 (red) only label cells that have successfully completed the cell cycle

3.4.2 Fixing

1. To prepare the 4% PFA fixation buffer, dilute 16% PFA 1:4 in PBS. This solution can be stored for up to 4 weeks at 4 °C protected from light.
2. Gently remove cell medium. Gently rinse with 300 μL of PBS.
3. Remove PBS, add 150–300 μL of cold 4% PFA buffer per well, and incubate at room temperature for 10 min.
4. Stop fixation by aspirating the PFA and wash cultures twice with 500 μL of PBS at room temperature (*see* **Note 11**).

3.4.3 Immunostaining

1. Calculate the required volume of complete blocking buffer: for an 8-well chamber slide: 120 μL/well × blocking step, primary and secondary antibody steps.
2. Prepare required volume of blocking buffer by adding fresh NGS to the blocking stock buffer.
3. Incubate cultures with 120 μL/well of blocking buffer at room temperature for 1 h.
4. Prepare Rb-anti-Ki67 and Gp-anti-insulin primary antibody solution in blocking buffer, and mix well by vortexing.
5. Remove blocking buffer and add 120 μL of primary antibody solution to each well. Incubate for 2 h at room temperature on a slow shaker.
6. After 2 h, wash 3× 5 min with 500 μL of PBS (*see* **Notes 12** and **13**).
7. Prepare appropriate secondary antibody solution: Alexa Fluor-conjugated Goat-anti Rb-594 and Goat-anti-Gp-488 secondary antibodies in blocking buffer, and mix well by vortex.
8. Add 120 μL of secondary antibody solution to each well. Incubate for 1 h at room temperature in the dark on a slow shaker.

9. Wash 3× 5 min with PBS.
10. Prepare a 2 μg/mL solution of DAPI in water. Add 100–300 μL of DAPI solution to each well for 10 min. Wash twice with PBS.
11. Remove the gasket and seal glue from chamber slides (*see* **Note 14**).
12. Coat samples with mounting medium and coverslip; use anti-fade reagent. Seal coverslip with nail polish if necessary.

3.4.4 BrdU Incorporation

To measure BrdU incorporation, the Amersham Cell Proliferation Labelling Reagent must be added to each treatment 18 h prior to fixation.

1. In order not to dilute drug concentrations in each treatment condition, only add 75 μL of medium containing a final effective 1:100 dilution of the labeling reagent.
2. Incubate overnight.
3. To prepare the 4% PFA fixation buffer, dilute 16% PFA 1:4 in PBS. This solution can be stored for up to 4 weeks at 4 °C protected from light.
4. Gently remove cell medium. Gently rinse with 300 μL of PBS. When working with many samples remove medium and wash cells in small groups to prevent cells from drying out.
5. Remove PBS and add 150–300 μL of cold 4% PFA buffer per well, and incubate at room temperature for 10 min.
6. Stop fixation by aspirating the PFA and wash cultures twice with 500 μL of PBS at room temperature (*see* **Note 11**).

3.4.5 Immunostaining

1. BrdU immunostaining requires antigen retrieval. After fixation, add 200 μL 1 N HCl to each condition and incubate at 37 °C for 30 min.
2. Wash twice with 500 μL PBS.
3. Calculate the required volume of complete blocking buffer: for an 8-well chamber slide: 120 μL/well × blocking step, primary and secondary antibody steps.
4. Prepare required volume of blocking buffer by adding fresh NGS to the blocking stock buffer.
5. Incubate cultures with 120 μL/well of blocking buffer at room temperature for 1 h.
6. Prepare Rat-anti-BrdU and Gp-anti-Insulin primary antibody solution in blocking buffer, and mix well by vortex.
7. Remove blocking buffer and add 120 μL of primary antibody solution to each well. Incubate for 2 h at room temperature on a slow shaker.

8. After 2 h, wash 3× 5 min with 500 μL of PBS (*see* **Notes 12** and **13**).
9. Prepare appropriate secondary antibody solution, Alexa-conjugated Goat-anti-Rat-594 and Goat-anti-Gp-488 secondary antibodies in blocking buffer, and mix well by vortex.
10. Add 120 μL of secondary antibody solution to each well. Incubate for 1 h at room temperature in the dark on a shaker.
11. Wash 3× 5 min with PBS.
12. Prepare a 2 μg/mL DAPI solution in water. Add 100–300 μL of the DAPI solution to each well, and incubate for 10 min. Wash twice with PBS.
13. Remove the gasket and seal glue from chamber slides (*see* **Note 13**).
14. Cover samples with mounting medium and coverslip; use anti-fade reagent.
15. Seal coverslip with nail polish if necessary.

3.5 Microscopy Image Acquisition and Proliferation Data Analysis

A fluorescence or confocal microscope with the ability to capture ×20 magnification images of the fluorophores used is required for image acquisition. Although automated microscopy systems can be used to image each condition, manual imaging each condition tends to provide more accuracy. To establish imaging parameters for each experiment, use negative and positive control samples to set each channel exposure time. For each condition, image 5–10 random subfields at ×20 magnification (Figs. 1, 2, and 3). Count the total number of beta cells and total number of proliferating beta cells for each condition. A minimum of 1000 beta cells per condition should be counted. With these data, calculate the percent proliferation as a function of Ki67-, BrdU-, or pHH3-positive

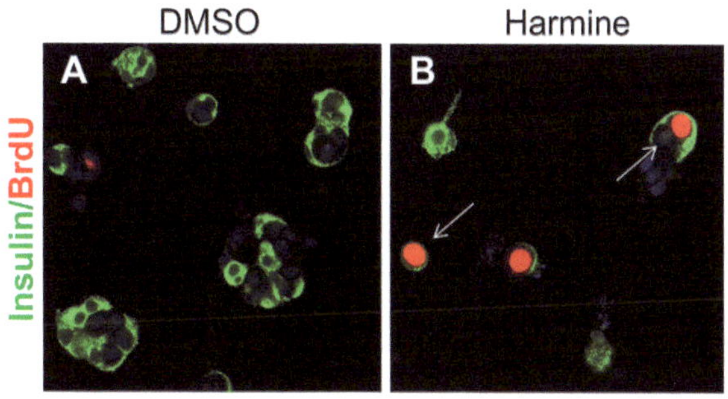

Fig. 3 Immunolabeling for Insulin and BrdU incorporation allows for quantitation and visualization of beta-cell replication. Dispersed human islets were treated with the (**a**) vehicle control (DMSO) and (**b**) positive control (10 μM harmine) for 96 h. Following incubation with BrdU labeling reagent, cells co-immunostained for insulin (green) and BrdU (red) can label proliferating beta cells

beta cells. After testing multiple donors, proper statistical analysis can determine significance of a treatment's induction of proliferation.

4 Notes

1. Prepare compounds or adenovirus to working concentrations. Thaw compounds reconstituted in DMSO at room temperature; all other compounds, peptides, or adenoviruses should be thawed on ice while preparing treatments. Adenoviruses should be diluted with serum-free medium and stored on ice. Warm complete islet culture medium and serum-free islet medium.
2. Make stock blocking buffer containing 1% BSA and 0.5% Triton X-100, in PBS, filtered. Store at 4 °C for up to 4 months. Store NGS at −20 °C. Thaw small-volume NGS and add required volume to stock blocking buffer fresh for each staining.
3. Coat slides under sterile conditions; all solutions and equipment coming into contact with cells must be sterile.
4. When adding the coating solution, make sure that the coating solution fully covers the surface of the slides and coats the surface evenly.
5. Coated slides should be air-dried completely before being stored at 4 °C. This significantly affects the ability to seed cells in small volume.
6. All culture incubations are performed in a humidified (constant 95% humidity), 37 °C, 5% CO_2, incubator.
7. Human islets should be used as soon as possible; extended culture time results in decreased beta-cell survival and increased resistance to proliferation.
8. Use p2 or p10 pipets when required for preparing compound solutions for maximum accuracy.
9. Calculate virus volume per the following equation:
   (a) $\mu\text{L virus} = \dfrac{(\text{IEQ})\left(\frac{1000\ \text{cells}}{\text{islet}}\right)\left(\frac{\text{pfu}}{\text{cell}}\right)}{\frac{\text{pfu}}{\mu l}}$
   (b) Example: $\mu\text{L virus} = \dfrac{(\text{XIEQ})\left(\frac{1000\ \text{cells}}{\text{islet}}\right)\left(250\ \frac{\text{pfu}}{\text{cell}}\right)}{1.5\times10^9\ \frac{\text{pfu}}{\mu l}}$
   (c) Or, if counting cells: $\mu\text{L virus} = \dfrac{(\text{total cells})\left(\frac{\text{pfu}}{\text{cell}}\right)}{\frac{\text{pfu}}{\mu l}}$
10. Infect dispersed islet cells in individual tubes, and then plate entire suspension volume to the chamber slides.
11. After cell fixation, cells can be stored in PBS at 4 °C for a few weeks.
12. Be sure not to overstain. High background can occur particularly with the insulin antibody and can cause false positives in

the proliferation assay. For this reason, incubating the insulin antibody overnight at 4 °C is not recommended; for best results, incubate primary antibodies at room temperature for 2 h. If overnight incubation is required, reduce the insulin antibody concentration to 1:1000 dilution factor.

13. For each wash and incubation, do not allow cells to dry out, as this will increase the background dramatically.
14. When removing the chamber slide gasket, follow the manufacturer's instructions carefully and be sure not to provide too much pressure and break the slide. In addition, be sure to remove all adhesive glue on the slide as it will interfere in imaging.

Acknowledgements

The authors thank Dr. Andrew Stewart and the members of the Stewart lab, Dr. Adolfo Garcia-Ocaña (all at Icahn School of Medicine at Mount Sinai, New York, NY), for invaluable and continuous insight. The authors thank the National Institutes of Health/National Institute of Diabetes and Digestive and Kidney Diseases Integrated Islet Distribution Program and Dr. Tatsuya Kin (Alberta Diabetes Institute, Edmonton, Alberta, Canada) for providing human islets. This work was supported by JDRF grant 2-SRA-2015-62-Q-R; National Institutes of Health grants R-01 DK105015 and T32 GM 062754; and the Human Islet and Adenovirus Core of the Einstein-Mount Sinai Diabetes Research Center P30 DK020541-38.

References

1. Butler AE, Janson J, Bonner-Weir S, Ritzel R, Rizza RA, Buter PC (2003) Beta-cell deficit and increased beta-cell apoptosis in humans with type 2 diabetes. Diabetes 52:102–110
2. Eisenbarth GS (1986) Type I diabetes mellitus. A chronic autoimmune disease. N Engl J Med 314:1360–1368
3. Gregg BE, Moore PC, Demozay D, Hall BA, Li M, Husain A et al (2012) Formation of a human beta-cell population within pancreatic islets is set early in life. J Clin Endocrinol Metab 97:3197–3206
4. Matveyenko AV, Butler PC (2008) Relationship between beta-cell mass and diabetes onset. Diab. Obes Metab 10:23–31
5. Meier JJ, Bhushan A, Butler AE, Rizza RA, Butler PC (2005) Sustained beta cell apoptosis in patients with long-standing type 1 diabetes: indirect evidence for islet regeneration? Diabetologia 48:2221–2228
6. Bleich D, Jackson RA, Soeldner JS, Eisenbarth GS (1990) Analysis of metabolic progression to type I diabetes in ICA+ relatives of patients with type I diabetes. Diabetes Care 13:111–118
7. Lindgren CM, McCarthy MI (2008) Mechanisms of disease: genetic insights into the etiology of type 2 diabetes and obesity. Nat Clin Pract Endocrinol Metab 4:156–163
8. Pepper AR, Gala-Lopez B, Ziff O, Shapiro AJ (2013) Current status of clinical islet transplantation. World J Transplant 3:48–53
9. Shapiro AM, Lakey JR, Ryan EA, Korbutt GS, Toth E, Warnock GL et al (2000) Islet transplantation in seven patients with type 1 diabetes mellitus using a glucocorticoid-free

immunosuppressive regimen. N Engl J Med 343:230–238
10. Fiaschi-Taesch NM, Kleinberger JW, Salim FG, Troxell R, Wills R, Tanwir M et al (2013) Human pancreatic beta-cell G1/S molecule cell cycle atlas. Diabetes 62:2450–2459
11. Heit JJ, Karnik SK, Kim SK (2006) Intrinsic regulators of pancreatic beta-cell proliferation. Annu Rev Cell Dev Biol 22:311–338
12. Takane KK, Kleinberger JW, Salim FG, Fiaschi-Taesch NM, Stewart AF (2012) Regulated and reversible induction of adult human beta-cell replication. Diabetes 61:418–424
13. Salic A, Mitchison TJ (2008) A chemical method for fast and sensitive detection of DNA synthesis in vivo. Proc Natl Acad Sci U S A 105:2415–2420
14. Preusser M, Heinzl H, Gelpi E, Höftberger R, Fischer I, Pipp I et al (2008) Ki67 index in intracranial ependymoma: a promising histopathological candidate biomarker. Histopathology 53:39–47
15. Endl E, Steinbach P, Knüchel R, Hofstädter F (1997) Analysis of cell cycle-related Ki-67 and p120 expression by flow cytometric BrdUrd-Hoechst/7AAD and immunolabeling technique. Cytometry 29:233–241
16. Healy S, Khan P, He S, Davie JR (2012) Histone H3 phosphorylation, immediate-early gene expression, and the nucleosomal response: a historical perspective. Biochem Cell Biol 90:39–54
17. Aamodt KI, Aramandla R, Brown JJ, Fiaschi-Taesch NM, Wang P, Stewart AF et al (2016) Development of a reliable automated screening system to identify small molecules and biologics that promote human β-cell regeneration. Am J Phys Endocrinol Metab 311: E859–E868
18. Dirice E, Walpita D, Vetere A, Meier BC, Kahraman S, Hu J et al (2016) Inhibition of DYRK1A stimulates human beta-cell proliferation. Diabetes 65:1660–1671
19. Shen W, Taylor B, Jin Q, Nguyen-Tran V, Meeusen S, Zhang YQ et al (2015) Inhibition of DYRK1A and GSK3B induces human β-cell proliferation. Nat Commun 6:8372
20. Wang P, Alvarez-Perez J-C, Felsenfeld DP, Liu H, Sivendran S, Bender A et al (2015) A high-throughput chemical screen reveals that harmine-mediated inhibition of DYRK1A increases human pancreatic beta cell replication. Nat Med 21:383–388
21. Ricordi C, Rastellini C (2000) Methods in pancreatic islet separation. In: Methods in cell transplantation. RG Landes, Austin, pp 433–438
22. Walpita D, Wagner BK (2014) Evaluation of compounds in primary human islet cell culture. Curr Prot. Chem Biol 6:157–168

Chapter 8

High-Content Imaging Phenotypic Screen for Neurogenesis Using Primary Neural Progenitor Cells

Li Sharon Wu and Jingjun Li

Abstract

Neurogenesis phenotypic screen of small-molecule library enables the discovery of small-molecule inducers, and identification of associated biological targets and pathways that control neuronal formation from neural progenitor cells (NPCs). Here, we describe protocols for preparing mouse embryonic NPCs, setting up a high-content imaging assay that quantifies the production of Tuj1-labeled neurons, and analysis of high-throughput screens.

Key words Neurogenesis, High-content imaging, Phenotypic screen, Neural progenitor cells, Neuronal differentiation, Small-molecule chemical library

1 Introduction

Regenerative medicine that targets neural progenitor cells (NPCs) for neuronal replacement holds great promise to treat brain injuries and neurodegenerative diseases [1–3]. Phenotypic screening for neurogenesis is an effective approach to discover new chemical entities and targets for potential therapeutics [4, 5]. Identification of the new targets and associated pathways through phenotypic screening has greatly enhanced our understanding of biological processes in health and disease [6, 7].

NPCs exist in both developing and adult mammalian central nervous system (CNS). In the embryonic brain, neuroepithelial cells and cortical radial glia appear sequentially, and are the ancestors of NPCs in both late developmental stage and adulthood [8]. NPCs proliferate in vitro when cultured in serum-free basal medium supplemented with fibroblast growth factor (FGF). Upon removal of FGF, they undergo differentiation, giving rise to neurons and glial cells. Adherent monolayer culture of NPCs provides a uniform culture environment. NPCs and their progeny can be readily monitored at single-cell resolution—a prerequisite for imaging-based cellular assays [9].

Bridget Wagner (ed.), *Phenotypic Screening: Methods and Protocols*, Methods in Molecular Biology, vol. 1787, https://doi.org/10.1007/978-1-4939-7847-2_8,

We have established a high-content imaging assay to quantify the formation of neurons upon NPC differentiation using immunofluorescent staining of the neuronal marker Tuj1 [10]. The assay has been optimized for high-throughput screen (HTS) of small-molecule libraries in 384-well format [4]. Here we provide protocols for the isolation, culture, and passage of NPCs prepared from mouse embryonic cortices, the process of the HTS assay, and the post-HTS analysis and validation of active hits.

2 Materials

2.1 Dissection of Cortical NPCs from Mouse Embryos

1. Animals: Time-mated pregnant CD1 or ICR mice carrying embryonic day 12.5 (E12.5) embryos (*see* **Note 1**).
2. Disinfectant: 75% ethanol in water.
3. Dissection instruments: Fine dissecting forceps (two pairs), delicate dissecting scissors (one pair), regular forceps and scissors (one pair each), and a small spatula. All instruments are autoclaved.
4. Dissecting microscope with light source.
5. Dissection buffer: Calcium- and magnesium-free Hanks' balanced salt solution (CMF-HBSS), supplemented with penicillin-streptomycin (P/S).

2.2 Cell Culture, Passage, and Cryopreservation

1. Coating solutions for adherent culture: 15 μg/mL poly-L-ornithine (PO) and 10 μg/mL fibronectin (FN). Dissolve 7.5 mg PO in 5.0 mL of distilled water (dH_2O), and sterilize with a 0.22 μm pore size syringe filter to make 100× stock solution. To make 100× FN stock solution, add 5 mL of sterile Dulbecco's phosphate-buffered saline (DPBS) directly into the vial containing 5 mg of FN. Gently swirl and invert the vial for several times. Let FN dissolve for ~1 h at room temperature (*see* **Note 2**). Make small aliquots of 100× PO and 100× FN stock solutions, respectively. Store at −20 °C, protected from light.
2. Cell culture vessels and microtiter assay plates: 100 mm tissue culture petri dishes, microtiter plates with clear and flat bottom but black walls for cell culture and screening, e.g., ViewPlate-384 (PerkinElmer), or Corning® 96-well plates.
3. Plasticware and glassware: 15 and 50 mL polypropylene conical tubes, 70 μm mesh size cell strainer, Pasteur pipettes (borosilicate type I clear glass, 150 mm) autoclaved and fire-polished.
4. Basic fibroblast growth factor (bFGF) stock solution (10,000×): 100 μg/mL bFGF. Dissolve 100 μg bFGF in 1 mL of 10 mM Tris–HCl (pH 7.6) containing 0.1% bovine serum albumin and sterilize with a 0.22 μm pore size filter. Make 10–20 μL aliquots and store at −20 °C.

5. Basic culture medium (BCM): Dulbecco's modified Eagle medium (DMEM)/F12 containing 2% B-27 supplement (*see* **Note 3**) and 1% P/S.
6. Complete BCM: BCM supplemented with 10 ng/mL bFGF.
7. Detaching cells for passaging: TrypLE™ Express (GIBCO, *see* **Note 4**).
8. Cryogenic vials and cell freezing container (e.g., Nalgene® Cryo 1 °C "Mr. Frosty").
9. Cell freezing medium: DMEM/F12 containing 40% fetal bovine serum (FBS) and 20% dimethyl sulfoxide (DMSO) (*see* **Note 5**).

2.3 Small-Molecule Chemical Library

1. High-quality collections of compounds: Small-molecule compounds, either acquired through vendor or internally generated, are usually supplied as pre-dissolved DMSO solutions at 10 mM in 96- or 384-well format (*see* **Notes 6–8**).
2. Positive reference compound (*see* **Note 9**), e.g., SB-216763, dissolved in DMSO at 10 mM.

2.4 Immunocytochemistry (ICC) Fluorescence Staining

1. Fixation solution: 4% Formaldehyde in DPBS pH 7.4, freshly diluted from 16% formaldehyde (w/v, ethanol free) (*see* **Note 10**).
2. Blocking solution: 3% Normal donkey serum in DPBS containing 0.1% Triton X-100.
3. Primary antibody: Neuron-specific class III β-tubulin (Tuj1) mouse monoclonal antibody (TU-20, *see* **Note 11**).
4. Secondary antibody: Appropriate fluorophore-conjugated donkey antibody (e.g., donkey anti-mouse Alexa Fluor 488, *see* **Note 11**).
5. DNA staining solution: 1 μg/mL DAPI.

2.5 Equipment for Automated HTS (See Note 12)

1. Automated reagent dispenser, e.g., Multidrop Combi Reagent Dispenser (Thermo Fisher Scientific) for seeding cells or dispensing reagents during ICC staining in 96- or 384-well format.
2. Automated liquid handler, e.g., Echo 555 Liquid Handler (Labcyte) for dispensing compounds in 96- or 384-well format.
3. Microplate washer or washing system for fluorescent immunoassays, e.g., ELx405 (BioTek).
4. High-content imaging system, e.g., Acumen (TTP Labtech), Cellomics (Thermo Fisher) or Operetta (PerkinElmer) for high-throughput imaging and analysis.

3 Methods

3.1 Basic Setup and Preparations

1. Coat culture dishes and microtiter plates with PO/FN at least 2 days prior to use (*see* **Note 13**). Make 15 μg/mL PO from 100× stock solution with sterile dH_2O, and add to petri dishes or assay plates to cover the culture surface.
2. Incubate overnight at 37 °C.
3. Aspirate the PO and wash 3× with dH_2O. Allow to air-dry.
4. Add 10 μg/mL FN solution freshly diluted from 100× stock solution with cold dH_2O, and incubate at 37 °C for 3–24 h.
5. Remove the FN, and wash once with dH2O. Use immediately, or allow to air-dry for storage. PO/FN-coated culture vessels can be stored at 4 °C for up to 2 weeks in a closed sterile container or sterile sealable bag.
6. On the day of dissection, prepare an appropriate amount of complete BCM and pre-warm to 37 °C in a water bath. Place a small amount of BCM and the dissection buffer (CMF-HBSS) on ice.
7. Disinfect dissection microscope and bench with 75% ethanol.

3.2 Harvesting E12.5 Mouse Brain and Micro-dissection

Follow dissection procedure as reported [11] (*see* **Note 14**).

1. Anesthetize a time-mated pregnant mouse on day 12 of gestation according to individual's institutional approved animal protocol. Perform cervical dislocation.
2. Spray the skin of the abdomen with 75% ethanol for sterilization. Grasp the skin with forceps and cut through the skin and the underlying fascia to open the entire peritoneal cavity. Remove the uterine horn with scissors and forceps. Wash with cold dissection buffer in 100 mm petri dish.
3. Open the uterine horn with fine scissors, remove the embryos from amniotic sacs using fine forceps, and transfer the embryos to a 100 mm petri dish containing cold dissection buffer with a spatula.
4. Under the microscope, decapitate the embryos with forceps, and transfer the heads to a new petri dish filled with cold dissection buffer. Free the brain from the skin and meninges using a pair of fine forceps.
5. Using micro-dissection scissors, on each hemisphere perform a dorsal horizontal cut from the rostral to caudal through the entire cortex. Repeat this procedure until all the brains are micro-dissected. Mince the cortices into small pieces using micro scissors.
6. Collect all the cortical pieces into a 50 mL conical tube containing 3–5 mL of cold BCM.

3.3 Dissociation of Cortices and Primary Culture of NPCs (See Note 15)

1. Dissociate the tissue mechanically using a fire-polished glass Pasteur pipette or P1000 pipette tip. Triturate the tissue gently by pressing the pipetting tip to the bottom of the tube and pipetting the suspension up and down for ~20 times to dissociate into a single-cell suspension.
2. Add another 3–5 mL of BCM to the cell suspension, and pass the entire cell suspension through a cell strainer (mesh size 70 μm) to a new 50 mL tube.
3. Pellet cell suspension by centrifugation at 200 × *g* for 7 min, room temperature.
4. Carefully discard the supernatant, resuspend the cell pellet in 10 mL of pre-warmed complete BCM supplemented with 10 ng/mL bFGF, and gently dissociate the pellet.
5. Perform a cell viability count using trypan blue stain. Adjust the viable cell concentration to 2×10^5 cells/mL with complete BCM.
6. Plate 10 mL of dissociated cell suspension into a 100 mm culture dish coated with PO/FN (i.e., 2×10^6 cells per dish).
7. Incubate at 37 °C in a humidified incubator with 5% CO_2.
8. Feed cultures with fresh bFGF every day at the final concentration of 10 ng/mL. Change half medium every other day. It takes ~6 days for the primary cells (P0) to reach 70–80% confluence.

3.4 Passaging and Expansion of NPCs

1. Passage P0 cells when reaching ~80% confluence (*see* **Note 16**). Aspirate the medium from the culture dish, and rinse cells with 10 mL of DPBS. Add 3 mL of TrypLE™ Express and incubate at 37 °C for 3–5 min, or until the cells become rounded and detached from the surface.
2. Remove the dish from incubator, and tap side of the dish with hand for several times to dislodge cells. Immediately add 10 mL of BCM, resuspend cells smoothly by pipetting up and down for two or three times, and then transfer cells to a 50 mL conical tube.
3. Rinse the dish once with BCM and transfer the remaining cells to the same tube.
4. Centrifuge at 200 × *g* for 7 min, room temperature.
5. Resuspend the cell pellet in 30 mL of complete BCM, count cells, and plate the cells in PO/FN pre-coated petri dishes at the density of 2×10^5/mL (~2×10^6 cells per dish). Culture cells to ~80% confluence (P1 cells).
6. Passage and re-plate P1 cells following **steps 1–5** to generate P2 cells for cryopreservation.

3.5 Cryopreservation of NPCs

1. Follow standard cryopreservation procedure as reported [12]. Briefly prepare cell freezing medium and keep at 4 °C.
2. Harvest P2 cells, resuspend the pellets in complete BDM, perform cell count, and adjust the cell density to 4×10^6/mL.
3. Transfer 0.5 mL of cell suspension to each cryogenic vial, and add an equal volume of cold cell freezing medium in drops while gently swirling the vial.
4. Transfer vials into a Nalgene® Cryo 1 °C freezing container (*see* **Note 17**). Place it in a −80 °C freezer overnight, and transfer the cryovials into a liquid nitrogen tank for long-term storage the following morning.

3.6 Thawing Cryopreserved NPCs and Expanding Cells for Assay

1. Estimate the number of vials to be thawed. Follow standard thawing procedure as reported [12]. Briefly warm complete BCM to room temperature. Remove the cryovial(s) containing frozen NPCs (P2) from liquid nitrogen storage and immediately place it into a 37 °C water bath.
2. Quickly thaw the cells (<1 min) by gently swirling the vial in the water bath until there is just a small bit of ice left in the vial (*see* **Note 18**).
3. Wipe the outside of entire cryovial with 75% ethanol, and then transfer the cell suspension from each cryovial to a 15 mL tube containing 10 mL of pre-warmed BCM.
4. Centrifuge at $200 \times g$ for 7 min at room temperature.
5. Gently resuspend the cell pellet from each cryovial in 10 mL of complete BDM, and plate the cells in a PO/FN pre-coated 100 mm petri dish.
6. Culture and passage P2 cells as described in Subheadings 3.3 and 3.4.
7. When P3 culture reaches ~80% confluence, harvest the cells for the screening assay (*see* **Note 19**).

3.7 Primary Screen

1. Set up plate map: The 384-well format is preferred for high- or medium-throughput screen using this assay. To prevent "edge effect" in microplates, an issue due to the increased evaporation rate of circumferential wells compared to the centrally located wells, the wells at the outer edges are filled with BCM but not used for compounds.
2. For each microplate, at least two columns of 384-well plate are arrayed with negative and positive control, respectively (Fig. 1). Negative control is the vehicle, i.e., BCM containing 0.01% DMSO, while positive control is a reference tool compound at the optimal concentration that can generate the largest neurogenic effect (*see* **Note 9**).

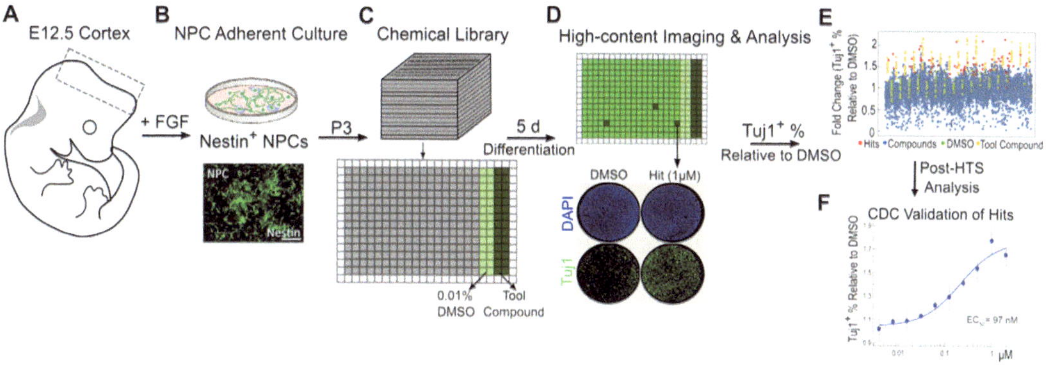

Fig. 1 Workflow of neurogenesis phenotypic screen using mouse primary NPCs. (**a**) Cortices of mouse E12.5 embryos are harvested for NPC adherent culture in the presence of FGF. (**b**) Nestin^{+} NPCs are enriched and expanded through multiple passages. (**c**) Passage 3 NPCs are added to 384-well microplates with arrayed small-molecule library at single dose, together with negative (0.01% DMSO) and positive (tool compound) reference control. (**d**) Following 5-d differentiation, all the plates are subject to ICC, high-content imaging, and analysis. The percentage of TuJ1-labeled neuron is quantified by each compound. All compounds are normalized to the controls on each plate. (**e**) Positive hits from primary screen are scored and illustrated. (**f**) CDC is used for validation of active hits (reproduced from [4, 10] with permission from Elsevier and SAGE Science Press, respectively)

3. A separate plate containing only positive and negative controls will be used to determine the Z'-factor [13]. A barcode with a unique ID is preferred to label each assay plate.
4. Prepare compound plates for single-dose primary screen: Compounds are properly dispensed from a chemical library into destination plates using a liquid handler (*see* **Note 12**) according to user's manual. The final concentration of each compound for the single dose usually ranges from 1 to 10 μM, taking into consideration of both expected efficacy and potential cytotoxicity associated with the dose.
5. Due to the considerable variability in cell-based assays in general, triplicates for each compound are preferred in the primary screen.
6. Add NPCs to microplates: Harvest P3 NPCs as described in Subheading 3.6. Resuspend the cells in BCM without FGF.
7. Using an automated reagent dispenser, dispense cell suspension into 384-well destination plates with the compound library, positive and negative controls at a density of 5×10^3/well. The final volume is 50 μL/well for 384-well microplates (*see* **Note 20**). The final concentration of DMSO in all wells is 0.01% in BCM.
8. Gently shake plates for 8–10 s to fully mix compounds with cells. Allow cells to settle down and attach to the bottom of plates before moving to the incubator.
9. Incubate NPCs at 37 °C with 5% CO_2 for 5 days for differentiation.

3.8 ICC Fluorescence Staining

1. Microplate washer or dispenser is set at the slowest speed for washing or dispensing during the process of staining.
2. Aspirate culture medium with a microplate washer.
3. Immediately dispense 50 μL of freshly prepared 4% PFA per well and fix cells for 20 min.
4. Rinse with 50 μL DPBS per well three times. Then aspirate DPBS using the microplate washer.
5. Dispense 50 μL of blocking buffer. Incubate cells for 30–60 min at room temperature.
6. Aspirate blocking buffer, and add 30 μL of class III β-tubulin (Tuj1) mouse monoclonal antibody diluted in blocking buffer at an optimal working concentration.
7. Incubate for 2 h at room temperature or overnight at 4 °C.
8. Rinse cells in 50 μL DPBS three times, and aspirate.
9. Add 30 μL of blocking buffer containing both appropriate fluorescence-conjugated secondary antibody at an optimal working concentration and 1 μg/mL DAPI.
10. Incubate for 2 h at room temperature in the dark.
11. Rinse cells in 50 μL DPBS three times.
12. Plates can be immediately scanned using high-content imager or kept from light at 4 °C overnight.

3.9 High-Content Imaging and Analysis

1. Load plate (plates) in position, and allow the registration of plate identity including user information and plate ID/barcode.
2. Choose appropriate imaging fields within sample wells excluding the edge wells, define pixel resolution, magnification, autofocus, exposure time, etc., and adjust image acquisition settings using positive and negative reference controls to acquire quality images (*see* **Note 21**).
3. To identify the primary objects (nuclear DAPI stain) and score secondary objects (Tuj1+ neurons), use positive and negative reference controls to determine the optimal image analysis algorithms, and to adjust features or parameters, such as segmentation, area, mask, and intensity threshold. The images with these parameters visible should be reviewed to vet the images and ensure that the objects appropriately scored from the background or separated from the artifacts [14].
4. Scan all the plates. With built-in software, high-content imager can run image analysis in real time. The image data can also be reanalyzed with modified measurement parameters. Export the data, including the number of $DAPI^+$ objects and $Tuj1^+$ objects.

5. The percentage of $Tuj1^+$ cells in total cell number ($DAPI^+$ cells), i.e., $Tuj1^+$%, is quantified as the readout for neurogenesis. The average of $Tuj1^+$% for 0.01% DMSO vehicle controls in each plate is used as the baseline reference to normalize the data by each plate (*see* **Note 22**). All compound data are normalized and shown as the fold change of $Tuj1^+$ % relative to DMSO controls. Positive hits are defined as the compounds that enhanced $Tuj1^+$ cell formation by at least three standard deviations above the mean of DMSO controls.
6. Assay validation and robustness are assessed by *Z'*-factor calculated as below, where σ is the standard deviation, μ the mean, *p* the positive control, and *n* the negative control:

$$Z' - \text{factor} = 1 - 3^* \frac{(\sigma_{\text{p}} + \sigma_{\text{n}})}{|\mu_{\text{p}} - \mu_{\text{n}}|}$$

7. Assays with a *Z'*-factor more than 0.4 are considered excellent (*see* **Note 23**).
8. The primary screen data can be visualized and reported using software such as Tibco Spotfire (Fig. 1).

3.10 Validation of Primary Hits

1. To confirm the activity of the hits in the primary screen, individual samples of active compounds from a different source—either repurchased from the original vendor or requested from a different synthesis lot—are arrayed in 384-well plates to generate concentration-response curves (CRC).
2. Dispense 8- or 10-point serial dilutions (at a ratio 1:3) of a compound (usually starting from 10 μM) in triplicates into destination plates using a liquid handler.
3. For each 384-well plate, use the similar plate map as the primary screen including negative and positive reference controls, add NPCs, and perform ICC, high-content imaging, and analysis as described in Subsections 3.7, 3.8 and 3.9.
4. Plot and analyze concentration-response data using a four-parameter logistic regression fit formula in Excel add-in XL fit or GraphPad Prism. Calculate EC_{50} values (Fig. 1).
5. Based on the objective of the screen, rank reproducible, concentration-dependent, active compounds by their efficacy, potency, and other available information such as physicochemical and ADME (absorption, distribution, metabolism, and excretion) properties of compound.
6. Prioritized active compounds are subject to target identification/validation [15], or further in vitro and in vivo assays for the further development of lead compound.

4 Notes

1. To generate a large batch of NPCs for assay characterization and high-throughput screening, 15–20 brains from 2 to 3 pregnant mice are usually required. The accuracy of gestation time is critical to isolate NPCs at desirable developmental stage for in vitro proliferation and differentiation, and to achieve reliable and reproducible assay results.
2. Do not vertex or agitate when dissolving or thawing FN. Vortex or rapid thaw to room temperature may cause proteins to aggregate and become sticky. Slowly thaw FN stock solution on ice, and dilute it in cold sterile dH2O.
3. Different batches of B27 supplement may exhibit large variability in their capabilities [16] to support NPCs in culture, which may result in considerable variations manifested in the rate of NPC proliferation and differentiation, the morphological characteristics of differentiated cells, and the cellular response to compounds. We recommend testing several batches of B27 supplement, and selecting the best batch for bulk storage and usage.
4. TrypLE is gentler on cells compared to trypsin.
5. Mix DMEM/F12 and FBS first before adding DMSO to prevent protein aggregation from serum.
6. The selection of compound library is based on (1) the objectives of the screen—identification of new chemistry entity for a novel target, or repositioning of drug-like molecules or approved drugs; (2) the biological targets of interests, e.g., kinases, G-protein-coupled receptors, ion channels, or CNS target-focused library; and/or (3) automation screen capabilities and cost. In addition to proprietary compound libraries generated by individual organization, there are collections of libraries commercially available, e.g., FDA-approved Drug Library [17].
7. High-quality compound library and compound management are vital to the successful outcome of the screen. The creators or vendors of compound library should enforce greater quality standards to guarantee the integrity of libraries, and the identity and purity of individual members of each library. The quality of all compounds should be greater than 90% pure with provided quality control procedures and data. It can be verified by 5% random sampling. A good collection of compound library selects the compounds that have lead-like qualities and desired extent of molecular complexity. Well-practiced procedures on compound management—the receipt, partition, storage, registration, QC, and tracking of all compounds—are

critical to the integrity and stability of each sample, the life span of a library, and the success of screen follow-up efforts.

8. The composition and cheminformatics information of a library should be provided to facilitate hit follow-up and prioritization, i.e., unique ID of each compound, structure, and if possible target annotation, physicochemical, and ADME (absorption, distribution, metabolism, and excretion) properties.
9. In addition to GSK3β inhibitors, BET inhibitors, e.g., (+)-JQ1, can be used as positive reference compounds for neurogenesis phenotypic screen [4].
10. Appropriate fixation of cultured cells is a critical step to the success of ICC staining and subsequent high-content imaging. Incomplete fixation may lead to reduced specific immunoreactivity and cell detachment from culture surface. Over-fixation may result in masking of the epitope or strong background staining. To ensure the reproducibility of the assay, we recommend prepare fresh 4% formaldehyde fixative in DPBS only before fixation.
11. Antibodies, especially polyclonal primary antibody from different batches, may result in altered signal-to-noise ratio for immunostaining. Prior to using a new batch of primary or secondary antibody for phenotypic screen assay, characterization of the antibody and optimization of antibody concentration is required to produce optimal signal-to-noise ratio for ICC staining.
12. Automated reagent dispenser or liquid handler is an essential tool for HTS. Not only does it enable lab automation and high speed, but also deliver high-performance and precision—a key to ensure data quality. Multidrop Combi Reagent Dispenser (Thermo Fisher Scientific) is often chosen for dispensing cell suspension into 96- or 384-well microplates owing to its easy-to-use protocol and quick operation process. To ensure reproducible cell dispensing results, maintain the evenness of cell suspension in the reservoir by gently swirling the conical tube for several times during dispensing intervals. For dispensing compounds, taking advantage of an acoustic non-contact dispensing technology, Echo 555 Liquid Handler (Labcyte) avoids the risk of compound loss and cross-contamination associated with traditional tip-based liquid handling. It also supports and simplifies a broad range of applications in HTS primary and secondary assays with a dispensing volume as low as 2.5 nL.
13. Appropriate coating is critical to maintaining a healthy adherent culture and ensuring the success of ICC staining and subsequent high-content imaging. The sequential coating

procedure with PO/FN described here is optimized for NPC in vitro proliferation and differentiation.

14. To have a consistent primary NPC preparation with high viability and high quality, we recommend the following: (1) prepare all materials and reagents required for tissue dissection, cell isolation, and culture before starting to harvest embryos; (2) keep the embryos, brains, and cortices in cold dissection buffer or BCM throughout the dissection; (3) perform the procedure as quickly as possible; (4) teamwork is an effective way to reduce individual's workload and stress associated with micro-dissection, and to speed up the whole process for NPC isolation.
15. To avoid over-triturating cells, we recommend using pre-coated fire-polished Pasteur glass pipette or polypropylene P1000 tips by pipetting BCM up and down for several times before trituration. Then triturate slowly by hand to avoid getting bubbles and shearing cells.
16. Do not allow cells to exceed 90% confluence, because it will lead to premature differentiation of NPCs and reduced yields with passages.
17. Isopropanol should not be subject to $\geq$5 freeze-thaw cycles; otherwise it will decrease cell viability during cryopreservation.
18. To avoid contamination, immerse the vial in the water without submerging the cap in the 37 °C water bath.
19. With three passages, NPC culture is enriched selectively with Nestin$^+$ progenitors (~90% population, Fig. 1) and expanded exponentially from a limited number of starting cells, which is sufficient for high-throughput screening. Given the reliability and reproducibility of P3 cells on cellular response in our hands, P3 NPCs are recommended to use in HTS. Further passaging the cells is not recommended.
20. For the 96-well format, the cell density is 1×10^4/well and the final volume is 200 μL/well.
21. It is imperative that a good image is captured for subsequent analysis. However, compromises may need to be made when adjusting the mage acquisition settings to ensure that all images within one experiment are taken using exactly the same parameters.
22. If the performance of positive controls is reproducibly robust, the data for each plate can also be normalized to negative and positive controls by each plate as 0% and100%, respectively, which allows normalization across plates and batches.
23. The Z′ factor is a measurement of statistical effect size using particular positive and negative control. It informs the user whether the response in an assay manifested with particular

positive and negative control is large enough to warrant a full-scale HTS. If the controls, especially the positive control, are not appropriately behaved, to achieve a favorable Z′ factor (≥0.4) could be challenging. However, unfavorable Z′ factor may not necessarily inform the reliability of assay [14]. Primary hits can be further validated by CDC.

References

1. Gage FH, Temple S (2013) Neural stem cells: generating and regenerating the brain. Neuron 80:588–601
2. Goldman SA, Nedergaard M, Windrem MS (2012) Glial progenitor cell-based treatment and modeling of neurological disease. Science 338:491–495
3. Lu P, Wang Y, Graham L, McHale K, Gao M, Wu D et al (2012) Long-distance growth and connectivity of neural stem cells after severe spinal cord injury. Cell 150:1264–1273
4. Li J, Ma J, Meng G, Lin H, Wu S, Wang J et al (2016) BET bromodomain inhibition promotes neurogenesis while inhibiting gliogenesis in neural progenitor cells. Stem Cell Res 17:212–221
5. Schneider JW, Gao Z, Li S, Farooqi M, Tang TS, Bezprozvanny I et al (2008) Small-molecule activation of neuronal cell fate. Nat Chem Biol 4:408–410
6. Filippakopoulos P, Qi J, Picaud S, Shen Y, Smith WB, Fedorov O et al (2010) Selective inhibition of BET bromodomains. Nature 468:1067–1073
7. Nicodeme E, Jeffrey KL, Schaefer U, Beinke S, Dewell S, Chung CW et al (2010) Suppression of inflammation by a synthetic histone mimic. Nature 468:1119–1123
8. Kriegstein A, Alvarez-Buylla A (2009) The glial nature of embryonic and adult neural stem cells. Annu Rev Neurosci 32:149–184
9. Danovi D, Folarin AA, Baranowski B, Pollard SM (2012) High content screening of defined chemical libraries using normal and glioma-derived neural stem cell lines. Methods Enzymol 506:311–329
10. Kim KJ, Wang J, Xu X, Wu S, Zhang W, Qin Z et al (2012) A chemical genomics screen to discover genes that modulate neural stem cell differentiation. J Biomol Screen 17:129–139
11. Azari H, Sharififar S, Rahman M, Ansari S, Reynolds BA (2011) Establishing embryonic mouse neural stem cell culture using the neurosphere assay. J Vis Exp 47:2457
12. Ricardo R, Phelan K (2008) Freezing, thawing, and packaging cells for transport. J Vis Exp 17:757
13. Zhang JH, Chung TD, Oldenburg KR (1999) A simple statistical parameter for use in evaluation and validation of high throughput screening assays. J Biomol Screen 4:67–73
14. Buchser W, Collins M, Garyantes T, Guha R, Haney S, Lemmon V et al (2004) Assay development guidelines for image-based high content screening, high content analysis and high content imaging. In: Sittampalam GS et al (eds) Assay guidance manual. Eli Lilly & Company and the National Center for Advancing Translational Sciences, Bethesda
15. Schenone M, Dancik V, Wagner BK, Clemons PA (2013) Target identification and mechanism of action in chemical biology and drug discovery. Nat Chem Biol 9:232–240
16. Chen Y, Stevens B, Chang J, Milbrandt J, Barres BA, Hell JW (2008) NS21: re-defined and modified supplement B27 for neuronal cultures. J Neurosci Methods 171:239–247
17. Dandapani S, Rosse G, Southall N, Salvino JM, Thomas CJ (2012) Selecting, acquiring, and using small molecule libraries for high-throughput screening. Curr Protoc Chem Biol 4:177–191

Chapter 9

Discovery of Modulators of Adipocyte Physiology Using Fully Functionalized Fragments

Andrea Galmozzi, Christopher G. Parker, Bernard P. Kok, Benjamin F. Cravatt, and Enrique Saez

Abstract

Defects in adipocyte function associated with obesity drive the development of systemic insulin resistance and type 2 diabetes. Agents that correct obesity-linked adipocyte dysfunction serve as useful insulin sensitizers in humans, as is exemplified by the thiazolidinediones (TZDs). We have developed a new platform that integrates advanced chemoproteomics with phenotypic screening to identify small molecules that promote differentiation and lipid storage in adipocytes, and, in tandem, their molecular target(s). These molecules mimic the activity of TZDs in culture and thus may also serve as insulin sensitizers in vivo. Central to this platform is the use of fully functionalized fragment (FFF) probes that consist of a variable, fragment-like recognition element linked to an alkyne-diazirine group that enables the photoactivated capture of probe-bound proteins directly in living cells and subsequent copper-catalyzed azide-alkyne cycloaddition to reporter tags for enrichment and identification of these probe-bound proteins by mass spectrometry. This platform, which can be adapted to diverse screens and cell types beyond adipocytes, has the potential to uncover new biological pathways amenable to pharmacological modulation that may impact human disease.

Key words Phenotypic screening, Adipose tissue, Adipogenesis, Photoreactive small molecules, Chemical proteomics, Target identification

1 Introduction

As the incidence of obesity and obesity-associated type 2 diabetes continues to increase, there is a critical need to develop new diabetes drugs with a mechanism of action distinct from that of existing treatments. Advances in our understanding of adipose tissue highlight the central role that adipocytes play in systemic glucose homeostasis and in the development of obesity-linked diabetes [1, 2]. The ability of obese, hypertrophied adipocytes to dispose of glucose, store lipids, and secrete insulin-sensitizing adipokines is severely

Andrea Galmozzi and Christopher G. Parker are co-first authors.

Bridget Wagner (ed.), *Phenotypic Screening: Methods and Protocols*, Methods in Molecular Biology, vol. 1787, https://doi.org/10.1007/978-1-4939-7847-2_9, © Springer Science+Business Media, LLC, part of Springer Nature 2018 compromised, and this contributes to insulin resistance, hyperglycemia, and deposition of lipids in other tissues (liver, muscle) that impairs whole-body insulin action [3]. Agents such as TZDs that can revert these defects and restore adipocyte function (e.g., lipid retention, secretion of beneficial adipokines) and normal lipid partitioning among tissues are used as diabetes medications in humans.

Phenotypic screens in adipocytes are an attractive strategy to discover molecular pathways that may be pharmacologically modulated to correct obesity-linked adipocyte dysfunction. It has been shown that molecules that mimic the effects of TZDs in cultured preadipocytes can have therapeutic effects in diabetic mice [4–6]. However, technical hurdles intrinsic to phenotypic screens have restricted wider application of this approach. Primary among these is the fact that identification of the molecular target(s) of bioactive compounds is often an arduous task [7, 8]. Without a target, understanding the mode of action of hits and optimizing leads can be difficult.

To address this limitation of cell-based screens, we recently described an integrated platform for phenotypic screening and chemoproteomics-driven target identification based on the use of libraries of fully functionalized fragment (FFF) probes [9] that enable broad exploration of the proteome and direct progression from phenotypic screen to target identification in living cells. Each member of the FFF library consists of a variable drug-like low-molecular-weight (MW $<$ 300 Da) small molecule that is affixed to a conserved element comprised of a photoreactive diazirine group to enable UV light-induced cross-linking of probe to interacting proteins, and a "clickable" alkyne handle for reporter tag conjugation to visualize, enrich, and identify cross-linked proteins (Fig. 1). The use of FFF libraries in phenotypic screens offers important advantages over classical approaches, such as affinity-based chromatography, where a hit compound is immobilized on a solid support for target enrichment from cell lysates. First, unlike affinity chromatography, FFFs enable direct identification of phenotype-relevant target proteins without requiring chemical modification of the screening hit(s). Second, treatment of cells with FFFs, followed by UV light-induced covalent trapping of small molecule-protein interactions, provides a way to enrich and identify FFF-interacting proteins directly from intact cells. Finally, because the identification of FFF probe-interacting proteins is performed in living cells, lower affinity or labile interactions prone to disruption by cell lysis/processing protocols used in biochemical affinity enrichment strategies are preserved, thus facilitating the discovery of context-dependent and/or short-lived probe-protein interactions.

These notable strengths of FFF-based screens are balanced by the need to perform careful target deconvolution studies to discern phenotypically relevant target(s) of FFF hits from proteins that may interact with these compounds, but not contribute to the phenotypic

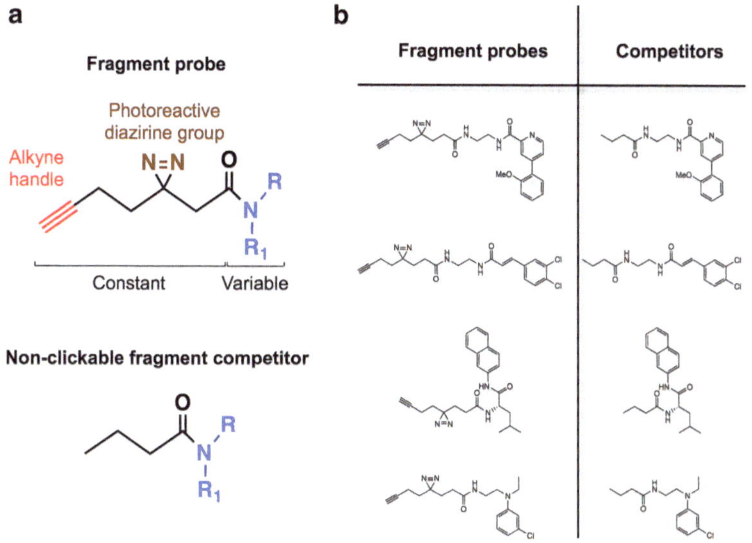

Fig. 1 Basic structure of fully functionalized fragment probes (FFFs). (**a**) Schematic of FFF probes showing their basic features: (1) a variable head group that confers specificity toward target proteins, (2) a photoreactive diazirine group for UV light-induced cross-linking of probes to their interacting proteins, and (3) an alkyne handle for copper(I)-catalyzed alkyne-azide cycloaddition (CuAAC) of reporter tags (e.g., fluorophores or biotin) for downstream applications. (**b**) Examples of FFF probes are shown, along with their paired nonfunctionalized competitors used in target identification studies

effect. Accordingly, both inactive control compounds and paired competitors lacking the photoreactive/clickable element serve as critical components of the workflow for target deconvolution, and we outline below how these reagents can be used in quantitative mass spectrometry (MS)-based competitive assays to facilitate target identification of FFF probes that promote adipocyte differentiation [9].

In a typical workflow, a phenotypic screen is performed in preadipocytes or adipocytes and hit FFFs are validated and prioritized for target identification. Protein target(s) of prioritized bioactive FFFs are identified using quantitative MS-based chemical proteomics methods, such as stable isotope labeling with amino acids in cell culture (SILAC) [10] or isotopic reductive dimethylation (ReDiMe) [11] coupled to multidimensional protein identification technology (MudPIT) [12]. The relevant protein target of the prioritized FFF for the elicited biological response is then confirmed using genetic tools (e.g., RNAi). Here, we describe the general procedure to identify positive regulators of adipocyte differentiation and lipid storage using the preadipocyte line 3T3-L1, although other cell lines or primary preadipocytes can also be used. The sections on FFF hit selection and protein target identification also apply to screens in adipocytes, or in any other cell type, in which a different phenotype is desired (e.g., glucose-stimulated insulin secretion).

2 Materials

2.1 Phenotypic Screen

1. 3T3-L1 preadipocyte cell line.
2. Culture medium: DMEM supplemented with 10% bovine calf serum (BCS), 1% penicillin and streptomycin.
3. Adipogenesis induction medium: DMEM supplemented with 10% fetal bovine serum (FBS), 1% penicillin and streptomycin, 1 μg/mL insulin, 0.4 μg/mL dexamethasone, 0.5 mM 3-isobutyl-1-methylxantine (IBMX).
4. Maintenance medium: DMEM supplemented with 10% FBS and 1 μg/mL insulin.
5. Rosiglitazone (a thiazolidinedione that is a direct synthetic ligand of the proadipogenic transcription factor PPARγ).
6. Library of fully functionalized fragment probes (FFFs) [9].
7. Nile red stain (AdipoRed): Working solution is 1:250 dilution of AdipoRed reagent in DPBS. BODIPY 493/503 can also be used for neutral lipid staining instead of Nile red.
8. Hoechst 33342: Working concentration: 0.1 μg/mL.
9. 4% Paraformaldehyde (PFA) solution in DPBS.
10. DPBS, no calcium, no magnesium.
11. High-content imaging system
12. 0.05% Trypsin-EDTA.
13. 96-Well black-walled, clear-bottom plates.

2.2 Gel- and Mass Spectrometry-Based Studies

All solutions are prepared with mass spectrometry (MS)-grade water/DPBS.

1. DMEM medium for SILAC.
2. Fetal bovine serum, dialyzed, heat inactivated.
3. [^{12}C, ^{14}N]-L-lysine, [^{12}C, ^{14}N]-L-arginine ("light") and [^{13}C, ^{15}N]-L-lysine, [^{13}C, ^{15}N]-L-arginine ("heavy") amino acids.
4. UV cross-linker for 365 nm UV light exposure.
5. Probe sonicator.
6. Click-chemistry reaction mix (10×): 10 mM Tris-(2-carboxyethyl)-phosphine HCl (TCEP, freshly made), 1 mM tris-(benzyltriazolylmethyl)-amine (TBTA), 10 mM $CuSO_4$, and 250 μM tetramethylrhodamine (TAMRA) azide (for gel-based visualization) or 1 mM biotin-PEG-azide (for MS analysis).
7. 4–12% Bis-Tris protein gels.
8. Methanol/chloroform 4:1 solution.
9. 6 M Urea (proteomics grade) in DPBS.

10. 10% SDS solution in DPBS.
11. 600 mM Potassium carbonate solution in DPBS.
12. 400 mM Iodoacetamide solution in DPBS.
13. Streptavidin agarose resin.
14. Calcium chloride.
15. Sequencing-grade modified trypsin.
16. Formic acid.

3 Methods

3.1 Phenotypic Screen

1. Seed 10,000 3T3-L1 cells in culture medium on gelatin-coated 96-well plates (final volume 100 μL/well). Replace medium every 2 days until preadipocytes reach confluence (*see* **Note 1**).
2. After cells reach confluence (day -2), wait for 2 days before starting differentiation regimen.
3. At day 0, gently aspirate/remove medium and replace with adipogenesis induction medium supplemented with vehicle (DMSO), rosiglitazone 2 μM (positive control), or FFFs (10–50 μM). Make sure to include multiple control wells per 96-well plate at different spots. The screen should be performed at least in duplicate.
4. On day 2, carefully aspirate/remove adipogenesis induction medium and replace with maintenance medium supplemented with vehicle, rosiglitazone, or FFFs.
5. Refresh maintenance medium every 2 days with minimal agitation (i.e., on days 4 and 6).
6. On day 8, gently aspirate medium, wash cells with DPBS, and add 80 μL of staining solution containing the fluorescent neutral lipid dye Nile red and Hoechst 33342 (for nuclei counterstain). Incubate cells for 10 min at 37 °C. Wash with DPBS and add 50 μL of DPBS to each well.
7. Measure fluorescence and image all wells for lipid (red channel) and nuclei (blue channel) staining using a high-content imager (*see* **Note 2**). A general workflow scheme is shown in Fig. 2.
8. (Optional) Alternatively, cells can be fixed without staining and plates stored at 4 °C for a short time (up to 2 weeks) before staining and imaging.
9. On day 8, wash cells with DPBS, and add 80 μL of 4% paraformaldehyde solution. Incubate cells for 10 min at room temperature, remove fixative, wash with DPBS, and add 80 μL of DPBS to each well.

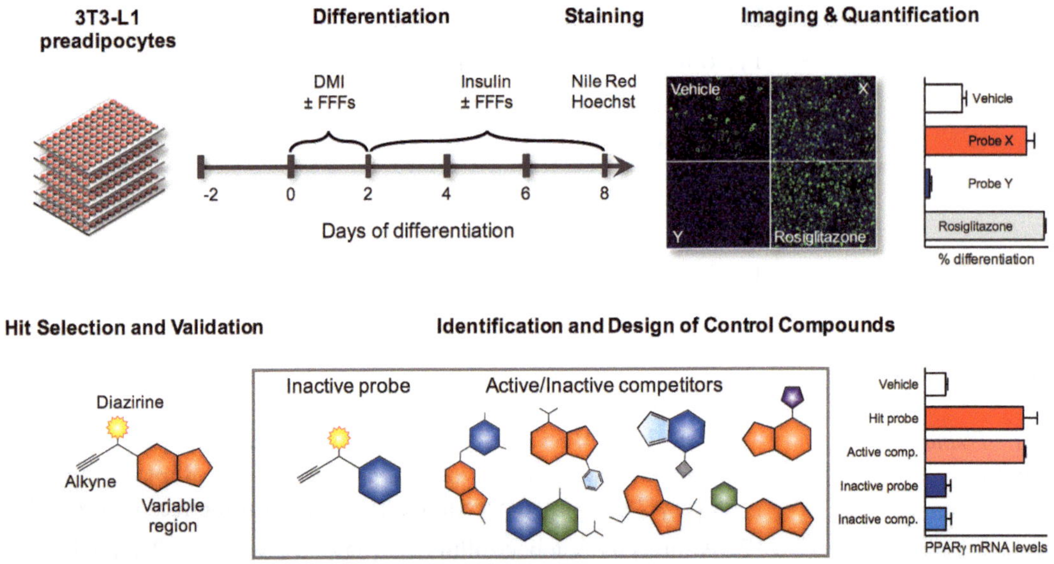

Fig. 2 Phenotypic screen to isolate proadipogenic FFFs. (**a**) Preadipocytes are seeded and induced to differentiate into adipocytes in the presence of FFFs. On day 8, cells are stained with the lipid dye Nile red and imaged and quantified using a high-content imager. FFFs are ranked based on their ability to promote differentiation and lipid accumulation. (**b**) Following validation of hit FFF probes, control FFF probes and nonfunctionalized active and inactive competitors are selected to facilitate target deconvolution

10. Seal plates with parafilm and aluminum foil. Store at 4 °C. Perform lipid and nuclei staining right before imaging as described in **step 6** of Subheading 3.1.

3.2 Hit Selection and Prioritization for Target Identification

In this screen, FFFs that induce lipid accumulation >3-fold relative to vehicle are selected as positive hits. Hit FFF probes are then retested at different concentrations (at least 2, the screening concentration and a lower dose) in larger formats (e.g., 24-well) and with additional biological replicates. Lipid staining, as well as gene expression analysis of adipocyte markers evaluated using real-time quantitative PCR (e.g., PPARγ, adiponectin, FABP4, GLUT4, CD36), is used to rank order FFF hits based on their potency (*see* **Note 3**). To prioritize FFF probes that work via novel mechanisms of action, validated hits are also tested for their ability to directly activate PPARγ in a luciferase reporter assay [4, 9]. Compounds that act as direct PPARγ agonists, an established mechanism for their proadipogenic activity, are eliminated.

3.3 Design and Identification of Control Compounds

The availability of control molecules is critical for successful target identification. FFF probes that are structurally similar to active FFFs may share many, but not all, interactions with protein targets. However, if these structurally related compounds exhibit little or no ability to promote differentiation and lipid accumulation, it is likely that the protein(s) responsible for these effects are exclusive

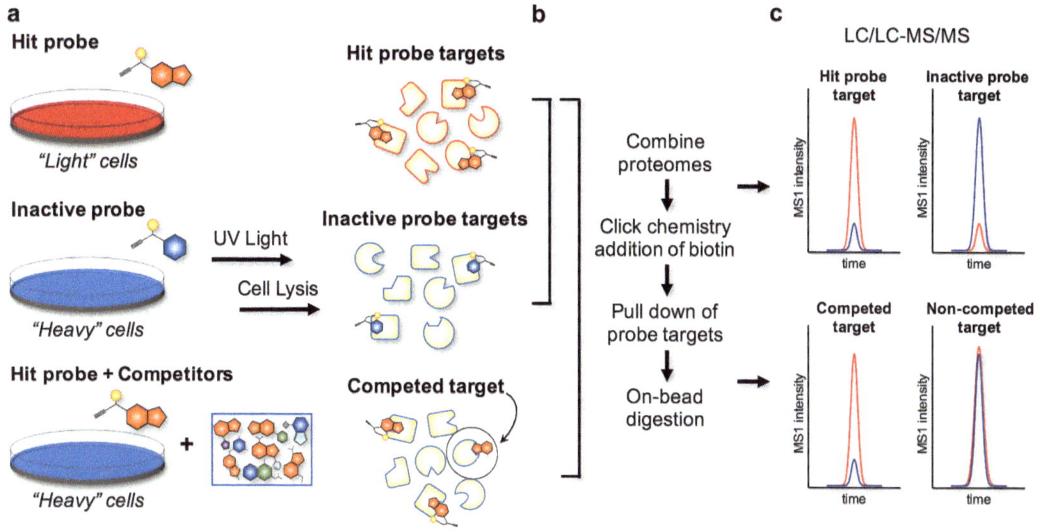

Fig. 3 SILAC-MudPIT analysis to identify relevant protein targets of FFF probes. Experimental workflow used to identify protein target(s) of active FFF probes. (**a**) Light and heavy SILAC cells are treated with the indicated FFF probes in the presence or absence of excess active or inactive competitor. After 30 min, cells are irradiated with UV light for 10 min to cross-link probe-interacting proteins. (**b**) Heavy and light proteomes are mixed in equal proportions, and click chemistry is used to add a biotin tag to the FFF probe to enable enrichment of target proteins. After streptavidin pull down, FFF targets are digested on-bead with trypsin and the resulting peptides analyzed by liquid chromatography and tandem mass spectrometry. (**c**) Relevant targets are defined as proteins with >3-fold enrichment by the active FFF probe over the inactive FFF probe, competed >3-fold with the active nonfunctionalized competitor, *and* competed <2-fold by any inactive nonfunctionalized competitor

to the active FFF probe. As such, comparisons of interacting proteins of active versus inactive control FFF probes allow for the exclusion of proteins shared between related molecules but unrelated to the observed bioactivity, as well as the identification of phenotypically relevant target(s) specific to the active FFF probe (Fig. 3). Further, to identify high-occupancy, saturable targets specific to the active FFF probe, it is also essential to perform competitive assays using paired nonfunctionalized versions of both the phenotypically active and inactive FFF probes (i.e., molecules lacking the photoreactive/clickable element). Analysis of primary screening data should ideally furnish a set of control compounds that include the following:

- Inactive probes: FFF probes that structurally resemble active hits, but show no effect in the phenotypic screen. Inactive probes can assist in the sorting of targets that are more selective for the active FFF probe. They are expected to share some, but not all, protein targets of active FFF probes.
- Active competitors: Nonfunctionalized derivatives of the hit FFF probe. Active competitors show similar performance in the phenotypic assay. When incubated in excess with an active

FFF probe, active competitors allow the identification of high-occupancy targets of the active FFF probe and therefore potentially relevant to the observed activity.

- Inactive competitors: Nonfunctionalized derivatives of inactive FFF probes. Similar to their functionalized counterparts, they have no effect on the phenotypic screen. Inactive competitors are used to identify high-occupancy targets of the active FFF probe that are unrelated to the observed phenotype (*see* **Note 4**).
- Background FFF probe: This probe has a methyl group substituted at the recognition element and is designed to identify nonspecific interactions of the cross-linking enrichment tag that is shared by all FFF probes. Proteins enriched with this control FFF can be considered as the "background noise" of the method.

3.4 Preparation of Probe-Labeled SILAC Samples for MS- and Gel-Based Protein Analysis

1. Grow cells in which target identification is to be performed (e.g., 3T3-L1 preadipocytes) for five passages in SILAC DMEM—10% dialyzed FBS supplemented with isotopically "light" and "heavy" amino acids, to allow their complete incorporation into proteomes. For gel-based analyses, cells can be grown in standard label-free medium.
2. Seed approximately two million heavy and light 3T3-L1 cells in separate 10 cm dishes. Grow cells to confluence.
3. Wash cells with DPBS.
4. Add serum-free DMEM containing FFF hits to be tested. For active-versus-inactive probe experiments, treat light and heavy cells with active and inactive FFF probes, respectively. For competition experiments, treat light cells with serum-free DMEM containing the active FFF probe and vehicle (DMSO), and heavy cells with the active FFF probe in the presence of a 10× excess of active or inactive competitor. Incubate at 37 °C for 30 min (*see* **Note 5**).
5. Remove medium and expose cells to 365 nm UV light for 10 min at 4 °C. Include a no-UV condition in which cells are incubated at 4 °C for 10 min under ambient light (*see* **Note 6**).
6. Wash cells once with ice-cold DPBS (5 mL).
7. Harvest cells with cold DPBS and spin down to pellet cells. Cell pellets can be stored at −80 °C until analysis.
8. Resuspend cell pellets in 500 μL of ice-cold DPBS.
9. Sonicate with a probe sonicator (if using a Branson Analog Sonifier 250, five pulses, 30% duty cycle, output setting = 4).
10. Adjust cell lysates to a concentration of 1.5 mg/mL.

3.5 Proteome Processing for Mass Spectrometry Analysis

1. Combine light and heavy proteomes in equal proportions to obtain 1 mL of 1.5 mg/mL (50% light, 50% heavy).
2. Add 110 μL of freshly prepared 10× click chemistry reaction to each sample to conjugate biotin to FFF probe-labeled proteins.
3. Rotate at room temperature for 1 h.
4. Transfer to a 15 mL tube. Incubate on ice and add 2.5 mL of cold 4:1 methanol (MeOH)/chloroform ($CHCl_3$) mixture, and 1 mL of cold DPBS.
5. Vortex until solution appears homogeneously cloudy.
6. Centrifuge samples at 5000 × *g* for 10 min at 4°C to fractionate the protein interphase from the organic and aqueous solvent layers.
7. Wash the protein disk carefully three times with 1 mL cold 1:1 MeOH:$CHCl_3$.
8. Add 3 mL of cold 4:1 MeOH:$CHCl_3$ and sonicate samples to ensure that unreacted click chemistry reagents are efficiently removed.
9. Pellet the remaining precipitate by centrifugation (5000 × *g*, 10 min, 4°C).
10. Discard supernatant and add 500 μL of 6 M MS-grade urea solution (prepared fresh in DPBS) containing 10 μL of 10% SDS.
11. Resuspend pellet by sonication.
12. Add 50 μL of a 1:1 mixture containing TCEP (200 mM in DPBS) and potassium carbonate (600 mM in DPBS) to reduce disulfide bonds. Incubate for 30 min at 37°C.
13. Alkylate reduced thiols by adding 70 μL of 400 mM iodoacetamide. Incubate for 30 min at room temperature, protected from light.
14. Add 130 μL of 10% SDS (in DPBS) and 5.5 mL DPBS and incubate with pre-equilibrated streptavidin agarose resin (100 μL 1:1 slurry) for 1.5 h at ambient temperature on a rotator.
15. Pellet streptavidin beads by centrifugation (1400 × *g*, 2 min).
16. Wash with 5 mL of 0.2% SDS in DPBS. Pellet beads by centrifugation.
17. Wash two times with 5 mL of detergent-free DPBS. Pellet beads by centrifugation.
18. Wash two times with 5 mL of dH_2O. Pellet beads by centrifugation.
19. Transfer streptavidin beads to a Protein LoBind tube (Eppendorf).

20. Digest bound proteins on-bead overnight at 37 °C under constant shaking in 200 μL of DPBS containing 2 μg sequencing-grade modified trypsin, 2 M urea, and 1 mM $CaCl_2$.
21. Transfer proteolyzed supernatant to a new Protein LoBind tube and acidify with formic acid (5% final).
22. Analyze samples by liquid chromatography-tandem mass spectrometry (LC/LC-MS/MS) or store samples at −20 °C until analysis. *See* [9, 13] for examples of instrument settings, quantitation methods, and data analysis.

3.6 Identification of Relevant Active FFF Probe Protein Target(s)

To be classified as active FFF probe targets, proteins must be:

1. Labeled in a UV-dependent manner (>5-fold enrichment in UV versus no-UV experiments; *see* **Note 6**).
2. Selectively enriched by the active FFF probe over the inactive FFF probe (>3-fold enrichment; *see* **Note 7**).
3. Competed by active nonfunctionalized competitor (>3-fold), but not competed by inactive nonfunctionalized competitors (<2-fold).

3.7 Gel-Based Analysis

Direct hit FFF probe labeling of proteins that respect the above criteria is confirmed by overexpression of putative targets in HEK293T cells in 6-well plates. Mock-transfected and target-overexpressing cells are exposed to active FFF probe and labeling is competed with increasing concentrations of active or inactive nonfunctionalized competitors (Fig. 4). Treatments are performed as described above (*see* Subheading 3.4, **step 4**), but cells do not need to be metabolically labeled unless subsequent MS analysis is desired. The following steps are performed after UV light exposure:

1. Wash cells with ice-cold DPBS (1 mL).
2. Harvest cells with 200 μL of ice-cold DPBS and keep on ice.

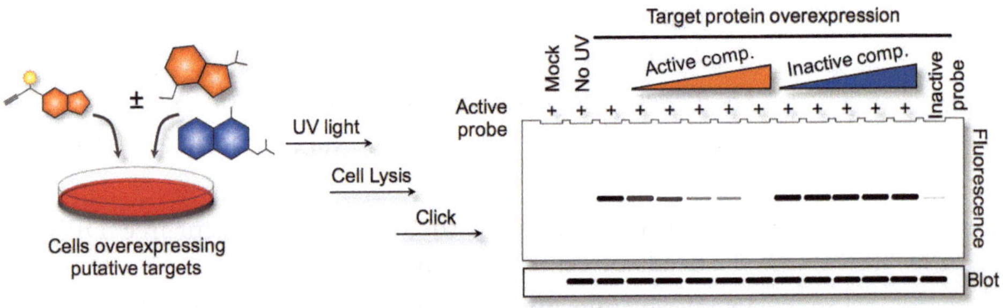

Fig. 4 Visualization of FFF probe target proteins by SDS-PAGE. Cells overexpressing the putative relevant protein target of an active FFF are exposed to the FFF probe in the presence and absence of nonfunctionalized competitors for 30 min, followed by UV light-induced photocross-linking, cell lysis, click chemistry conjugation of a fluorescent tag, SDS-PAGE separation, and visualization by in-gel fluorescence scanning and western blot

3. Lyse cells by sonication.
4. Quantify protein concentration.
5. Aliquot 50 μg of proteomes into new tubes and adjust volume to 50 μL with DPBS.
6. Add 6 μL of 10× click chemistry reaction mix to each sample to conjugate rhodamine to FFF probe-labeled proteins.
7. Mix vigorously and incubate for 1 h at room temperature.
8. Quench click chemistry reaction by adding 17 μL of 4× SDS loading buffer.
9. Load 25 μL on a 4–12% Bis-Tris acrylamide gel.
10. Run SDS-PAGE for 45 min at 165 V.
11. Visualize in-gel fluorescence using a fluorescence scanner.
12. Transfer proteins to PVDF or nitrocellulose membrane and perform western blot analysis to confirm overexpression and normalize fluorescent signals.

3.8 Genetic Validation of Relevant Active FFF Probe Target Protein

This phenotypic screening strategy can identify compounds that either inhibit or act as gain-of-function ligands on their targets. To discern the relevant protein target for a compound's effects, lentiviruses (e.g., pLKO.1 and/or regulated versions if needed) can be used to express shRNA against each putative target identified in Subheading 3.6 (four constructs/target) in the cells in which the compound exhibits its effects. The infected cells are then phenotyped in the presence and absence of active FFF probe. For FFFs behaving as inhibitors, knockdown of the relevant protein target will mimic the effect of the FFF. For gain-of-function ligands, knockdown of the relevant protein target will abolish the effect of the FFF. In this case, complementation studies can be used to further establish the identity of the relevant target.

4 Notes

1. Edge effects due to uneven evaporation and humidity amid wells can be common in 96-well plates, especially when cells are cultured for several days. This usually results in reduced adipocyte differentiation in wells along the edges and at corners. To avoid these problems and minimize variability, external wells should not be used and should instead be filled with 200 μL of medium or DPBS only. In addition, to verify that differentiation is comparable between plates, several wells of vehicle- and rosiglitazone-treated cells should be present in each plate and located in different positions.
2. The adipogenic potential of 3T3-L1 cells can vary significantly from batch to batch and must be considered carefully before

setting up the phenotypic screen. If the chief goal is to find positive modulators of adipogenesis, vehicle-treated cells should display a low-medium level of differentiation, such that the assay window is optimal to identify compounds that promote differentiation. On the other hand, if negative modulators of adipogenesis are sought, a batch of cells showing a medium-high level of basal differentiation is preferred.

3. As described, this phenotypic screen constitutes an intrinsic filter of toxicity, given that cells are treated for several days with compounds and hits are identified using a positive readout. Therefore, hit FFF probes are likely to be well tolerated at the screening concentration. If the screen is run to identify compounds that block differentiation, hits must be carefully filtered in secondary cell viability assays (e.g., CellTiter-Glo) to exclude compounds that show significant toxicity and can thus be considered false-positive hits.
4. Control FFF probes can be sought within the library of FFFs that has been screened. A set of structurally similar compounds that do not affect differentiation/lipid accumulation can be selected for initial validation. Small, second-generation libraries can also be synthetically generated around hit FFF probes to identify additional active and inactive analogs.
5. Prior to SILAC MS analysis, it is important to define the temporal window of action of active FFF probes (the time at which the relevant protein target is present in the cells), as well as the minimal concentration at which active FFF probes exert their effects (to reduce the number of nonspecific interactions). In initial MS-based target deconvolution studies, concentrations between the apparent EC_{50}/IC_{50} and EC_{90}/IC_{90} should be used. Using lower concentrations may run the risk of failure to detect lower abundance targets, while using significantly higher concentrations may result in the enrichment of low-affinity off-targets that are unlikely to be relevant for the observed bioactivity. Cells can be induced to differentiate and be exposed to different concentrations of active FFFs at different times for varying duration. This will help to characterize the effects of hit FFF probes and select optimal experimental conditions for successful target identification.
6. Proteome profiling of FFF probes and target deconvolution can be challenging to interpret. FFF probes can potentially interact with a large number of proteins with varying affinity and selectivity. Similarly, some proteins can be very promiscuous and may interact with many different FFF probes. Having several structurally similar, but inactive analogs greatly assists in the identification of the FFF target responsible for the phenotype observed. Further, it is important to confirm that FFF

probe targets are enriched in a UV-dependent manner. This can be accomplished by comparing FFF probe-treated cells that are exposed to UV irradiation to those treated with probe but not exposed to UV irradiation.

7. Active and inactive FFF probes may bind the same interacting proteins but with notably different affinity. By directly comparing relative target enrichment between active and inactive probes, differential affinity toward those targets can be directly assessed and higher affinity targets of the active probe identified.

Acknowledgements

This work was supported by NIH DK099810, DK114785, CA132630, and 1S10OD16357.

References

1. Rosen ED, Spiegelman BM (2014) What we talk about when we talk about fat. Cell 156:20–44
2. Kusminski CM, Bickel PE, Scherer PE (2016) Targeting adipose tissue in the treatment of obesity-associated diabetes. Nat Rev Drug Discov 15:639–660
3. Samuel VT, Petersen KF, Shulman GI (2010) Lipid-induced insulin resistance: unravelling the mechanism. Lancet 375:2267–2277
4. Dominguez E, Galmozzi A, Chang JW, Hsu KL, Pawlak J, Li W et al (2014) Integrated phenotypic and activity-based profiling links Ces3 to obesity and diabetes. Nat Chem Biol 10:113–121
5. Waki H, Park KW, Mitro N, Pei L, Damoiseaux R, Wilpitz DC et al (2007) The small molecule harmine is an antidiabetic cell-type-specific regulator of PPARgamma expression. Cell Metab 5:357–370
6. Galmozzi A, Sonne SB, Altshuler-Keylin S, Hasegawa Y, Shinoda K, Luijten IH et al (2014) ThermoMouse: an in vivo model to identify modulators of UCP1 expression in brown adipose tissue. Cell Rep 9:1584–1593
7. Schenone M, Dancik V, Wagner BK, Clemons PA (2013) Target identification and mechanism of action in chemical biology and drug discovery. Nat Chem Biol 9:232–240
8. Swinney DC, Anthony J (2011) How were new medicines discovered? Nat Rev Drug Discov 10:507–519
9. Parker CG, Galmozzi A, Wang Y, Correia BE, Sasaki K, Joslyn CM et al (2017) Ligand and target discovery by fragment-based screening in human cells. Cell 168:527–541
10. Ong SE, Blagoev B, Kratchmarova I, Kristensen DB, Steen H, Pandey A et al (2002) Stable isotope labeling by amino acids in cell culture, SILAC, as a simple and accurate approach to expression proteomics. Mol Cell Proteomics 1:376–386
11. Hsu JL, Huang SY, Chow NH, Chen SH (2003) Stable-isotope dimethyl labeling for quantitative proteomics. Anal Chem 75:6843–6852
12. Washburn MP, Wolters D, Yates JR 3rd (2001) Large-scale analysis of the yeast proteome by multidimensional protein identification technology. Nat Biotechnol 19:242–247
13. Cisar JS, Cravatt BF (2012) Fully functionalized small-molecule probes for integrated phenotypic screening and target identification. J Am Chem Soc 134:10385–10388

Chapter 10

The Ancient Genetic Networks of Obesity: Whole-Animal Automated Screening for Conserved Fat Regulators

Wenfan Ke, Anna Drangowska-Way, Daniel Katz, Karsten Siller, and Eyleen J. O'Rourke

Abstract

Caenorhabditis elegans is the first and only metazoan model that enables whole-body gene knockdown by simply feeding their standard laboratory diet, *E. coli*, carrying RNA interference (RNAi)-expressing constructs. The simplicity of the RNAi treatment, small size, and fast reproduction rate of *C. elegans* allow us to perform whole-animal high-throughput genetic screens in wild-type, mutant, or otherwise genetically modified *C. elegans*. In addition, more than 65% of *C. elegans* genes are conserved in mammals including human. In particular, *C. elegans* metabolic pathways are highly conserved, which supports the study of complex diseases such as obesity in this genetically tractable model system. In this chapter, we present a detailed protocol for automated high-throughput whole-animal RNAi screening to identify the pathways promoting obesity in diet-induced and genetically driven obese *C. elegans*. We describe an optimized high-content screening protocol to score fat mass and body fat distribution in whole animals at large scale. We provide optimized pipelines to automatically score phenotypes using the open-source CellProfiler platform within the context of supercomputer clusters. Further, we present a guideline to optimize information workflow from the automated microscope to a searchable database. The approaches described here enable unveiling the whole network of gene-gene and gene-environment interactions that define metabolic health or disease status in this proven model of human disease, but similar principles can be applied to other disease models.

Key words Whole-animal screening, *C. elegans*, Genetics of obesity, Automated microscopy, Live screening, Fat screening

1 Introduction

Caenorhabditis elegans has become a leading model organism to tackle important biomedical questions. It is a powerful system due to its short life span (less than 30 days), high reproduction rate (300–500 eggs per animal in total with peak around 150 eggs/day), fast development (90 h from hatching egg to adult at 15 °C), transparency of the organism, and fully sequenced and annotated genome [1]. Furthermore, *C. elegans* has a stereotypical distribution of cells identical from worm to worm and its neuronal

Bridget Wagner (ed.), *Phenotypic Screening: Methods and Protocols*, Methods in Molecular Biology, vol. 1787,
https://doi.org/10.1007/978-1-4939-7847-2_10,

connectome has been completely traced, providing great advantages for studying biological processes that rely on proper cell-to-cell, tissue-to-tissue, or whole-body communication [2].

The most basic biological processes and their effectors (genes) are conserved from *C. elegans* to human, which makes research in *C. elegans* translatable to mammals and even humans. Although there are important differences in terms of tissue and organ function, and some key mammalian metabolic players are missing (i.e., leptin) most of the core lipid, sugar, and protein metabolism pathways are highly conserved between worms and mammals. For example, adipose triglyceride lipase, hormone-sensitive lipase, and lysosomal lipases play essential roles in fat storage and mobilization in *C. elegans* and mammals [3, 4]. TOR kinase, AMPK, sterol response element-binding protein, and many other transcription factors similarly control metabolism-gene regulation and cell responses to nutrients in *C. elegans* and mammals. Loss of function of such regulators causes severe metabolic defects such as obesity in worms [5–7]. In addition, several pathways systemically regulating metabolism are also conserved. Importantly, the major components of the insulin-signaling pathway are conserved and chronically defective insulin signaling leads to symptoms associated with insulin resistance including obesity in worms [8]. These and other examples suggest that unwinding *C. elegans* metabolic players and how they are modulated by changing environmental factors, including food quantity and quality or genetic defects, would help us better understand metabolic disease in humans.

C. elegans is also the first model system enabling whole-genome systemic RNA interference in vivo. This capability is particularly critical to understand metabolic disease since metabolic status is the product of coordinated action of multiple cells and organ systems. In *C. elegans*, RNAi can be delivered by simply feeding worms with bacteria that overproduces double-stranded RNA against a specific worm gene upon isopropylthiogalactoside (IPTG) induction [9]. RNAi libraries developed by Julie Ahringer's group cover close to 87% of the worm genome and these feeding constructs are readily available from Source BioScience [10].

The fast development and reproduction rate imply that *C. elegans* can be used as an in vivo model to perform high-throughput (HT) screening. Automated and quantitative high-throughput screening methods have been developed and improved throughout the past years, and these HT assays allow us to identify lead compounds in in vivo HT chemical screens and genes in RNAi-based screens [11].

Previously we described an improved oil red O staining technique to provide an image-based quantitative measurement of fat mass that correlated to biochemically measured triglyceride mass [12]. This method uses isopropanol to preserve worm-tissue structure, which is otherwise fragile (i.e., by paraformaldehyde-based fixation). This method can be used as a powerful quantitative technique to identify genes that alter body fat mass in *C. elegans* [12]. Here we present several advances to the processing and imaging of the worm populations, and we delineate considerations and solutions to the handling of the massive amount of data generated by image-based HT screening of *C. elegans.*

Comparing to our previous method, we optimized conditions in preparation, RNAi treatment, and data analysis. In this chapter, we describe detailed step-by-step procedures with improvement on reducing non-RNAi variants such as position of the wells, growth conditions, and precipitation of ORO dyes, and moreover we incorporate the data processing power from supercomputer to enable analyses of high-resolution composite images from high-content screening experiments.

Studying metabolic diseases such as obesity is a challenge due to the complex genetic interactions between multiple pathways throughout development, between cells and organs, and the modifier effect of the environment, in particular the diet. Our whole-animal automated high-throughput genetic screening method provides an approach to unveil the genetic pathways underlying genetic- and diet-induced obesity in intact animals (Fig. 1).

2 Materials

2.1 Bacteria and Worm Strains

1. RNAi Library from Ahringer laboratory can be acquired from Source BioScience: http://www.sourcebioscience.com/products/life-science-research/clones/rnai-resources/c-elegans-rnai-collection-ahringer/ (*see* **Note 1**).
2. *C. elegans* strains NL2099 (*rrf-3*(pk1426)II (*see* **Note 2**) and GMW004 (*rrf-3*(pk1426) II; *daf-2*(e1368) III) can be acquired from Caenorhabditis Genetics Center (CGC) [13]: https://cbs.umn.edu/cgc/home (*see* **Note 3**).
3. L4440 empty vector transformed bacteria can be obtained from Addgene: https://www.addgene.org/1654/ (*see* **Note 4**).
4. *daf-16* RNAi bacteria can be found in Ahringer RNAi library, number 1717, plate 18, well H1 (*see* **Note 5**).

2.2 Reagents

1. Potassium phosphate buffer: 108.3 g KH_2PO_4, 35.6 g K_2HPO_4, and water to 1 L, pH 6.0. Sterilize by autoclaving.
2. NGM RNAi agar: 3 g NaCl, 2.5 g peptone, 17 g agar, and water to 1 L. Sterilize by autoclave and cool to 55 °C. After cooling to 55 °C add in the following order: 1 mL of 5 mg/mL cholesterol dissolved in ethanol, 1 mL of 1 M $CaCl_2$, 1 mL of 1 M $MgSO_4$, 25 mL of 1 M potassium phosphate, pH 6.0,

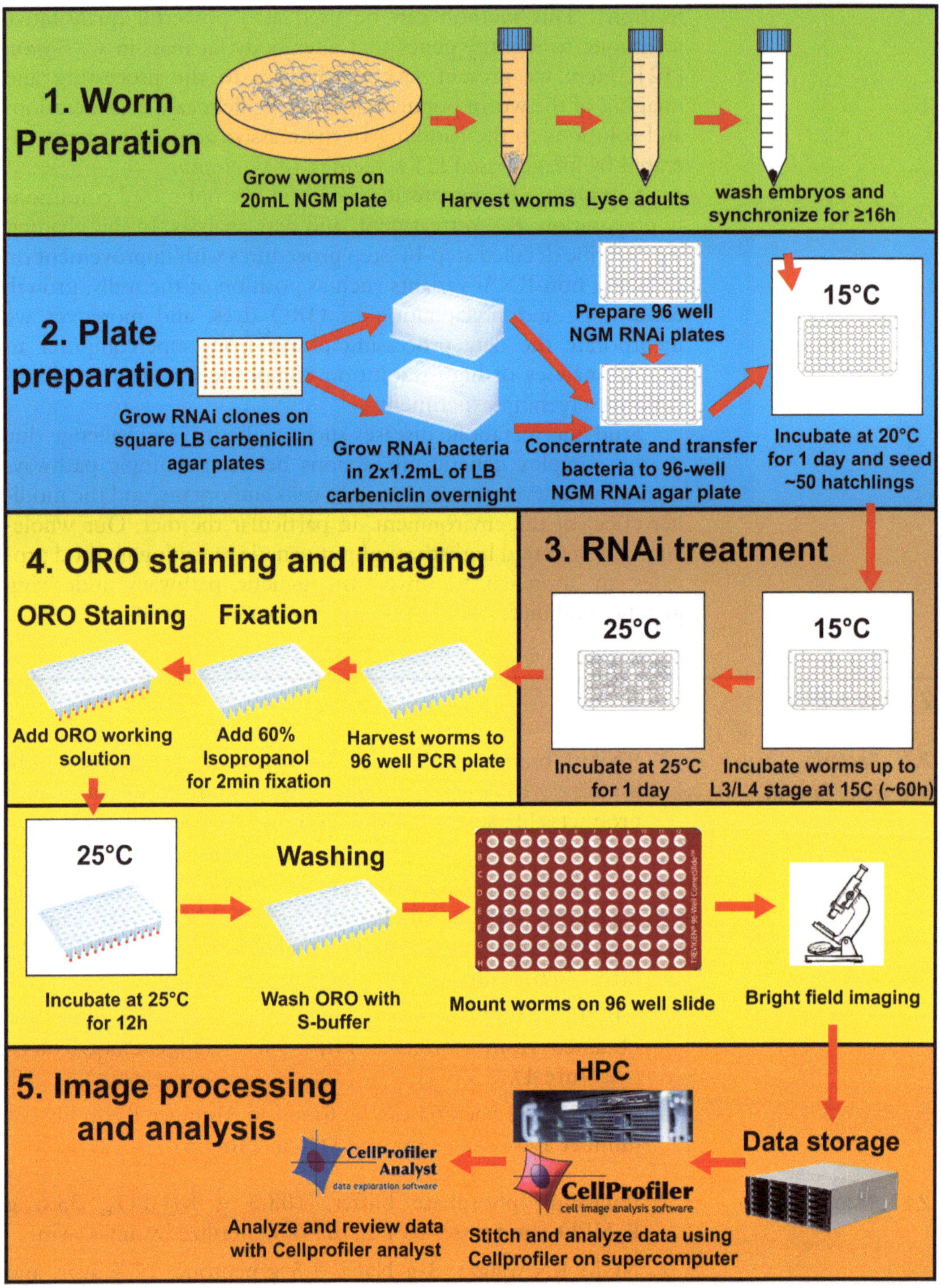

Fig. 1 RNAi screening workflow. Worm preparation is shown in green (*see* Subheading 3.3); bacterial RNAi preparation is shown in blue (*see* Subheading 3.2); worm RNAi treatment is shown in brown (*see* Subheading 3.4); ORO staining and imaging procedures are shown in yellow (*see* Subheadings 3.5 and 3.6); and image analysis and data transfer and processing are shown in orange (*see* Subheading 4)

IPTG (1 mM final concentration), and 1 mL of 50 mg/mL carbenicillin.

3. LB broth: 10 g Bacto-tryptone, 5 g yeast extract, 10 g NaCl, and water to 1 L, pH to 7.5 with NaOH, and sterilize by autoclaving. Before use, add 1 mL of 50 mg/mL carbenicillin.
4. LB agar: 10 g Bacto-tryptone, 5 g yeast extract, 10 g NaCl, 17 g agar, and water to 1 L, pH to 7.5 with NaOH, and sterilize by autoclaving. Cool to around 55 °C and add 1 mL of 50 mg/mL carbenicillin.
5. S-basal without cholesterol buffer (S-buffer): 5.85 g NaCl, 1 g K_2HPO_4, 6 g KH_2PO_4, water to 1 L. Sterilize by autoclaving.
6. Lysis buffer: 195 μL 10 *N* NaOH, 600 μL sodium hypochlorite, and water to 3 mL
7. S-buffer with 0.01% Triton-X: Add 100 μL of Triton-X 100% to 1 L of S-buffer (resuspend with stirring bar for ≥1 h).
8. 60% Isopropanol: Mix 30 mL of 100% isopropanol with 20 mL of water in a 50 mL conical tube.
9. Oil red O stock solution (0.5%): Resuspend 0.25 g of oil red O in 50 mL of 100% isopropanol by overnight shaking at room temperature. Prepare at least 2 days before use. Can be stored for ≥4 weeks at room temperature.
10. Diluted S-buffer (20%): 20 mL S-buffer and 80 mL of distilled water.

2.3 Supplies and Equipment

1. Omnitray single-well square plates.
2. 96-pin replicator.
3. 96-Well clear V-bottom 2 mL polypropylene deep-well plate.
4. 96-Well cell culture plates.
5. Breathe-Easy sealing membrane.
6. HT115 *E. coli* strain.
7. AlumaSeal 96^{TM} sealing film.
8. Sealing mats for 96-well PCR plates.
9. 96-Well channel vacuum manifold.
10. 96-Well PCR plates.
11. Petri dish (100 mm × 15 mm).
12. Sterile reservoirs.
13. 37 °C Shaking incubator with holders for 96-well plates.
14. Benchtop centrifuge with 96-well plate adaptors.
15. Vertical flow biological hood.
16. Comet assay 96-well slides.

17. Upright or inverted microscope with an automated stage capable of doing high-throughput tile capture and stitching of images of 96-well slides (*see* **Note 6**).
18. Digital Triple Heat Block.
19. Multichannel pipettes and regular pipettes.

3 Methods

3.1 Reagent Preparation

3.1.1 Preparing Agar Plates (1 Week Before Experiment)

1. Make and autoclave 1 L LB agar as described in Subheading 2.
2. Place LB agar in a 60 °C water bath for 5–10 min to cool down.
3. Add 1 mL of 50 mg/mL carbenicillin.
4. Pour LB agar into plates in a hood. 20 mL of LB agar can be poured into 100 mm Petri dish or omnitray single-well square plates.
5. Place the plates in the hood for 30 min to solidify.
6. Store the plates in a sealed box at 4 °C. LB agar plates can be stored at 4 °C for up to 1 month (*see* **Note 7**).

3.1.2 Preparing NGM RNAi Plates (1 Week Before Experiment)

1. Make and autoclave 1 L NGM agar as described in Subheading 2.
2. Place NGM agar in a 60 °C water bath for 5–10 min to cool down.
3. In a biological hood, add in the following order and thoroughly mix after each ingredient: cholesterol, $CaCl_2$, $MgSO_4$, potassium phosphate, IPTG, and carbenicillin (volumes and stock concentrations listed in materials). Keep molten medium in water bath until pouring plates in the hood.
4. For a 100 mm petri dish, pour 20 mL of NGM agar in the hood and go to **step 9**. For pouring 96-well cell culture plates, go to **step 5**.
5. Right after **step 3**, take a heat block pre-warmed to 70 °C into the hood.
6. Bed clean aluminum foil in the heat block, and place a sterile reservoir and a 96-well cell culture plate on top of the foil.
7. Pour NGM agar into the sterile reservoir.
8. Use multichannel pipettes to transfer 150 μL of NGM agar to every well of the 96-well plate (*see* **Note 8**).
9. Place the plates in the hood for 30 min to solidify.
10. Store the plates in a sealed box in 4 °C. NGM RNAi agar plates can be stored at 4 °C for up to 1 month.

3.1.3 Preparing Oil Red O Working Solution (First Day of Experiment)

1. Prepare 0.5% oil red O stock solution several days beforehand.
2. Filter the oil red O stock solution with a 0.45 μm pore size filter.
3. Prepare a 60% working solution with filtered sterile water and shake for ≥2 h at room temperature.
4. Filter the solution with a 0.45 μm filter and shake for another 2 days at room temperature prior to usage.

3.2 Bacteria RNAi Strain Preparation

3.2.1 Growing Bacteria Colonies on Square LB Agar Plate (3 Days Before Experiment)

1. Clean the bench area with Cavicide™ and turn on the gas and fire on the bench (this step can also be performed in a hood).
2. Prepare sterilized water, 95% ethanol, and 10% bleach in three separate containers (e.g., empty 200 μL pipette boxes) next to the fire.
3. Briefly dip the 96-pin replicator into 10% bleach, rinse thoroughly with water, and wet in 95% ethanol. Flame every side of the replicator to make sure that it is sterile.
4. Place the replicator back into 95% ethanol.
5. Transport RNAi library to the bench with dry ice from −80 °C.
6. Remove the seal of the RNAi library plates next to the flame.
7. Flame the replicator quickly to evaporate the excess ethanol and wait for 5–15 s for it to cool down (*see* **Note 9**).
8. Dip the tips of the replicator into the library plate and make sure that they touch the bacteria in every single well.
9. Quickly transfer and stamp the tips of replicator on the LB square plate, and make smooth rounded movements to increase bacterial growth area without overlapping RNAi clones (*see* **Note 10**).
10. Briefly dip the replicator into 10% bleach and repeat **steps 3** and **4** for each successive plate.
11. Re-seal the library plates with AlumaSeal 96™ Sealing Film and store back at −80 °C (do not ever let the library thaw!).
12. Incubate the RNAi clones seeded in the LB square plate at 37 °C overnight.
13. For control RNAi clones, use sterile toothpicks or wire loops to streak bacteria on the 10 cm LB agar carbenicillin 50 μg/mL plates from the glycerol stock and incubate at 37 °C overnight.
14. After incubation, record the RNAi clones that have no colonies.
15. The LB agar plates can be parafilmed and stored at 4 °C for up to 2 weeks.

3.2.2 Growing Bacteria in 96 Deep-Well Plates (2 Days Before Experiment)

1. In the biological hood, fill the 96 deep-well plate with 1200 μL/well of LB broth with 50 μg/mL carbenicillin using a multichannel pipette.
2. Sterilize the bench with Cavicide™ and turn on the gas and fire on the bench (this step can also be performed in hood).
3. Prepare sterilized water, 95% ethanol, and 10% bleach in three separate containers (e.g., empty 200 μL pipette boxes) next to the fire.
4. Dip the 96-pin replicator into 10% bleach, thoroughly rinse with water, then briefly dip in 95% ethanol, and flame every side of the replicator to make sure that it is sterile.
5. Dip replicator back in 95% ethanol, and quickly flame again to evaporate excess ethanol. Wait for 5–15 s for replicator to cool down.
6. Touch the replicator pins to the mini RNAi bacterial lawns.
7. Dip the tips of the replicator into the 1.2 mL LB broth carbenicillin in two 96 deep-well plates (*see* **Note 11**).
8. Re-sterilize replicator by briefly dipping it into 10% bleach and repeat **steps 4** and **5** sequentially for the next plates.
9. Use sterile toothpicks or wire loops to pick and transfer colonies from positive and negative control 10 cm LB agar carbenicillin 50 μg/mL plates to all empty wells of a 96 deep-well plate and record these wells (*see* **Note 12**).
10. Seal the 96 deep-well plates with Breathe-Easy sealing membrane and shake at 1000 rpm at 37 °C overnight.

3.2.3 Seeding RNAi Bacteria in 96-Well NGM RNAi Plates (1 Day Before Experiment)

1. Take out 96 deep-well plates from the shaker and centrifuge for 10 min at 4415 × *g*.
2. Thoroughly clean the sink area with Cavicide™. Next to the sink, remove the seal, and discard the supernatant into a bucket containing bleach by quick inversion of the plate. Place plates upside down on a stack of clean paper towels and move to the hood.
3. Using a 12-channel pipette, add 100 μL of diluted S-buffer (20%) to each well of *just one* of the plates with RNAi bacterial pellets, and resuspend the bacterial pellets by vigorously pipetting.
4. Transfer bacterial suspension from plate *replicate* 1 to plate *replicate* 2 (we had grown two 96 deep-well plates of culture for each RNAi clone set). Use the transferred suspension to resuspend pellets of replicate plate 2. Do not add S buffer to the second plate (*see* **Note 13**).
5. Add 20 μL of distilled water to the wells located in the four edges of the plate. Then, transfer 100 μL of the bacterial

suspension into the wells of the 96-well NGM RNAi agar plate using multichannel pipette (*see* **Note 14**).

6. Dry the bacterial suspensions in the hood for around 5 h, until no liquid remains (do not overdry the plates, covering faster drying wells with Breath-Easy if necessary).
7. Place the lids on the NGM plates and incubate the plates overnight in a 20 °C incubator for induction of expression of double-stranded RNAs.

3.3 Worm Preparation

3.3.1 Egg Preparation and Synchronization (1 Day Before Experiment)

1. Sterilize the bench with Cavicide™ and carry out the next steps next to flame.
2. Add 10–15 mL of S-buffer to a worm plate with ≥2000 gravid adult worms.
3. Using S-buffer, transfer worm/egg suspension to a 15 mL conical tube.
4. Centrifuge at ≤2000 × *g* for 30 s.
5. Discard supernatant above 3 mL using sterilized glass Pasteur pipettes connected to a vacuum line.
6. Add 3 mL of lysis buffer.
7. Shake vigorously for 1 min and then add 9 mL of S-buffer.
8. Centrifuge at 2000 × *g* for 30 s.
9. Discard supernatant above 3 mL using sterilized glass Pasteur pipettes connected to a vacuum line.
10. Add 3 mL of lysis buffer.
11. Shake vigorously for 1 min and add 9 mL of S-buffer.
12. Centrifuge at 2000 × *g* for 30 s.
13. Using sterilized glass Pasteur pipettes connected to the vacuum line, discard as much supernatant as you feel comfortable with, without disturbing the pellet. The volume remaining should be ≤500 μL; otherwise two additional washing steps are necessary.
14. Add 14 mL of S-buffer.
15. Centrifuge at 2000 × *g* for 30 s.
16. Wash by repeating **steps 13–15** three times.
17. Discard as much supernatant as you feel comfortable with, without disturbing the pellet using sterilized glass Pasteur pipettes connected to the vacuum line, and fill up to the 10 mL mark with S-buffer.
18. Rotate embryos in a 20 °C incubator for ≥16 h.
19. The next day, estimate worm density (*see* **Note 15**).
20. Dilute the worm suspension with S-buffer, or concentrate by centrifugation, to bring the density to 10 worms/μL.

3.3.2 Seeding Worms (First Day of Experiment)

1. Take out the NGM plates with RNAi bacteria from the 20 °C incubator and move them to the hood.
2. In the hood, pour the worm suspension into a 50 mL sterile reservoir.
3. Use a multichannel pipette to seed 5 μL of worms from the reservoir to each well (constantly move the reservoir to keep the worms suspended).
4. Dry the plates in the hood for around 15 min.

3.4 RNAi Incubation (First Day of Experiment)

1. Move the plates to a 15 °C incubator and incubate for 60 h (*see* **Note 16**).
2. Transfer the plates to a 25 °C incubator for 24 h.
3. Check the worms growing in the wells to make sure that they are at the gravid stage.

3.5 Oil Red O (ORO) Staining

1. Filter the ORO working solution with a 0.45 μm filter.
2. Pour S-buffer into a reservoir (*see* **Note 17**).
3. Using a 12-channel pipette, add 100 μL S-buffer to each well.
4. Using a 12-channel pipette, transfer the whole-worm suspension to a 96-well PCR plate.
5. Connect the 96-channel vacuum manifold to a vacuum source and calibrate the height for aspiration using a mock PCR plate with water. Calibrate to leave ~25 μL of water behind (*see* **Note 18**).
6. Once aspirator is calibrated, aspirate the S-buffer from the 96-well PCR plates containing worms using the 96-channel vacuum manifold.
7. Wash twice by adding and aspirating 100 μL of S-buffer using a multichannel pipette and 96-channel vacuum manifold.
8. Using a 12-channel pipette, add 200 μL of 60% isopropanol to each well, intentionally disrupting the worm pellet while adding the isopropanol.
9. Let worms sink for about 2 min.
10. Aspirate 60% isopropanol using 96-channel vacuum manifold.
11. Using a 12-channel pipette, add 200 μL of ORO working solution to each well. Intentionally disrupt the worm pellet while adding the ORO solution or resuspend by pipetting but *do not* vortex, invert, or flick the plates. Avoid ORO droplets from reaching the walls of the tubes to prevent precipitation of the dye.
12. Seal the 96-well PCR plates using sealing mats, and put plates into a sealed box with a wet paper towel (*see* **Note 19**).
13. Incubate the box at 25 °C for 12–16 h (*see* **Note 20**).

14. Remove the seal and aspirate the ORO working solution using the 96-channel vacuum manifold.
15. Wash the wells twice with 100 μL of S-buffer.
16. Add 100 μL of S-buffer with 0.01% Triton.
17. Use a 12-channel pipette to mount 8 μL of worms on the Comet assay 96-well slides and carefully place the coverslips for imaging (*see* **Note 21**).

3.6 Image Acquisition and Processing

1. To capture images from the 96-well slide, we recommend using an automated microscope or high-content screening platform. In our case, we use a Nikon Eclipse Ti microscope with an automated stage.
2. The optics include a 10× 0.45NA objective and a DsRi2 camera. Initial storage into local SSD drives (≥10 TB) renders fastest capturing rates (*see* **Note 22**).
3. It is recommended that all data processing is completed on a high-processing computer (HPC) cluster. In the HPC context, image processing and analysis can be done in parallel (*see* **Notes 23–31**, Fig. 2).

4 Notes

1. Two major sets of RNAi feeding libraries are currently available, the initial Ahringer library and the supplementary library. The initial Ahringer library covers 72% of the worm genome, and the supplementary library covers another 15% of all genes, so the combination of both libraries covers 87% of the worm genome. All the feeding library sets are delivered as glycerol stocks and should be stored at −80 °C.
2. NL2099 strain carries a 3015 base-pair deletion between exons 4 and 11 of *rrf-3* gene, and this allele(pk1426) can be detected by PCR using an internal forward primer: AATTGGAAGAATGAGTCACG, an external forward primer: AAATCATACGTCATCGATGC, and an external reverse primer: GCCACGAAATACCATTGCC. Amplification of genomic preps from wild-type N2 strain with internal forward primer and external reverse primer results in a 0.7 kb band, while amplification of genomic prep from NL2099 results in a 0.85 kb band. In addition, no product can be amplified on N2 strain with external forward primer and external reverse primer due to the size-limiting condition of this PCR. No product can be amplified on NL2099 strain using internal forward primer and external reverse primer because the internal forward primer sits in the deleted sequence. We use NL2099 strain for screen because *rrf-3* is an RNA-dependent RNA

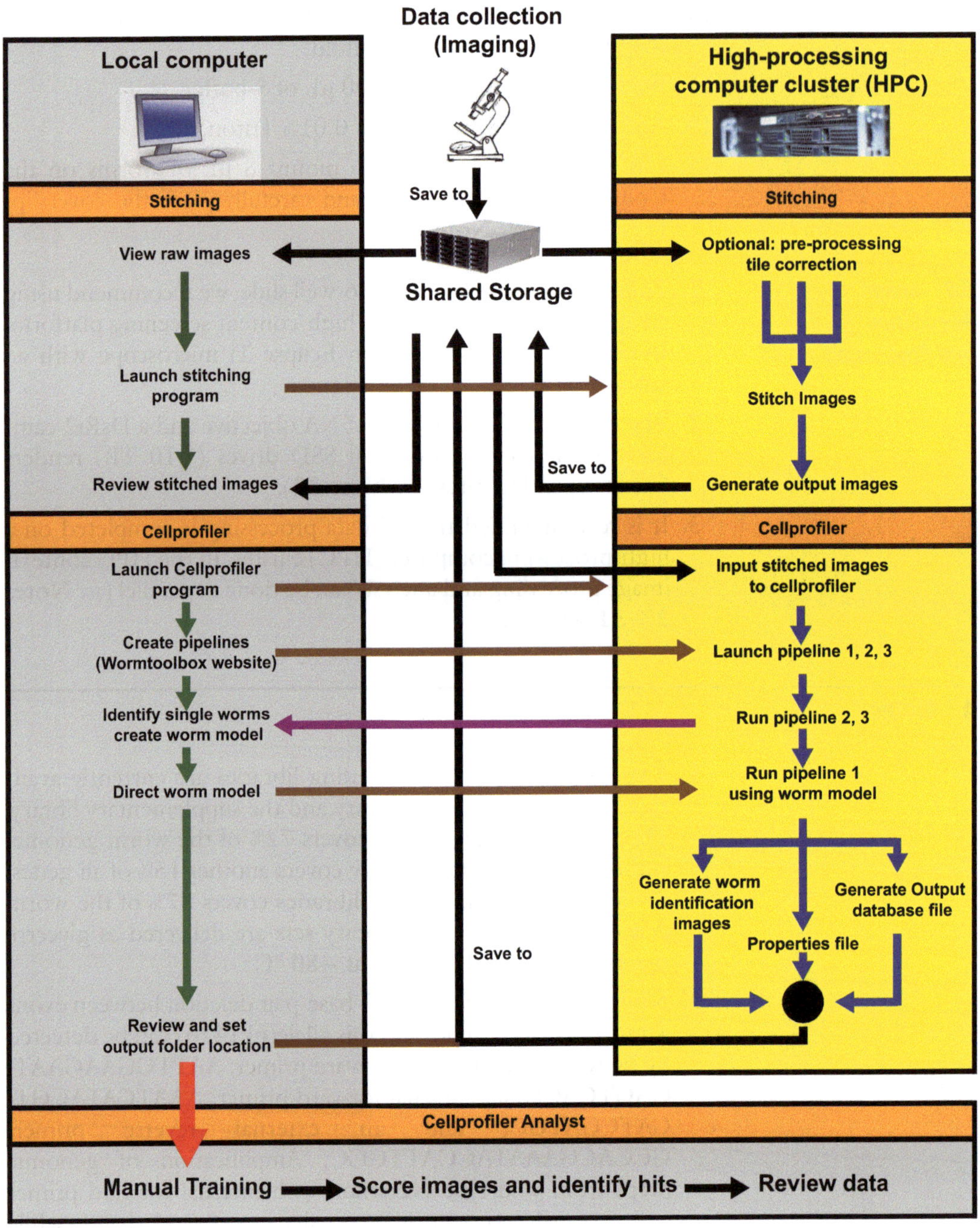

Fig. 2 Post-HCS image processing workflow. Imaging data may be saved in a shared storage positioned between local computers and a local or cloud-based HPC cluster. Both homemade encoded image stitching (available at http:/orourkeeyleenlab.wix.com/obesity-aging-lab/outreach) and CellProfiler data analysis can be run on an HPC cluster. In this diagram, black arrows indicate saving and loading data to and from shared storage; green arrows indicate data processing steps within a local computer; brown arrows indicate control of programs on the HPC cluster; blue arrows indicate automated (and often parallelized) programs running in the HPC cluster; purple arrows indicate direct communication from the HPC cluster to a local computer in the pipeline 2 and 3 to optimize worm models manually; and red arrow indicates that the output of CellProfiler will input into CellProfiler Analyst on the local computer for final data analysis and hit detection

polymerase and the *rrf-3*(pk1426) allele makes this strain hypersensitive to RNAi. Also, worms with this allele become sterile at 25 °C, which allows us to image and quantitate fat in adult worms without the interfering signal of the progeny.

3. GMW004 is a double-*rrf-3*(pk1426);*daf-2* (e1368) mutant. DAF-2 is the worm insulin receptor and the e*1368* allele of *daf-2* is a hypomorphic thermosensitive mutation that leads to an obesity phenotype at 25 °C. *daf-2*(e1368) is a single-nucleotide substitution from G to A, causing a change of protein sequence from S573 to L573. This allele can be detected by PCR amplification of genomic DNA using forward primer CTCACCATTTGTCCCTTC and reverse primer CAATCGTCACCGTTTATCTC, followed by incubation with restriction enzyme TSP45I, which only cut wild-type *daf-2* allele on the mutation site.
4. L4440 is the 2790 bp empty plasmidic vector of the RNAi constructs. It is shipped from Addgene as transformed bacteria and generally used as negative control for *C. elegans* RNAi screens.
5. DAF-16 is the *C. elegans* homolog of human FOXO, which is a transcription factor inhibited by insulin signaling. Reduced insulin signaling causes activation of DAF-16. Loss of DAF-16 is sufficient to rescue the obesity phenotype of daf-2 mutant worms. Hence, daf-16 RNAi can be used as a positive control in a daf-2 fat-suppressor screen.
6. The Nikon Eclipse Ti microscope with 10 × 0.45 NA objective and DsRi2 camera is suitable for this purpose.
7. These LB carbenicillin agar square plates need to be prepared well ahead or thoroughly dried in the hood. Otherwise, moisture will cross-contaminate stamped RNAi clones.
8. Pipette tips need to be prewarmed by pipetting up and down molten NGM RNAi agar several times before transferring. While aspirating the 96-well plates, place the tips on the side and the bottom of the well, and aspirate slowly to prevent bubbles from forming. It is very important to avoid bubbles in the wells because worms will crawl into the agar through bubbles, leading to irregular feeding, and will make it difficult to harvest worms for further processing.
9. It is important to make sure that the replicator is not too hot or it will kill the bacteria. There are two extra tips on the replicator that allow the user to get a sense of the pins' temperature; simply touch the two extra tips with your hand while making sure not to touch any other tip of the replicator.
10. While stamping the replicator, it is important to be gentle and not poke into the agar. Also, every tip of the replicator has to

touch the LB agar to secure the successful transfer of all RNAi clones.

11. One 96 deep-well plate worth of bacterial culture is insufficient food to sustain the growth of ~50 worms into gravid adulthood. Two or three 96 deep-well plates are preferred per RNAi library plate. In our experimental setup, two strains of worms (GMW004 and NL2099) are tested, so 4–6 96 deep-well plates are prepared for each RNAi library plate.
12. There are some RNAi clones that would not grow on the LB agar omnitray (not represented in the original library or experimental error). Take advantage of these empty wells and seed positive and negative control RNAi clones in these wells.
13. As referred in **Note 11**, two or three 96 deep-well plates of samples are grown for each RNAi library plate to be tested. These replicates need to be combined to support the growth of ~50 worms into gravid adulthood. If there are two deep-well plates per original plate, add 100 μL of 20% S-buffer to the first deep-well plate and resuspend the bacteria by pipetting. Then transfer the sample to the second plate and resuspend the bacteria of the second plate.
14. While transferring bacteria to the 96-well NGM RNAi agar plate, it is important to be careful not to poke into the agar. After carefully transferring 100 μL of bacteria, it is necessary to add 20 μL water to the edge wells because these wells dry much faster than the wells at the center of the plate. If these wells become too dry, the agar may crack, worms will crawl into the cracks, and that well will not be scorable.
15. To estimate the density of the worms, thoroughly resuspend the hatchlings by inverting 15 mL conical several times, take three or more 2 μL aliquots of worms, place them in a clean microscope slide, count the number of live hatchlings under the dissecting microscope, and average the hatchlings/microliter.
16. The 15 °C incubation can take longer or shorter than 60 h. The development of worms needs to be checked every 2 h starting from 50 h to 70 h, to ensure that the worms are at late L3 or early L4 stage for transferring to 25 °C incubator. It is critical not to transfer daf-2 mutant worms earlier than L3 because they will enter the dauer program and arrest, and also no later than L4 stage to observe full penetrance of the *rrf-3* sterility phenotype.
17. 0.01% of Triton in the S-buffer helps prevent the worms from sticking to the pipette tips, and reduce worm loss.
18. The amount of liquid remaining in the 96-well PCR plates is determined by the height of the 96-channel manifold. This can

be adjusted and calibrated by filling water in a 96-well PCR plates and testing the water level after aspiration. It is important to keep around 25 μL of solution to minimize the chances of accidental loss of worms by aspiration.

19. Keeping the plate completely sealed and incubating in a humid box are necessary because the oil red O solution is a saturated solution, and loss of water causes precipitation of black crystals that prevent quantification of ORO levels.
20. It is important to restrict oil red O staining to ≤16 h at 25 °C because longer incubation may change the tone of the oil red O staining and affect the quality of the images.
21. It requires some practice to be able to pipette up most of the worms in such a small amount of liquid. To achieve this, it is important to place the tip at the very bottom of the well and aspirate as fast as possible (i.e., use a "bad" pipetting procedure). Check the slide and make sure that you have worms in every well. If the worms in some wells are not transferred, use a single-channel pipette to transfer again.
22. The images are most likely to be stored on the local computer connected to microscope. To analyze them on an HPC cluster you will need to transfer them to a fast-transfer storage unit locally, remotely, or in the cloud. Your local administrator should advise you on the best available option.
23. When performing a high-throughput image-based genetic screening, capturing, storing, and processing approximately thirty 25 MB images per channel per well become bottlenecks due to the limited storage and processing power of regular lab computers or even workstations. For example, for the screening setup described here, each stitched RGB image of a whole well is 9001 × 8928 pixels, with each whole-well tiff image sized at around 250 MB, so one 96-well plate represents approximately 25 GB of data. Analyzing these data is an even bigger challenge than storage. Our single-worm identification algorithms, although highly accurate, are computationally demanding. Therefore, it can take from a day to a week to run a 0.25–1 TB dataset on a conventional desktop. A previous paper from our lab used compressed images to achieve analysis (5 Mb *.bmp file/well). However, this is not suboptimal due to the reduction in image resolution and consequent loss of information [14].
24. When naming folders and images avoid spaces. This will be useful when using a command line. If you are unable to change the names to avoid spaces you will need to use quotation marks at the beginning and at the end of the path to the file while using the command line.

25. As introduced in Subheading 3.6, data acquisition begins with automated imaging of the experimental plates. These data will be written directly to a mounted PC, often the one used to set up and begin microscopic data collection. A few considerations help ensure fast data collection and transfer, including local writing space, naming conventions, and an automated transfer framework. For local storage, it is recommended to have at least 10 TB of solid-state drive (SSD) space. SSD is much faster than its alternative, hard drive disk (HDD), and should be used to prevent bottlenecks. Data acquisition is also the optimal step to enforce naming conventions. Since there are no built-in frameworks for naming, it is up to the screening lab to provide and enforce easy-to-read naming conventions that will ensure smooth data processing. Finally, to prevent any impedance to data collection, it is recommended to have a second mounted PC that can separately upload data to the desired server. This allows for a slower upload speed, which is often unavoidable, while still maintaining scope operation. A detailed list of our setup and examples of file naming can be found at http://orourkeeyleenlab.wix.com/obesity-aging-lab/outreach.
26. While some of the processing steps can be executed on a personal computer, utilization of an HPC cluster will dramatically increase computational speeds and allow for use of full-resolution raw screening images. Interactive processing steps (i.e., refining worm identification) will require extra considerations, and a rudimentary knowledge of command-line language may be necessary to operate some of the programs described herein within the context of an HPC cluster. First, several systems and software tools must be installed, created, or optimized. These include CellProfiler (CP), CellProfiler Analyst (CPA), a MySQL database (or an alternative, but CP/CPA-compatible, database of your choice), and appropriate CellProfiler pipelines [15]. It is important to confer with a local information technologist to ensure proper installation and setup of these programs and communication with HPC cluster.
27. The setup and optimization of CellProfiler pipelines for new screens may be done outside of the HPC cluster environment, which facilitates the use of CP interactive optimization tools, as long as special attention is paid to any and all naming and PATH conventions required for CP and CPA to generate and retrieve data. CellProfiler's export and import tools work within a HPC cluster environment as long as the final modifications to the pipeline—including image folder location and PATH names—are correctly assigned in the HPC environment. At this stage pay special attention to the MySQL database used to ensure that files are sent to the correct locations.

Depending on the preferred CP analysis output, CSV data sheets can be generated instead of databases with CellProfiler, which can then be used to manually analyze data. This approach is less robust and may result in errors due to CellProfiler Analyst's particular file location requirements.

28. A comprehensive analysis of the images is possible after the initial steps are completed, and can be easily completed using the scripts provided at http://orourkeeyleenlab.wix.com/obesity-aging-lab/outreach.
29. This data analysis process will require work on the HPC cluster through programs such as FastX or similar. FastX provides a graphical user interface (GUI)-enabled program for HPC cluster operation, and is an easy solution for users unfamiliar with linux or remote systems. If your institution does not have access to this program, or prefers a different one, alternatives should work just as well. Note that most alternatives may require basic command-line knowledge.
30. After initial data have been collected, post-processing can be done. This may include image review, and further image analysis if necessary. It is recommended the user conducts analysis within the HPC cluster to ensure that all image data can be retrieved without permission issues. Due to the preestablished database, it should be easy to locate and retrieve any files for further use. Another consideration is kinds of storage, in particular, the differences between private user storage and shared storage. It is recommended that both be set up in all cases: the shared storage for the organization and maintenance of all raw image data sets (terabytes of information), and the private user storage for processing images, outputting data, and linking of relevant experimental metadata or information fro databases. In this way, private data folders would beoptimized for processing, and commonly used and large image files are not copied to all users, but instead located in a central location. Nevertheless, naming conventions for files, subfolders, and folders should be established and strictly followed.
31. For users that want to compile an easy-to-use image viewer, programs such as Omero exist. Omero is an open-source program that allows for broad image organization and utilization. It may be possible to set the HPC cluster to send data outputs directly to Omero. Alternatively, integration of the MySQL database into Omero may provide the most success. These types of programs are not strictly necessary for post-processing, but may allow for additional organization and integration of screen results with data banks (i.e., wormbase, ensemble). Optimal data processing and organization solutions would vary with type and scale of the screens, as well as with lab resources.

Acknowledgments

We want to thank Carolina Wählby, Anne Carpenter, Jonah Larkins-For, and Annie Lee-Conery for their contributions to development of the earlier versions of these protocols. We would like to thank Wei Ma for practical advice on developing the assays. We also thank the University of Virginia Advanced Research Computing Services for support on Rivanna HPC cluster usage. This work would not have been possible without the generous support of the W. M. Keck Foundation.

References

1. Muschiol D, Schroeder F, Traunspurger W (2009) Life cycle and population growth rate of Caenorhabditis elegans studied by a new method. BMC Ecol 9:14
2. Sulston JE, Horvitz HR (1977) Post-embryonic cell lineages of the nematode, Caenorhabditis elegans. Dev Biol 56:110–156
3. Wang MC, O'Rourke EJ, Ruvkun G (2008) Fat metabolism links germline stem cells and longevity in *C. elegans*. Science 322:957–960
4. Narbonne P, Roy R (2009) Caenorhabditis elegans dauers need LKB1/AMPK to ration lipid reserves and ensure long-term survival. Nature 457:210–214
5. Long X, Spycher C, Han ZS, Rose AM, Muller F, Avruch J (2002) TOR deficiency in *C. elegans* causes developmental arrest and intestinal atrophy by inhibition of mRNx translation. Curr Biol 12:1448–1461
6. McKay RM, McKay JP, Avery L, Graff JM (2003) C. elegans: a model for exploring the genetics of fat storage. Dev Cell 4:131–142
7. Sze JY, Victor M, Loer C, Shi Y, Ruvkun G (2000) Food and metabolic signalling defects in a Caenorhabditis elegans serotonin-synthesis mutant. Nature 403:560–564
8. Garcia AM, Ladage ML, Dumesnil DR, Zaman K, Shulaev V, Azad RK et al (2015) Glucose induces sensitivity to oxygen deprivation and modulates insulin/IGF-1 signaling and lipid biosynthesis in Caenorhabditis elegans. Genetics 200:167–184
9. Tabara H, Grishok A, Mello CC (1998) RNAi in *C. elegans*: soaking in the genome sequence. Science 282:430–431
10. Kamath RS, Fraser AG, Dong Y, Poulin G, Durbin R, Gotta M et al (2003) Systematic functional analysis of the Caenorhabditis elegans genome using RNAi. Nature 421:231–237
11. O'Rourke EJ, Conery AL, Moy TI (2009) Whole-animal high-throughput screens: the *C. elegans* model. Methods Mol Biol 486:57–75
12. O'Rourke EJ, Soukas AA, Carr CE, Ruvkun G (2009) *C. elegans* major fats are stored in vesicles distinct from lysosome-related organelles. Cell Metab 10:430–435
13. Simmer F, Tijsterman M, Parrish S, Koushika SP, Nonet ML, Fire A et al (2002) Loss of the putative RNA-directed RNA polymerase RRF-3 makes C. elegans hypersensitive to RNAi. Curr Biol 12:1317–1319
14. Wählby C, Conery AL, Bray MA, Kamentsky L, Larkins-Ford J, Sokolnicki KL et al (2014) High- and low-throughput scoring of fat mass and body fat distribution in *C. elegans*. Methods 68:492–499
15. Kamentsky L, Jones TR, Fraser A, Bray M, Logan D, Madden K et al (2011) Improved structure, function, and compatibility for CellProfiler: modular high-throughput image analysis software. Bioinformatics 27: 1179–1180

Chapter 11

Modulation of Threat Response in Larval Zebrafish

Andrew J. Rennekamp

Abstract

High-throughput, whole-organism phenotypic drug screening is made possible using live zebrafish larvae. Many human drugs have now been shown to affect zebrafish larvae in similar ways, through homologous molecular mechanisms. At this stage in life, zebrafish are small enough to fit in multi-well, microliter plates, yet developed enough to exhibit complex phenotypes, such as hunting behaviors and avoidance of predators. Importantly, zebrafish larvae can be easily dosed via automated pipetting of chemical compounds directly into their liquid medium, without injection. Only microgram amounts of small molecules are required, making animal husbandry and dosing regimens cost effective. This chapter describes how the stereotyped zebrafish larval responses to darkness and strobe light—which cause hyperactivity and freezing behavior, respectively—can be used to efficiently screen small molecules for brain and behavior-modulating activity.

Key words Fear, Freezing behavior, CNS, Whole-organism screening, Phenotypic screens, High-throughput drug discovery, Small molecules, Strobe light response

1 Introduction

Approximately two-thirds of all structurally new, small-molecule drugs approved by the US FDA in the past two decades (the so-called new molecular entities) were discovered using phenotypic chemical screening approaches [1]. In other words, twice as many screens predicated on assays that measured the effects of compounds on cells, tissues, or whole organisms were successful at producing translatable products than screens based on assays that measured the effect of compounds on a single-protein target. The advantage that phenotypic screens have had over target-based screens has been even more pronounced in specific therapeutic areas, most notably for CNS indications, where phenotypic discovery has beat target-based discovery seven to one [1].

Phenotypic screening permits unbiased, target-agnostic discovery in way that target-based screening does not. It also allows for the possibility of identifying polypharmacologic compounds, which require the engagement of more than one target for efficacy. The

Bridget Wagner (ed.), *Phenotypic Screening: Methods and Protocols*, Methods in Molecular Biology, vol. 1787,
https://doi.org/10.1007/978-1-4939-7847-2_11,

ideal phenotypic screen is a whole-organism screen. It allows for the maximum number of targets to be screened all at once. By starting the screening process at the whole-organism level, drug discovery scientists can focus their efforts on producing only hit compounds with certain favorable attributes. For example, we can begin to deal with acute toxicity issues, drug metabolism concerns, and biological hurdles, such as the blood–brain barrier during the initial screening phase. Screening at the organismal level gives us the ability to find systems-modulating compounds that would otherwise remain undetected and ignored by target-centric methods.

Most vertebrate animal models are not amenable to high-throughput screening. Their large size, low fecundity, inability to be dosed automatically, and ethical concerns surrounding their use make their use during the screening phases of drug discovery impractical if not impossible. Fortunately, zebrafish larvae have emerged as an ideal model organism for high-throughput screening [2]. They are fecund, small enough to fit into 96-well microliter plates, and can be dosed using automated equipment, and yet they share 90% protein identity with humans, respond to many drugs through the same molecular mechanisms, and have intact blood–-brain barriers.

This chapter provides the details required for conducting a high-throughput, whole-organism screen to discover new neuroactive small-molecule modulators of threat responses, using zebrafish larvae. The screen is predicated on the observation that zebrafish larvae reproducibly freeze in response to strobe light, a stimulus that is aversive to them [3]. During the strobe light response, larvae are exposed to alternating minutes of either darkness or strobe light for a total of 7 min. Fish movement is recorded as a video by an infrared camera. A software program tracks the changes in pixel intensity for each well within the 96-well plate over the course of the experiment. During the minutes of darkness, larvae activity will be high, but during the minutes of strobe light, larval activity will dramatically decrease. Some small molecules, such as the chemical compound finazine, will switch the behavioral response to strobe light from freezing to hyperactivity and escape behavior [3]. Using the strobe light response assay, high-throughput screens can be conducted to identify novel neuroactive compounds that further modulate threat responses in larval zebrafish.

2 Materials

2.1 Zebrafish Mating (Day 0)

1. One roll of plastic mesh with 1/16 in. holes (e.g., Pentair Aquatic Eco-Systems).
2. Pair of scissors.

3. Stapler with staples.
4. Binder clips (11/4 in. with 5/8 in. capacity): Two clips per 30 mating fish.
5. Aquaculture fish tanks (e.g., 10-L Aquatic Habitat, Pentair Aquatic Eco-Systems): One 10 L tank per 30 adult fish.
6. Adult zebrafish of breeding size and age (*see* **Note 1**).

2.2 Embryo Production (Day 1)

1. Nylon mesh tea strainer (3 in.).
2. Disposable polystyrene petri dishes (100 mm × 15 mm, sterile).
3. Wash bottle with cap and nozzle (e.g., polypropylene 32 oz., 1 L capacity).
4. HEPES-buffered E3 embryo medium (*see* **Note 2**).
5. Manual P1000 pipette with disposable tips.
6. Transfer pipettes (e.g., disposable polyethylene with 3.2 mL bulb volume).
7. Incubator (28 °C) or warm room with normal (14/10 h on/off) light cycle.

2.3 Larval Aquiculture (Days 2–6)

1. Transfer pipettes (e.g., disposable polyethylene with 3.2 mL bulb volume).
2. Wash bottle with cap and nozzle (e.g., polypropylene 32 oz., 1 L capacity).
3. HEPES-buffered E3 embryo medium (*see* **Note 2**).
4. Incubator (28 °C) or warm room with normal (14/10 h on/off) light cycle.

2.4 Zebrafish Plating and Chemical Treatment (Day 7)

1. Ice bucket with ice.
2. Transfer pipettes (e.g., disposable polyethylene with 3.2 mL bulb volume).
3. Manual P1000 pipette with disposable tips.
4. Clear 96-well plates, polystyrene (e.g., round-bottom 96-well plates, nonsterile).
5. Multichannel pipettes (P20, P200, P1000; 12 or 8 channels) with disposable tips.
6. Alternatively, robotic animation to dilute and dispense small molecules.
7. Chemical stock plates: Small molecules from library of interest, undiluted, arrayed in 96-well plates (*see* **Note 3**).
8. Chemical working plates, diluted to 1 mM, arrayed in 96-well plates.
9. Dimethylsulfoxide (DMSO) solvent.

10. Optional: Finazine, as a small-molecule positive control.
11. A centrifuge capable of spinning 96-well chemical stock plates.
12. Adhesive aluminum foil lids (e.g., "Seal & Sample" lids, Beckman Coulter).

2.5 Strobe Light Response Assay (Day 7)

1. ZebraBox Revolution (fourth Generation) video imaging system containing a computer-controlled light box and a video camera with an infrared filter (ViewPoint Life Sciences).
2. Accompanying computer system preloaded with ZebraLab Software (2014 version 1.3, ViewPoint).

3 Methods

3.1 Zebrafish Mating (Day 0)

1. Prepare a breeding basket to be used for adult group mating. Using a pair of scissors cut a plus sign (+) out of the roll of plastic mesh (with 1/16 in. holes) using a pair of scissors. The width of right and left arms of the plus sign should be slightly smaller than the length of one 10 L breeding tank, and the width of the top and bottom arms should be slightly less than the width of the 10 L tank. Staple together the adjacent arms of the mesh plus sign to create a lidless box-shaped basket using staples and a stapler. You will need one basket per group of 30 adult mating fish.
2. Fit the mesh basket snugly inside a 10 L tank. Use two binder clips (11/4 in. with 5/8 in. capacity) to hold the basket in place inside the tank so that the bottom of the basket is about 1 in. above the bottom of the tank (*see* **Note 4**). You will need one 10 L tank per group of 30 adult mating fish.
3. On the evening before embryo production, fill each 10 L tank containing a plastic mesh basket with enough clean zebrafish system water to cover the bottom of the mesh basket by about 2 in. (*see* **Note 5**).
4. Feed all adult fish to be used for mating an extra meal 1 h before setting them up for mating (*see* **Note 6**).
5. Transfer 30 adult fish (20 females and 10 males) into each mesh-lined tank (breeding apparatus) (*see* **Note** 7).
6. Store the breeding fish and apparatus in a large ventilated incubator or warm room (28 °C) with a light cycle overnight.

3.2 Embryo Production (Day 1)

1. Adult fish will automatically begin to breed at dawn (or artificial dawn, i.e., when the lights come on). Allow fish to breed for at least an hour.

2. Once the eggs have been laid and fertilized, return the adult fish to their home aquaculture tank by simply lifting the mesh insert.
3. Collect the embryos within 3 h of the start of mating by pouring the contents of the breeding tank through a nylon tea strainer (*see* **Note 8**).
4. Use a wash bottle containing embryo medium to rinse out the breeding tank and recover any remaining embryos.
5. While in the tea strainer, rinse embryos thoroughly to remove any accumulated waste or other debris.
6. Transfer the embryos to a single petri dish by turning the tea strainer upside-down over the dish and rinsing the strainer with the wash bottle (*see* **Note 9**).
7. Immediately aliquot embryos into fresh petri dishes each filled about 15–20 mL of embryo medium. Eight hundred microliter of concentrated embryos (about 200 embryos) can be pipetted per dish (*see* **Note 10**).
8. Using a transfer pipette, remove all dead embryos from the dish. Dead embryos turn white and opaque and can be easily visualized against a dark background (e.g., a black bench top).
9. Store plated embryos in a ventilated incubator or warm room (28 °C) with a light cycle overnight (*see* **Note 11**).

3.3 Larvae Aquiculture (Days 2–6)

1. Visually inspect embryo aquacultures daily. Remove any dead embryos using a transfer pipette (*see* **Note 12**).
2. Change the aquacultures medium as needed to maintain fish health. Carefully pour off most of the old medium without losing embryos, and add fresh embryo medium using a wash bottle.
3. Return plated embryos to ventilated incubator or warm room (28 °C) with a light cycle and store overnight. Do not feed the larvae (*see* **Note 13**).

3.4 Programming the ZebraLab Software (Before Day 7)

1. Turn on the computer controlling the ZebraBox Revolution behavioral imaging platform. This system is usually purchased together with the ZebraBox system and will be preloaded with the ZebraLab (2014 version 1.3) software.
2. Open the program ZebraLab program, typically by clicking on the related rocket ship icon. This will open the ViewPoint Launcher.
3. Within the "ZebraLab" tab, select "ZebraLab Quantitization," which will open a new unnamed window.
4. From the file menu toolbar select "File" and then "Generate protocol," which will open the "Protocol creation wizard"

window. Within the "Locations" section, set "Location count" to 96 and the "Animal root name" to z. Within the "Vignettes position" section, set the "First vignette position" top left corner coordinates (*x*, *y*) to (30, 30). Set "Vignettes per line" to 12, "Vignet [sic] count per column" to 8, and "Horizontal" and "Verticals space between vignettes" each to 40.

5. Click "OK" to save the information and close the window.
6. From the file menu toolbar select "Edit" and then "Protocol Creation Wizard," which will open the "Protocol creation wizard" window. Within the "Locations" section, set the "Location count" to 96. Within the "Common settings" section, set the "Animal root name" to z, the "Detection threshold" to 13, the "Scale" to $y = 0.1545$, the "Small/Large movement threshold" to 100, and the "Intact/Small movement threshold" to 50. Within the "Vignette positions" section, set the "Verticals space between vignettes" to 40, the "Top left corner" coordinates (*x*, *y*) to (30, 30), the "Vignettes count per line" to 12, and the "Horizontal" and "Verticals space between vignettes" each to 40. Within the "Areas position" section, set the "Areas per vignette count" to 1, the "Areas per line count" to 12, and the "Horizontal" and "Verticals space between areas" each to 74.
7. Within the "Areas position" section, click the "Add" button, which will open the "Area definition" box. Within the "Area definition" box, set the "Area type" to circle, the "Top left corner coordinates" (*x*, *y*) to (116, 90), and both the width and height of the "Area dimensions" to 72.
8. Click "OK" to save the information and close the window.
9. From the file menu toolbar select "Edit" and then "Animal List Creation wizard," which will open the "Animal list creation wizard" window. Within the "Locations" section there will be a list of 96 locations names "w001" to "w096"). Within the "Sessions" section, set the "Session count" to 1, the "Number of animals per session" to 96, and the "Animal count" to 0. Within the "Animals" section set the "Animal root name" to Zebrafish. Leave the fields within the "Custom parameters" section blank.
10. Click "OK" to save the information and close the window.
11. From the file menu toolbar select "Parameters" and then "Protocol parameters," which will open the "Protocol parameters" window. Leave the information in the "Locations tab" as is. Within the "Time" tab, in the "Time parameters" section set the "Experiment duration" to 0 days, 0 h, 7 min, and 0 s. Within the "Results period" section, check the box next to "Integration period" and set it to 1 s. Leave the box next to "No time bin" unchecked. Within the "Start" tab, in the "Start

type" section, select "All locations at a time" from the drop-down menu. Within the "Options" tab check the box next to "Numeriscope Video Recorders (to generates AVI files)" if you would like a video file saved for each experiment (recommended).

12. Click "OK" to save the information and close the window.
13. From the file menu toolbar select "Parameters" and then "Light driving," which will open the "Light Driver Settings" window. Select the "Use one of the 3 triggering methods below" radio button. Select the "Enhanced stimuli" radio button. In the "Enh. Stimuli editor" panel, within the "Starting method" section, select the "When session is starting" radio button. Set the "Initial light level" value to 0% and leave the box next to "Do not restore default at end" unchecked.
14. Within the "List of transitions to define stimuli" section, click the "Add" button to set up the transitions as follows:

    No = 1, Transition = Step, Duration = 0 ms, Final power = 0%

    No = 2, Transition = Step, Duration = 60,000 ms, Final power = 0%

    No = 3, Transition = Step, Duration = 100 ms, Final power = 100%

    No = 4, Transition = Step, Duration = 100 ms, Final power = 0%

    ... (Repeat alternating between 100% and 0%)

    No = 601, Transition = Step, Duration = 100 ms, Final power = 100%

    No = 602, Transition = Step, Duration = 100 ms, Final power = 0%
15. Within the "New Transition" section, select the radio button next to "Edge, then" and sent the "Duration" to 0 ms and the "End Power" to 0%. Within the "Repeat" section, select the radio button next to "Repeat the cycle." The "Total duration" time at the bottom of the panel should read "00:02:00:000."
16. From the file menu toolbar select "Options" and then "Customize." Within the "Raw Data Export Setup" tab set the "ASCII file extension" to txt, and check the boxes next to "Export key words" and "Export quantization data." Leave the other boxes unchecked.
17. Click "OK" to save the information and close the window.
18. From the file menu toolbar select "View" and then "View full screen." Within the "Activity" section, set the "Detection Sensitivity" to 13, check the box next to "Black," and click the "Apply to group" button. Within the "Activity threshold"

section, set the "Burst" to 100, the "Freezing" to 50, and the "Display scale" to 150 and click the "Apply to group" button.

19. From the file menu toolbar select "File" and then "Save Protocol As..." and save with a new file name (e.g., "SLR_Round_96well_Plate.vte").

3.5 Chemical Stock Preparation (Before Day 7)

1. Array all small molecules to be tested in 96-well polypropylene plate libraries (*see* **Note 3**) at high, undiluted concentrations (e.g., 10 mM). Be sure to leave a column (8 wells) free to be used as negative and positive control wells (4 of each).
2. Fill each negative control well with solvent only, and each positive control well with the drug finazine at half the concentration of the library compounds (e.g., if the library concentration is 10 mM, use 5 mM of finazine). These plates are referred to as the "chemical stock plates."
3. Make a similar library of "chemical working plates" by diluting compounds to 1 mM concentrations, in 96-well polypropylene plates, using DMSO solvent as the diluent. You may use robot automation to perform the dilutions or they may be carried out by hand using 8- or 12-channel hand pipettes (*see* **Note 14**).
4. Seal 96-well chemical plates with an adhesive aluminum foil lid.
5. Store both stock and working plate libraries in the dark at −80 °C, ideally in a desiccator.

3.6 Zebrafish Plating and Chemical Treatment (Day 7)

1. Thaw the chemical working plates (1 mM) to room temperature (*see* **Note 14**).
2. Visually inspect embryo aquacultures and discard any contaminated plates.
3. Completely fill an ice bucket with ice.
4. Anesthetize the zebrafish larvae by placing on more larvae-containing petri dish on top of the ice. Volumes from multiple plates may be pooled together to maximize efficiency. Increase the number of dishes on ice as needed.
5. Incubate on ice until the larvae are completely immobilized and sink to the bottom of the dish. This will take about 5–10 min. Do not leave larvae on ice longer than 30 min (*see* **Note 15**).
6. Place the petri dish on a white surface (e.g., a white sheet of paper) to aid in visualization of the fish. Using a transfer pipette, remove all dead, deformed, or floating larvae.
7. Cut about 4 mm off the pointed opening of a P1000 disposable pipette tip.
8. Attach the cut tip to a P1000 pipette, slowly draw ten larvae in embryo medium into the tip in a final volume of 250 μL, and

carefully aliquot the ten larvae into a single well of a clear, polystyrene 96-well round-sbottom plate (*see* **Note 16**). Be careful to pipette up the fish head or tail first to avoid damaging them.

9. Once the plates have been loaded with larvae, chemicals can be added directly to the fish water. Using either robot automation or a P20 8- or 12-channel hand pipette, transfer 2.5 μL of each small molecule from the 1 mM chemical working plates to the plates containing the zebrafish arrayed in 250 μL volumes. Use one chemical per well of ten fish. This will give a final concentration of 10 μM in each well (*see* **Note 17**).
10. After drug application, return plated larvae to 28 °C incubator or warm room and incubate the drug-treated larvae in ambient light for 1 h.

3.7 Strobe Light Response Assay (Day 7)

1. Load 96-well plates containing drug-treated larvae into the ZebraBox. Arrange the plate so that well A01 is positioned in the top left corner.
2. From the file menu toolbar select "File" and then "Open Protocol..." and select the previously saved protocol (e.g., "SLR_Round_96well_Plate.vte").
3. From the file menu toolbar select "View" and then "Draw areas" and line up the assay plate so that the wells are aligned with the circles drawn on the screen (*see* **Note 18**).
4. From the file menu toolbar select "Experiment" and then "Execute." Give the data file for the present experiment a new name (e.g., 20160530_SLR_Plate_1.vtr) and click "Save."
5. Pull down the door to enclose the plate in darkness and immediately click the "Start" button to begin. The clock display will turn red and begin to count up.
6. When the clock counter reaches 7 min (7:00), the experiment will be completed and the clock display will turn green. At this point data collection will stop. From the file menu toolbar select "Experiment" and then "Stop" to end the current experiment.
7. From the file menu toolbar select "Raw Data" and then "Export" to generate a text file containing the data points for all 96 wells across the entire 7-min experiment (*see* **Note 19**).
8. Freezing behavior can be quantified by simply calculating the "Freezing Index Score." To do this, first calculate the third-quartile fish motion values for all points during dark periods (D), second calculate the third-quartile fish motion values for all points during the strobe periods (S), and finally subtract

D from *S*. The more negative the Freezing Index Score, the more freezing occurs.

4 Notes

1. Breeding fish are typically more than 3 months old, but no more than 2 years of age. They ought to be from a wild-type strain, ideally the most fecund strain available (usually the most outbred strain) to obtain optimal embryo yields. The strains previously used successful include Ekkwill (EKW) and TuAB. For best results, the female-to-male ratio should be close to 2:1, and no less than 1:1. The number of fish required will depend on how many chemicals you will want to screen per week. Under ideal circumstances, 200 females and 100 males (300 fish total or ten 10 L tanks or 30 fish/tank) will produce about 20,000 embryos per week. This will be enough to screen at least twenty 96-well plates (at 10 fish/well), or about 1800 compounds in a day. Adult fish must rest 1 week before they can be mated again. This number can be easily scaled up as needed.
2. For best results, use 5 mM NaCl, 0.17 mM KCl, 0.33 mM $MgSO_4$, and 0.33 mM $CaCl_2$ buffered with 10 mM HEPES, and pH the solution to 7.8 with HCl. It is highly recommended to add at least 0.00002% methylene blue as an antifungal agent. Methylene blue at this concentration has no effect on zebrafish larval behavior in the strobe light assay.
3. Chemical stocks and working dilutions of compounds (typically dissolved in DMSO) can be stored in 96-well polypropylene plates with conical bottoms so that small compound volumes can be centrifuged to the center of the wells for easier pipetting. Plates should be stored in the dark at −80 °C, ideally in a desiccator. Ninety-six-well chemical stock plates can be stored with adhesive aluminum foil lids (e.g., "Seal & Sample" lids, Beckman Coulter).
4. When the eggs are laid and fertilized they will drop down through the mesh to the bottom of the tank. The mesh will provide a barrier between the embryos and the adult fish. Without this barrier, the adult fish will eat the embryos.
5. Zebrafish prefer to mate in shallow waters. The distance between the mesh bottom and the top of the water line is important. The water line should just cover the top of the freely swimming adult zebrafish. Zebrafish also prefer to mate on a shoreline. The bottom of the basket may be angled and secured with the binder clips in such a way that there is a deeper

and a shallow end of the basket. Creation of this artificial shoreline will greatly increase embryo yields.

6. This extra meal will increase embryo yield. The fish will take less than an hour to consume the food, but the wait will give them additional time to excrete waste. Setting up the fish immediately after feeding will increase the amount of excrement in the breeding tank. Too much excrement will reduce breeding efficiency.
7. For ease of setup, the 20 female and 10 male fish can be housed together in 10 L tanks and transferred all at once from their aquaculture tank to the breeding apparatus by net.
8. Uncollected embryos, left for more than 3 h, may begin to die from the accumulation of waste, crowding, and decrease in water oxygenation. The health of the adults may also decline.
9. Cutting off the tip of the water bottle can be useful to increase flow. Embryos from different clutches can be mixed. Group mating will provide diverse (outbred) genetic populations, and embryo mixing can further minimize possible clutch-to-clutch variations.
10. To pipette embryos using a standard P1000 pipette tip, you will need to cut off about 8 mm of the pointed end of the tip. This will create a larger opening. Pipetting through a smaller opening may cause damage to the embryos.
11. A physiological light/dark cycle provides the optimal environment to ensure normal development. This is especially important to ensure normal behavioral response in the strobe light assay [3].
12. This step will help maintain the health of the rest of the embryos in the dish. By day 4 (3 days postfertilization) the zebrafish embryos will have developed into larvae and hatched from their chorions. Under normal aquaculture conditions, chorion debris will not interfere with the development of the fish and the aquaculture medium should remain clear and colorless. If a plate becomes contaminated (e.g., with yeast, bacterial, paramecia, etc.) the plate will become cloudy and take on a yellowish color. This may negatively impact larval health. If evidence of contamination is visible by eye, the plate cannot be used and must be discarded. The addition of methylene blue to the medium will limit fungal growth.
13. During the first week of life, zebrafish obtain sustenance from their yolks rather than from external food. Feeding larvae is unnecessary at this stage and disadvantageous when such great numbers of fish are required. Feeding without medium change will lead to aquiculture plate contamination.
14. When thawing a multi-well microliter plate containing the small-molecule library, do so in the dark at room temperature,

in a desiccator when possible. After thawing, wipe off extra moisture from the outside of the plate, and then spin it in a centrifuge to remove droplets from the lid before removing the adhesive aluminum foil lid (e.g., "Seal & Sample" lids, Beckman Coulter). If the centrifuge step is omitted, and droplets remain on the lid, well-to-well cross contamination will occur when the plate is unsealed. Keep track of the number of thaws, as multiple freeze-thaw cycles may affect compound stability.

15. Larvae can tolerate incubation on ice for up to 30 min, depending on the volume of medium in the plates. Incubation times longer than 30 min may result in permanent damage to the larvae and eventually death.

16. The strobe light response assay has been previously optimized; fewer or more than 8–10 fish per well yielded smaller signal-to-noise ratios [3]. There are currently no automated methods available to plate this many fish per well. Manually pipetting has some advantages; it allows the larvae to be reliably counted and for their health and development to be visually assessed before plating. However, manually pipetting the fish can be a time-consuming step. For someone who is inexperienced, it could take up to 1 h to completely fill a single 96-well microliter plate with 10 larvae per well. An experienced technician will be able to complete a single plate within 10 min. All larvae (of both sexes) are used; zebrafish sex is indeterminable at this age.

17. There is no need to mix the compound in the well; the larvae motion will do this automatically. A final concentration of 1% DMSO solvent is used in this protocol. Larvae can tolerate this concentration of DMSO without exhibiting behavioral effects. Many chemicals administered to larvae in this way will easily penetrate larval fish. It is important that each plate contains vehicle (solvent only) control wells. This will provide a way for assessing the quality of the data, and account for any plate-to-plate variations. It is also helpful to include a positive control well(s) on each plate (e.g., 5 μM final concentration of finazine [3]), if available.

18. It is important that all wells of the assay plate are aligned with the circles drawn on the screen. If there is a misalignment, then some adjustments will need to be made to the "SLR_Round_96well_Plate.vte" program using the ZebraLab software. Specifically, the "Top left corner coordinates" (x, y), the "Area dimensions," and/or the Horizontal" and "Verticals space between areas" within the "Areas position" section of the "Protocol Creation Wizard" may need to be adjusted. The "Protocol Creation Wizard" can be found under the "Edit" option in the file menu toolbar.

19. The data will be exported as a .txt file. This file will be quite large.

References

1. Swinney DC, Anthony J (2011) How were new medicines discovered? Nat Rev Drug Discov 10:507–519
2. Rennekamp AJ, Peterson RT (2015) 15 years of zebrafish chemical screening. Curr Opin Chem Biol 24:58–70
3. Rennekamp AJ, Huang XP, Wang Y, Patel S, Lorello PJ, Cade L et al (2016) Sigma1 receptor ligands control a switch between passive and active threat responses. Nat Chem Biol 12:552–558

Chapter 12

A PLA-iRoCS Pipeline for the Localization of Protein–Protein Interactions In Situ

Taras Pasternak, William Teale, Thorsten Falk, Benedetto Ruperti, and Klaus Palme

Abstract

In plants as well as other organisms, protein localization alone is insufficient to provide a mechanistic link between stimulus and process regulation. This is because protein–protein interactions are central to the regulation of biological processes. However, they remain very difficult to detect in situ, with the choice of tools for the detection of protein–protein interaction in situ still in need of expansion. Here, we provide a protocol for the detection and accurate localization of protein interactions based on the combination of a whole-mount proximity ligation assay and iRoCS, a coordinate system able to standardize subtle differences between the architecture of individual *Arabidopsis* roots.

Key words Proximity ligation assay, iRoCS, Protein complex, Protein–protein interactions, In situ, In planta, 3D imaging, Root apical meristem, *Arabidopsis thaliana*

1 Introduction

Protein complexes regulate diverse cellular pathways. The analysis of protein–protein interactions is therefore important to understand biological processes. While many methods to investigate protein–protein interactions already exist, all have their respective drawbacks. In this chapter, we introduce a new method that enables the analysis of proteins based on their proximity in vivo.

Pull-down assays are widely used to isolate interacting proteins for analysis [1]. However, protein purification may affect a protein's conformation or charge and therefore potentially alter the range of interactions in which it participates. Furthermore, lysis disrupts subcellular compartmentalization of proteins, which may lead to aberrant interactions in vitro [2].

Bimolecular fluorescence complementation (BiFC) [3] and the yeast two-hybrid system [4] are commonly used to analyze protein–protein interactions in vivo. However, neither method

Bridget Wagner (ed.), *Phenotypic Screening: Methods and Protocols*, Methods in Molecular Biology, vol. 1787, https://doi.org/10.1007/978-1-4939-7847-2_12,

observes interactions under natural conditions. Therefore, protein complex formation observed with these methods may differ from interactions in situ.

A more recently developed technique, the proximity ligation assay (PLA) [5, 6], allows the detection of single-interaction events within protein complexes. Primary antibodies, raised in different animals, are used to bind target proteins in situ which, in turn, are bound by secondary antibodies conjugated to complementary oligonucleotides. A close proximity ($\leq$20 nm) results in hybridization and incorporation of fluorescently labeled oligonucleotides via a polymerase-mediated rolling circle reaction. The resulting fluorescence can be detected by fluorescence microscopy.

The method described here has been designed to combine the strengths of in vivo and in vitro analyses, for example, the direct detection of interactions as inferred from a close proximity, and the suppression of false-positive signals and map interactions within the three-dimensional context of the root apical meristem of *Arabidopsis thaliana*.

Representing three-dimensional structures with two-dimensional images captured by cameras and microscopes inevitably leads to information loss. To address this problem several software platforms for 3D imaging have been established including phytotyping4d [7], MARS-ALT [8], 3DCellAtlasOpen [9], and MorphoGraphX [10].

Phytotyping4D combines a focus image and a depth image to generate a 3D topological map. Similarly, MARS-ALT also combines images (this time after multi-angle acquisition) for 3D reconstruction and cell segmentation; developing cells may be traced with an associated automated lineage tracking function. MorphoGraphX uses curved surface images extracted from 3D data. This open-source software platform contains a collection of loadable modules such as 3DCellAtlasOpen, a rapid and modular computational approach to generate quantitative 3D cellular atlases.

The iRoCS (intrinsic root coordinate system) toolbox is an open-source software platform designed to compare root apical meristems among plant populations [11]. The program fits standardized coordinates to 3D images of roots gathered via confocal microscopy. It then detects nuclei or cell boundaries and classifies them into root tissue layers while only requiring minimal input from the user. The use of a machine-learning algorithm means it can be trained to recognize in situ PLA signals, allowing for fast and easy analysis. The following protocol can be easily divided into three parts: the proximity ligation assay, imaging with confocal microscopy, and image processing to detect PLA sites and analyze their distribution in the anatomical context of the root tip using the iRoCS.

2 Materials

All reagents should be made with high-purity water with a resistivity of 18 MΩ. Be sure to dispose of waste reagents safely.

2.1 Plant Material and Growth Conditions

1. Seeds of wild-type *Arabidopsis thaliana*.
2. Culture medium: ½ MS medium containing 1% sucrose and 1.3% agar.
3. Growth chamber capable of a light intensity of 80 μmol/m^2/s and a 16-h/8-h dark/light period.

2.2 Tissue Preparation

1. Microtubule-stabilizing buffer stock solution (2× MTSB): 15 g PIPES, 1.90 g EGTA, 1.22 g $MgSO_4 \cdot 7\ H_2O$, 2.5 g KOH in 500 mL water, adjust pH to 7.0 with 10 M KOH.
2. Methanol.
3. Fixative solution: 37% Formaldehyde stock solution diluted to a final concentration of 2% in MTSB supplemented with 0.1% Triton X-100.
4. Cell wall digestion solution: Weigh 20 mg Driselase and 15 mg Macerozyme R-10 into an Eppendorf tube and add 1 mL of water. Mix gently and centrifuge at 14000 × *g* for 1 min. Transfer the supernatant to a 15 mL Falcon tube and add 9 mL of 5.5 mM MES pH 5.2. The final solution contains 0.2% driselase and 0.15% macerozyme and can be stored at 4 °C for up to 2 weeks.
5. Permeabilization buffer: Prepare 3% IGEPAL CA-630 (IGEPAL CA-630 is similar to Nonidet P-40, which is no longer commercially available) and 10% dimethyl sulfoxide (DMSO) in 1× MTSB buffer.
6. Vacuum pump connected to a desiccator fitted with a sealable connection in order that it may be evacuated.
7. Benchtop rocking or orbital shaker for gentle shaking.

2.3 Proximity Ligation Assay and Mounting

1. Blocking solution: Dissolve 2% BSA fraction V in 1× MTSB.
2. Mouse primary antibody solution (*see* **Note 1**).
3. Rabbit primary antibody solution (*see* **Note 1**).
4. Proximity ligation PLUS solution: Duolink® In Situ PLA Probe Anti-Mouse PLUS, affinity-purified Donkey anti-Mouse IgG (H+L) (Invitrogen).
5. Proximity ligation MINUS solution: Duolink® In Situ PLA Probe Anti-Rabbit MINUS, affinity-purified Donkey anti-Rabbit IgG (H+L) (Invitrogen).
6. Proximity ligation reaction solution: Duolink® In Situ Detection Reagents Green (Invitrogen).

7. Nuclear stain: Dilute 4′,6-diamidino-2-phenylindole dihydrochloride (DAPI) in water to a final concentration of 0.2 mg/L from a 1 mg/mL stock, also prepared in water (*see* **Note 2**).
8. Antifade mounting medium: Fluoromount G (Southern Biotech, refractive index 1.39) or ProlongGold (Invitrogen, refractive index 1.47).

2.4 Confocal Microscopy

Root tips should be imaged with a confocal laser scanning microscope at an excitation wavelength of 405 nm for DAPI and 561 nm for PLA-labeled protein interactions with a 40×/1.2 W-corrected UV-VIS-IR objective.

2.5 Image Processing and iRoCS Assignment

Image processing requires a workstation (Linux preferred) equipped with at least 8 GB RAM running MATLAB (https://de.mathworks.com/products/matlab.html), Fiji (https://imagej.net/Fiji/), and the iRoCS Toolbox (https://lmb.informatik.uni-freiburg.de/lmbsoft/iRoCS/).

3 Methods

After each section, it is possible to pause the procedure and continue at a later date. Unless stated otherwise, all steps should be carried out at room temperature. An overview of the main steps is presented in Fig. 1. The volumes given have been calculated for the use of a 24-well plate and eight rubber frame slides.

3.1 Preparation of Plant Material

1. Imbibe seeds of wild-type *Arabidopsis thaliana* on plates of ½ MS medium containing 1% sucrose and 1.3% agar.
2. Incubate at room temperature for 5 h.
3. Transfer to a refrigerator at 4 °C and leave overnight (12–14 h) for stratification.
4. Induce germination by transfer to a growth chamber set at 22 °C with a light intensity of 80 μmol/m^2/s and a 16/8-h dark/light period. For all experiments, seedlings were used after 4 days in the growth chamber.

3.2 Proximity Ligation Assay

3.2.1 Fixation (50–60 min)

1. Carefully remove seedlings from their Petri dish and place in a 24-well plate containing at least 1.5 mL of fixative solution. Leave the lid off and transfer the 24-well plate to a desiccator connected to a vacuum pump (*see* **Note 3**).
2. Apply a vacuum for between 2 and 3 min before slowly releasing the vacuum (*see* **Note 4**).
3. Repeat **step 2**.

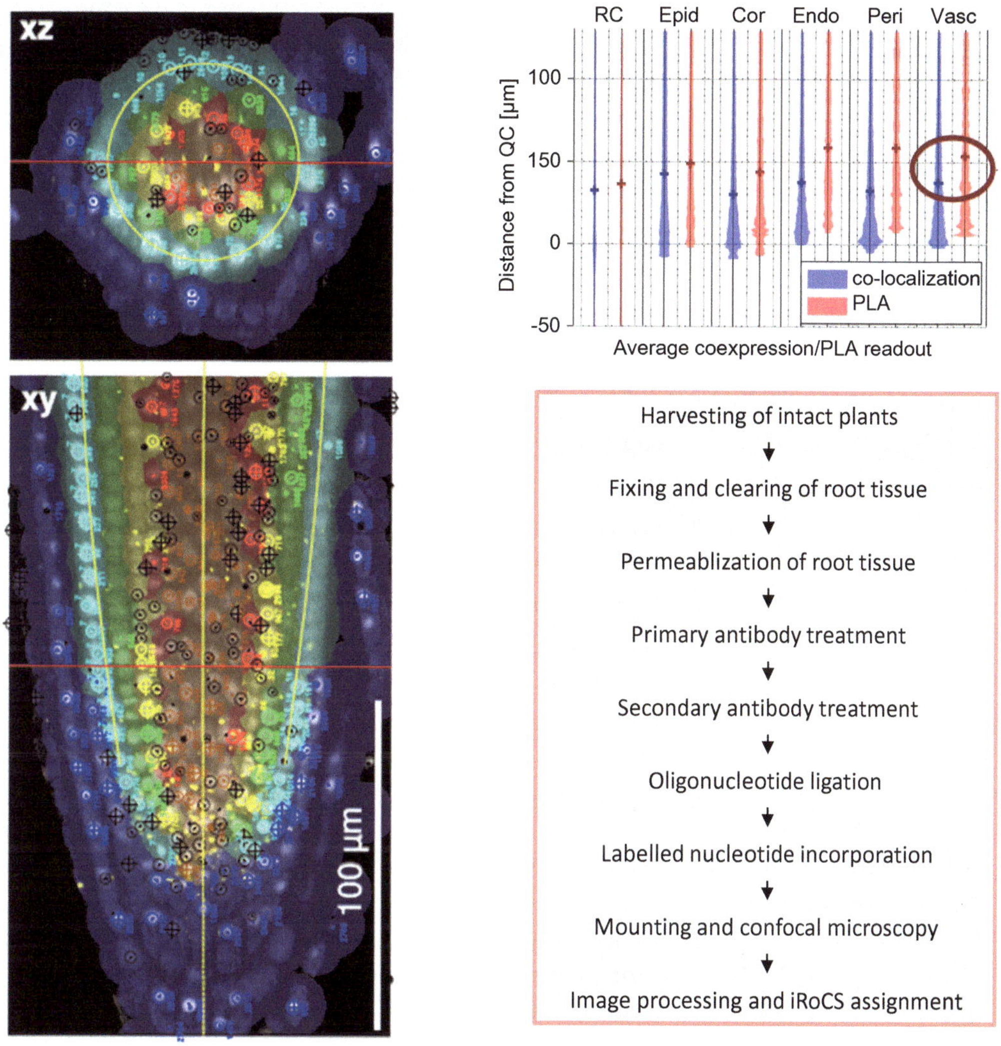

Fig. 1 *PLA-iRoCS in the Arabidopsis root apical meristem.* Left panel: Automatic annotation of nuclei superimposed on PLA signals (above, radial section; below, longitudinal section). Nuclei of cell layers are marked with circles and color coded as following: root cap, dark blue; epidermis, light blue; cortex, green; endodermis, yellow; pericycle, red; stele, brown. Right panel (above): Kite diagram showing normalized colocalization (blue) and PLA signal distribution (red) of two interacting target proteins as a function of distance from the quiescent center in two genotypes A and B. Each plot is the average of three experiments and the median signal is marked with a point. The root has been divided into its constituent cell layer for better visualization. Subtle differences between genotypes, in this case in the distribution of PLA signal in vascular cells, have been circled. Right panel (below): An overview of the steps which comprise the PLA-iRoCS pipeline

4. The seedlings should now have sunk to the bottom. Continue fixation for a further 40 min under atmospheric pressure with gentle shaking at 22 °C.
5. Replace the fixing solution with 2 mL of distilled water and incubate for 10 min (*see* **Note 5**).

3.2.2 Tissue Clearing (50–60 min)

1. Remove the water and replace with 800 μL of 100% methanol pre-warmed in a water bath to 60 °C. Incubate for 5–10 min.
2. Gradually decrease the methanol concentration by adding 150 μL of water every 2 min. Continue until the final methanol concentration reaches 20% (this corresponds to the addition of 3.2 mL of water) (*see* **Note 6**).
3. Remove the 20% methanol solution and replace with water. Incubate for 5 min.
4. Repeat **step 3**.
5. Transfer plants to rubber-framed slides, each pre-loaded with 60 μL of water.

3.2.3 Tissue Permeabilization (90 min)

1. Add 60 μL of cell wall digestion solution and incubate for 20–30 min at 37 °C (*see* **Note 7**).
2. Remove the cell wall digestion buffer and incubate for 4 min in 100 μL of 1× MTSB.
3. Repeat **step 2** a further three times.
4. Add 60 μL of permeabilization buffer and incubate for 15–20 min at 37 °C (*see* **Note 8**).
5. Remove the permeabilization buffer and incubate for 3 min in 100 μL of 1× MTSB.
6. Repeat **step 5** a further three times.

3.2.4 Blocking and Primary Antibody Incubation (150 min)

1. Remove the MTSB and add 60 μL of blocking buffer. Incubate for 20 min (*see* **Note 9**).
2. Remove the blocking buffer and replace with 60 μL of the primary antibody solution (*see* **Note 10**). Incubate for 1–2 h at 37 °C (*see* **Note 11**).
3. Remove the primary antibody solution and replace with 100 μL of 1× MTSB. Incubate for 5 min.
4. Repeat **step 3**.

3.2.5 Secondary Antibody Incubation (135 min)

1. Remove the MTSB, add 40 μL of the secondary antibody solution, and incubate for 1–2 h at 37 °C.
2. Remove the secondary antibody solution and replace with 100 μL of 1× MTSB. Incubate for 5 min.
3. Repeat **step 2** twice (*see* **Note 12**).

3.2.6 Ligation (120 min)

1. Thaw the 5× ligation buffer and add 8 μL to an Eppendorf tube.
2. Add 31 μL of water and 0.5 μL of ligase.
3. Mix well and incubate the tissue with 40 μL of the ligation solution at 37 °C for 2 h.

4. Remove the ligation solution and add 70 μL of buffer A.
5. Repeat **step 5**.

3.2.7 Amplification and Mounting (135 min)

1. Thaw the 5× amplification buffer and add 8 μL to an Eppendorf tube.
2. Add 31 μL water and 0.9 μL of polymerase. Mix well.
3. Remove buffer A and incubate the tissue with 40 μL of the ligation solution at 37 °C for 2 h.
4. Remove the ligation solution and add 70 μL of buffer B.
5. Repeat **step 4**.
6. Remove buffer B and incubate in 60 μL of water containing 1% buffer A and 1 mg/L DAPI for 5 min.
7. Gradually add 50% glycerol to a final concentration of 25%.
8. Transfer each seedling in turn to microscope slides with a jacket of antifade medium and cover each slide with a coverslip. Slides can be stored for months in a refrigerator (*see* **Note 13**).

3.3 Confocal Microscopy

1. Take serial optical sections with a spacing of 1 μm to a depth of 100 μm and reconstitute into a 3D image stacks, with an in-plane sampling of 0.15 μm.
2. Record three partially overlapping image stacks for each root.

3.4 Image Processing and iRoCS Assignment

1. Convert the recorded multichannel LSM stacks to HDF5 using the Fiji HDF5 plug-in.
2. Detect PLA spots using the function detectSpots in MATLAB (used noise levels: 0.2 for low and 0.5 for high noise level).

```
function spots = detectSpots(filename, dataset, outgroup,
noiselevel)
% function spots = detectSpots(filename, dataset, noiselevel)
%
% filename - hdf5 file containing the dataset to detect spots
in
% dataset - hdf5 dataset to detect spots in
% noiselevel - Expected noise level (between 0 and 1)
I = double(h5read(filename, dataset));
elSize = h5readatt(filename, dataset, 'element_size_um');
I = I / max(I(:));
I = imfilter(I, fspecial('gauss', 5, 0.7));
I2 = imhmax(I, noiselevel);
spotcenters = (I2 - I + noiselevel) < 0.00000001;
[X, Y, Z] = ind2sub(size(I), find(spotcenters));
spotpositions_px = [Z'; Y'; X'];
spotpositions_um = (double(spotpositions_px) - ...
ones(size(spotpositions_px))) .* ...
(elSize * ones(1, size(spotpositions_px,2)))
```

```
hdf5write(filename, [outgroup '/position_um'], spotposition-
s_um, ...
'WriteMode', 'append');
```

3. Attach the intrinsic root coordinate system (iRoCS).
   (a) Start the iRoCS Toolbox labeling GUI.
   (b) Run the following iRoCS plug-ins in sequence:
       01—Detect nuclei
       02—Label Epidermis
       03—Attach iRoCS
       04—Assign Layers
       Detailed information on the protocol can be found on the iRoCS project page https://lmb.informatik.uni-freiburg.de/lmbsoft/iRoCS/.
   (c) Load the detected PLA spots using "Channel → Import channels..." by selecting the HDF5 group used in the MATLAB detection script. After import, push "Update coordinates..." to compute their root-intrinsic coordinates. At this point you can review the detection result and manually correct detections if required.
4. Visualization of populations: To obtain per-layer statistics, assign each PLA spot to the cell layer of the closest nucleus. PLA heat maps can be generated by projecting the PLA spot positions to the z,r coordinates of iRoCS. To obtain smooth distributions, we recommend applying a Gaussian filter with a standard deviation of 2 μm to the resulting projections. Divide the resulting distribution by the number of samples in the population and normalize the intensity range of all populations to [0, 0.5]. For improved visibility, intensities may be shown using the jet color map of MATLAB.

4 Notes

1. The working concentration of antibody for PLA-iRoCS analysis can vary, depending on many diverse factors. In most cases, a concentration similar to that of immunolocalization should be used. Therefore, we advise an optimization procedure in which both primary antibodies are used to colocalize the target proteins in the root. This provides the added benefit of defining the colocalization domain inside which valid PLA signals should be confined.
2. Dissolve DAPI in water at a concentration of 1 mg/mL and dilute it before use to 2 μL in 10 mL. A 1 mg/mL solution is stable for at least 1 year at 4 °C.

3. Apply vacuum in a fume hood. Treat up to ten seedlings together in a single well.
4. Releasing the vacuum is a crucial step. Be sure to release the vacuum very gingerly to avoid the loss of your sample.
5. The integrity of the cells' structure must be constantly maintained throughout the fixation. For this, the rapid penetration of the fixative into cells is crucial. Freshly prepared 2% formaldehyde solution is used for this purpose, and gives best results. A 37% formaldehyde stock solution may also be used. However, formaldehyde is relatively unstable in aqueous solution and polymerizes during long-term storage. Both properties may deleteriously affect the assay and must be taken into account. Specimens should be fixed in a multi-well culture plate with as large a surface as practical to maximize efficiency of the fixation procedure.
6. The gradual addition of water is important to preserve the structure of the root tip.
7. The plant cells' rigid cell wall needs to be at least partially digested for efficient antibody penetration. This procedure is performed by incubation with a mixture of cell-wall-degrading enzymes. In the majority of published protocols, driselase is used. These conditions are suboptimal. Driselase is relatively poor at weakening the bonds among cells. Furthermore, its pectolytic and cellulolytic activities have different optimum pHs: between 4.0 and 6.0, and between 3.0 and 5.0, respectively. In order to increase cell wall digestion while maintaining the integrity of the tissue, we suggest using a mixture of Driselase and Macerozyme R-10 at pH 5.0.
8. Permeabilization creates pores in membranes that allow the antibody to penetrate. For this purpose, treatment with a mixture of DMSO and the detergent IGEPAL CA-630 should be used.
9. During the blocking step, nonspecific antibody binding is minimized. Block for at least 20 min. If background signal is high, you may extend this step up to 2 h.
10. This experiment requires an antibody raised against each target protein: one raised in rabbit, and the other in mouse. In our experience, polyclonal antibodies should always be affinity purified. If no steric hindrance occurs, dimerization of a single-target protein can be measured by using a single-mouse monoclonal antibody and an equal mixture of PLUS and MINUS anti-mouse secondary antibodies.
11. Do not mix the solution during this incubation step. Good results can be obtained with antibodies against translationally fused epitope tags or GFP. It is absolutely necessary to test

antibody specificity in Western blots. A loss of function mutant where the protein of interest is absent (or wild-type plants in the case of modified proteins) should be ideally used as a negative control.

12. Do not mix the solution during this incubation step.
13. To prepare samples for microscopy, embed in commercially available antifade solutions such as Fluoromount G (Southern Biotech) or Prolong® Gold (Invitrogen). These solutions are designed to protect samples from photobleaching. We highly recommend you match, as closely as possible, the refractive index of the mounting medium to the refractive index of the immersion medium used for the microscopic imaging to avoid optical artifacts and signal loss. One can also use homemade antifade solutions, containing glycerol (50%), N-propyl gallate (15 mg/mL final concentration), and H_2O (50%). For long-term storage of samples, addition of 0.1% sodium azide to the anti-bleaching solution is mandatory. In order not to damage the samples we suggest that you mount specimens after immunolocalization on microscopic slides pre-prepared with a 120 μm spacer made from TVC isolation tape. For whole-mounted seedlings, a 100 μm thick spacer gives enough height to avoid squashing the sample, maintaining its original 3D structure.

References

1. Uhlen M, Ponten F (2005) Antibody-based proteomics for human tissue profiling. Mol Cell Proteomics 4:384–393
2. Sprinzak E, Sattath S, Margalit H (2003) How reliable are experimental protein–protein interaction data? J Mol Biol 327:919–923
3. Hu CD, Chinenov Y, Kerppola TK (2002) Visualization of interactions among bZip and Rel family proteins in living cells using bimolecular fluorescence complementation. Mol Cell 9:789–798
4. Bartel PL, Fields S (1995) Analyzing protein--protein initeractions using yeast 2-hybrid system. Oncogene Tech 254:241–263
5. Soderberg O, Gullberg M, Jarvius M, Ridderstrale K, Leuchowius KJ, Jarvius J et al (2006) Direct observation of individual endogenous protein complexes in situ by proximity ligation. Nat Methods 3:995–1000
6. Soderberg O, Leuchowius KJ, Gullberg M, Jarvius M, Weibrecht I, Larsson LG et al (2008) Characterizing proteins and their interactions in cells and tissues using the in situ proximity ligation assay. Methods 45:227–232
7. Apelt F, Breuer D, Nikoloski Z, Stitt M, Kragler F (2015) Phytotyping(4D): a light-field imaging system for non-invasive and accurate monitoring of spatio-temporal plant growth. Plant J 82:693–706
8. Fernandez R, Das P, Mirabet V, Moscardi E, Traas J, Verdeil JL et al (2010) Imaging plant growth in 4D: robust tissue reconstruction and lineaging at cell resolution. Nat Methods 7:547–U594
9. Montenegro-Johnson TD, Stamm P, Strauss S, Topham AT, Tsagris M, Wood ATA et al (2015) Digital single-cell analysis of plant organ development using 3DCellAtlas. Plant Cell 27:1018–1033
10. de Reuille PB, Routier-Kierzkowska AL, Kierzkowski D, Bassel GW, Schupbach T, Tauriello G et al (2015) MorphoGraphX: a platform for quantifying morphogenesis in 4D. Elife 4:05864
11. Schmidt T, Pasternak T, Liu K, Blein T, Aubry-Hivet D, Dovzhenko A et al (2014) The iRoCS toolbox: 3D analysis of the plant root apical meristem at cellular resolution. Plant J 77:806–814

Chapter 13

High-Dimensional Profiling: The Theta Comparative Cell Scoring Method

Scott J. Warchal, John C. Dawson, and Neil O. Carragher

Abstract

Principal component analysis enables dimensional reduction of multivariate datasets that are typical in high-content screening. A common analysis utilizing principal components is a distance measurement between a perturbagen—such as small-molecule treatment or shRNA knockdown—and a negative control. This method works well to identify active perturbagens, though it cannot discern between distinct phenotypic responses. Here, we describe an extension of the principal component analysis approach to multivariate high-content screening data to enable quantification of differences in direction in principal component space. The theta comparative cell scoring method can identify and quantify differential phenotypic responses between panels of cell lines to small-molecule treatment to support in vitro pharmacogenomics and drug mechanism-of-action studies.

Key words Phenotypic screening, High-content analysis, Cell-based profiling

1 Introduction

Phenotypic screening allows the identification of treatments that modify a disease model without prior knowledge of the molecular target. This re-emerging method can generate hypotheses for the etiology behind poorly understood diseases, in addition to the discovery of potential therapeutics that act through novel biological mechanisms [1].

One form of phenotypic screening is high-content image-based screening which uses multiple measurements to create a detailed multivariate profile of a perturbagen. This can make screens less biased to prenominated target or therapeutic class hypothesis and also create a phenotypic fingerprint to inform mechanism of action [2–5].

A distinct phenotypic response between cell types which represent the broad heterogeneity of human disease and/or more defined clinical subtypes can highlight differences in cellular signaling, metabolic, and biochemical transporter mechanisms that

Bridget Wagner (ed.), *Phenotypic Screening: Methods and Protocols*, Methods in Molecular Biology, vol. 1787, https://doi.org/10.1007/978-1-4939-7847-2_13,

explain the variation of drug efficacy between patients often observed in the clinic. Correlation of distinct phenotypic response and drug sensitivity across genetically distinct cell types with genomic, transcriptomic, and proteomic data can help elucidate compound mechanism of action and identify molecular biomarkers which predict drug sensitivity and clinical outcomes [6, 7]. We can also use phenotypic similarity between different perturbagens to infer mechanistic similarities. One such example is that small molecules which elicit similar cellular phenotypes are likely to have similar mechanisms of action [8]. Phenotypes can also be used to model disease biology where the underlying signaling pathways and molecular targets associated with disease progression are lacking or poorly understood [9].

In order to quantify complex phenotypes, high-content screening generates multivariate datasets in which multiple phenotypic measurements are taken from a single cell or image. These datasets are usually subjected to some form of dimensionality reduction technique in order to make analysis more manageable. A common dimensional reduction method is principal component analysis, which creates new features (principal components) through orthogonal linear combinations of the original features in order to maximize variation. As principal components are ranked in order of variation, a subset of the principal components can be taken as a replacement for the original feature measurements—with the aim of reducing the number of variables while still retaining as much information as possible. This approach helps visualize complex multivariate data points by plotting them in 2D or 3D principal component space [10, 11].

A simple method used to identify active perturbagens in multivariate datasets is a distance measurement such as Euclidean or Mahalanobis distance between the perturbagen and the negative control in principal component space. This can be used to create a threshold distance to separate the active from inactive, as well as rank perturbations on phenotypic activity [11]. However, this distance metric cannot readily discern between different active phenotypes. Two perturbations may produce very different phenotypes and coordinates in principal component space, and yet have similar distances from the negative control.

In order to discern between perturbations such as these we need a measure of directionality. The idea behind the theta comparative cell scoring (TCCS) method is that different directions in phenotypic space indicate different phenotypes. Therefore measuring the angle between perturbagen-induced phenotypes can be used as a phenotypic similarity score independent of potency. This is very similar to the use of cosine similarity, though the TCCS method centers measurements on the negative control and removes inactive perturbagens that may otherwise produce inaccurate measures of directionality.

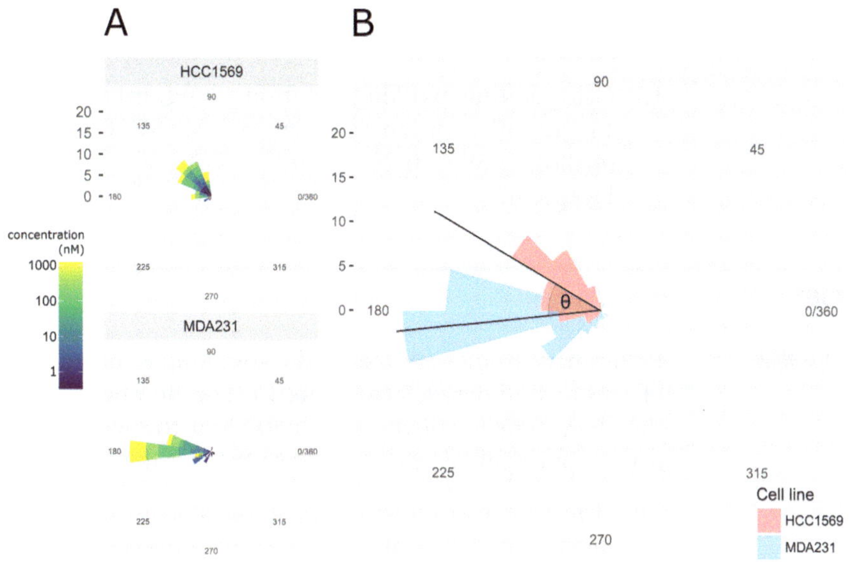

Fig. 1 Circular histograms showing the similar phenotypic direction of HCC1569 and MDA-MB-231 (MDA231) breast cancer cell lines treated with the aurora kinase inhibitor barasertib. (**a**) Theta values calculated from the first two principal components against a reference vector for both HCC1569 and MDA-MB-231 cell lines treated with barasertib at multiple concentrations. (**b**) Depiction of the θ value when calculated between a pair of cell lines representing the difference in phenotypic response

The idea of directionality can also be used to produce intuitive and quantitative figures such as circular histograms depicting the direction in phenotypic space or the difference in theta values between two perturbations or samples (Fig. 1).

2 Materials

1. Optical-bottom imaging plates (96- or 384-well).
2. Cell culture medium.
3. Trypsin.
4. Perturbagen Library.
5. Paraformaldehyde (PFA).
6. Triton X-100.
7. Wheat-germ agglutinin 594 (WGA), diluted in dH_2O.
8. SYTO14 green fluorescent nucleic acid stain.
9. Microtiter plate seals.
10. Aluminum foil.
11. Cell painting stock solution: 10 mg/mL Hoechst 33342, 1 mg/mL concanavalin A (diluted in 0.1 M $NaHCO_3$), 200 U/mL phalloidin-594 (diluted in methanol), 1 mg/mL WGA, 1 mM MitoTracker DeepRed.

12. Blocking buffer: 1% Bovine serum albumin (BSA) in PBS (w/v).
13. Cell painting working solution: 2 μg/mL Hoechst 33342, 11 μg/mL concanavalin A, 3 μM SYTO14, 2.5 U/mL phalloidin-594, 0.25 μg/mL WGA, 600 nM MitoTracker DeepRed.

3 Methods

3.1 Cell Seeding

Preliminary studies are required to determine the optimal number of cells to seed per well (*see* **Note 1**). This number is dependent on the characteristics of the cell line(s) and the well area in a given plate. Approximate values are provided in Table 1.

1. Using a sub-confluent population of cells, detach the cells by short-term incubation with trypsin and suspend to the desired concentration in cell culture medium.
2. Seed the cells into each well of an optical bottom microtiter 96- or 384-well plate. Make sure that the cells do not settle in the stock of cell suspension by frequently agitating the stock of cell suspension.
3. Incubate the plates containing cells for 24 h.

3.2 Compound Addition

1. Make up stock compound plates in DMSO at 1000× the final concentration.
2. Make an intermediate plate by diluting stock compound plate 1:50 in cell culture medium.
3. Remove cell plates from the incubator and transfer a volume from the intermediate plate to the cell plate in a 1:20 dilution.
4. Return cell plates to the humidified, 37 °C, 5% CO_2 incubator for an additional 48 h.

3.3 Fluorescent Labeling

3.3.1 Fixation

1. Make a solution of 8% paraformaldehyde (PFA) in phosphate-buffered saline (PBS).
2. Add an equal volume of PFA to each well, and incubate at room temperature for 30 min.
3. Wash wells three times with 50 μL of PBS.

Table 1
Approximate cell seeding densities for different plates

Plate	Cells/well	Volume/well (μL)
96	2000–3000	100
384	750–1500	50

3.3.2 Permeabilization

1. Add 30 μL of 0.1% Triton-X100 solution in PBS to each well, and incubate for 20 min at room temperature.
2. Wash wells three times with 50 μL of PBS.

3.3.3 Cell Labeling

Cell labeling protocol adapted from the cell painting protocol [12, 13].

1. Protect the staining solution from light sources by wrapping in aluminum foil.
2. Add 30 μL of cell painting solution and incubate in a dark place at room temperature for 30 min.
3. Wash plate three times with 50 μL of PBS. Do not aspirate the final volume.
4. Seal the plates. If the plates are not imaged immediately, then store them at 4 °C in the dark or wrapped in aluminum foil.

3.4 Imaging

1. Set up the microscope to image five channels at 20× magnification. *See* Table 2 for suggested filter settings.
2. Image multiple sites per well; we recommend a minimum of four.
3. Adjust the focus and exposure settings (*see* **Note 2**). These settings should be kept constant between batches and comparable experiments as intensity measurements are a function of exposure time.

3.5 Image Analysis

The following image analysis instructions use CellProfiler [14] nomenclature, though other image analysis software packages may be used to achieve similar results.

1. Extract metadata from either the image or the file path; record the date, plate number, plate name, well, site, and channel for each image.

Table 2
Cell painting reagents and suggested filters

		Filter wavelength	
Stain name	**Filter name**	**Excitation (nm)**	**Emission (nm)**
Hoechst 33342	DAPI	377 ± 40	447 ± 60
Con A	FITC	482 ± 35	536 ± 40
SYTO14	Cy3	531 ± 40	594 ± 40
Phalloidin & WGA	TxRed	562 ± 40	624 ± 40
MitoTracker DeepRed	Cy5	628 ± 40	692 ± 40

2. Add in external metadata from a .csv file such as compound labels or concentrations and match via plate name and well name/position.
3. Assign each image to a channel name using the extracted channel metadata.
4. Segment the nucleus using IdentifyPrimaryObjects.
5. Segment the cell body/cytoplasm using the nucleus object as a seed in the phalloidin/WGA channel with the IdentifySecondaryObjects module.
6. Measure image quality in the DAPI channel using MeasureImageQuality. Out-of-focus images and any debris can usually be detected in the DAPI channel. Image quality can also be measured in all the channels though the MeasureImageQuality module although this will increase analysis time.
7. Measure object size and shape of both the nucleus and cell body with MeasureObjectSizeShape.
8. Measure intensity of the nucleus in the DAPI channel and intensity of the cell body in the other four channels using MeasureObjectIntensity.
9. Measure texture in the channels for Golgi apparatus and actin staining (WGA channel) in the cell body objects, and the DAPI channel for the nuclei objects using MeasureTexture.
10. Measure object neighbors for both nuclei and cell bodies with MeasureObjectNeighbors.
11. Export measurement data as .csv files or to a database, excluding any feature measurements that may not be relevant such as object number or object *x*-*y* position.

3.6 Data Analysis

1. Check the data produced by the CellProfiler analysis for any missing rows or columns; these need to be removed as appropriate (*see* **Note 3**).
2. Using the ImageQuality measurements produced by CellProfiler, identify any images that may be out of focus or contain debris and after visually checking the images remove the data relating to that image if necessary (*see* **Note 4**).
3. If the data is at the object-level, i.e., measurements per cell, then aggregate this to a well median, so each measurement describes the median measurement per feature per well.
4. Remove non-informative features (any measurement columns that are not metadata) such as those with zero or very low variance.
5. Remove redundant features, such as one of a pair of features that are very highly correlated with each other. This can be performed by calculating a correlation matrix of the feature

dataset and finding groups of features that have Spearman's correlations greater than 0.95, and then removing all but one of these features from the dataset.

6. Normalize the data to the negative control values on each plate. This is performed by subtracting the median of the negative control for each feature, per plate (*see* **Note 5**).
7. Scale the features. For each feature: subtract the feature mean from each individual value, and then divide by the standard deviation of the feature. This standardizes the features to have a mean of zero and unit variance. This is done otherwise features with large values/small units—such as object area which is measured in pixels—will skew the subsequent statistical methods.
8. Calculate the principal components of the feature data and determine the number of principal components needed to account for a proportion of the variance in the dataset, typically 80–90% (*see* **Note 6**).
9. Remove those principal components that fall outside of this subset.
10. Calculate the negative control medoid, which is the median value for each feature of the negative controls.
11. Adjust the principal component values so that the negative control medoid is centered on the origin (*see* **Note 7**).
12. Calculate the l1-norm (AKA city-block or Manhattan distance) from the negative control medoid to each data point in principal component space.
13. Calculate the l1-norm of each negative control point from the origin and calculate a distance threshold as 2 standard deviations of these negative control distances. Any compound that has a distance less than this threshold from the medoid of the negative control can be labeled as inactive.
14. Once the inactive compounds have been removed, perturbagen similarity can be determined by the angle between perturbagen vectors (θ). In two dimensions—using the first two principal components—this can be visualized on a scatter plot. The θ value can be calculated in any number of dimensions, although visualization becomes more difficult. The similarity angle can be calculated by the cosine similarity converted to degrees (*see* Eq. (1)). Note that 180° is the value of maximum dissimilarity, where two perturbagens having completely different directions in phenotypic space, with values greater than 180° becoming increasingly similar as they approach 360°. Therefore θ values are constrained between 0 and 360 by subtracting from 360 any value greater than 180, i.e., $\theta > 180 \rightarrow \theta := 360 - \theta$:

$$\theta = \cos^{-1}\left(\frac{u \cdot v}{\|u\|\|v\|}\right) \times \frac{180}{\pi} \tag{1}$$

where u and v are the vectors of principal components for each compound.

15. If two principal components capture a large proportion of the variance in the dataset then a visualization can be made by calculating θ for every perturbagen against a common reference vector, and then plotting a circular histogram of the θ values (*see* **Note 8**).
16. Identify cell line pairs treated with the same compound that have significantly different theta values (*see* **Note 9**), indicating a distinct phenotypic response between cell lines to a compound treatment.
17. *See* **Notes 10** and **11** for additional troubleshooting steps.

4 Notes

1. Too few cells will provide fewer replicates and may run the risk of having no cells contained in an image if a perturbagen reduces the cell number. Seeding too many cells can mean cells do not form a single monolayer which makes image analysis more difficult. We advise seeding the number of cells to result in approximately 60–70% confluence.
2. After setting the focus for the first channel (DAPI/Hoechst), all additional channel's focus settings are based on these measurements. Therefore adjusting the focus settings for the first channel will also affect all of the other channels, so it is advised to set this first and check a few different wells to ensure that the settings are robust.
3. It is recommended to remove columns containing large amounts of missing numbers. This can often be caused by missing metadata in certain samples, or some features that remain constant between samples—such as Euler number—that may be transformed to missing data entries after scaling or aggregation. Once columns of largely missing data have been removed, rows containing missing values can be removed. Without first removing the missing data columns it is often possible to erroneously remove the entire or large proportions of the dataset when using missing rows as the first step.
4. Out-of-focus images can be detected using ImageQuality_PowerLogLogSlope measurements in the nuclei channel. Images with very low values are likely to be out of focus [15]. Debris such as dust or fibers typically show up in the nuclei channel,

and can be detected by identifying images with a large percentage of saturated pixels.

5. Normalizing to the negative control is a useful step in any plate-based screen to remove any batch effects between plates that may influence the results. It is especially important when comparing effects between cell lines as this converts the values to changes from the negative control for that particular plate; as we expected to have a single cell line per plate this also removes any inherent phenotypic differences between the cell lines, and allows the compound-induced changes to be comparable.
6. The number of principal components required to capture a specified proportion of the variance in the data can be calculated in *R* (assuming that data is numeric feature data), to calculate the value for 80% of the variance:

```
threshold <- 0.8
pca_output <- prcomp(data)
pc_variance <- pca_output$sdev^2
cumulative_proportion_variance <- cumsum(pc_variance) / sum(pc_variance)
n_components <- min(which(cumulative_proportion_variance >= threshold))
```

7. To center the principal component data so that the medoid of the negative control lies on the origin, find the medoid for the negative control values, which is the median value for each feature column for the negative control values; find how much this differs from the origin for each feature; shift all values for each feature by this difference, e.g., in R:

```
centre_control <- function(df, feature_cols, cmpd_col, neg_control = "DMSO") {
    # 1. the median value for the DMSO values for each measured feature
    medioid <- apply(df[df[, cmpd_col] == neg_control, feature_cols], 2, median)
    # 2. calculate the difference from the origin for each medioid position
    delta <- 0 - medioid
    # 3. iterate along columns and adjust to centre the DMSO data
    for (i in seq_along(feature_cols)) {
        feature <- feature_cols[i]
        df[, feature] <- df[, feature] + d[i]
    }
    return(df)
}
```

8. Creating circular histograms: If the principal component vector only contains information regarding two principal components, then we can calculate a θ value for each perturbagen against a common reference such as (0, 1). This generates a θ value for each perturbagen which can be plotted as a histogram. Without constraining them, the θ values are ranged between

0 and 360. As either end of this range is equivalent to the x-axis of this histogram can be wrapped round into a circle which can be used to visualize the phenotypic direction induced by a perturbagen (Fig. 1).

9. To identify distinct phenotypic responses between cell lines treated with a perturbagen, a theta value has to be calculated for each pair of cell lines per perturbagen. Cell lines that elicit a similar response to a given perturbagen will produce a low θ value, indicating that they produce similar phenotypic trajectories, whereas a θ value approaching 180 indicates opposite phenotypic directions. In our experience a histogram of all measured θ values produces a log-normal distribution, indicating that most perturbagens produce similar phenotypic response between cell lines.
10. Image analysis can take considerable time for large numbers of images. We recommended using either a computing cluster or a cloud computing service to process many images in parallel.
11. Large .csv files can also cause problems. If files exceed several GBs we recommend users switch to a database format such as SQLite.

Acknowledgments

This work was supported by a Cancer Research UK Ph.D. Studentship award to the Cancer Research UK Edinburgh Centre.

References

1. Swinney DC, Anthony J (2011) How were new medicines discovered? Nat Rev Drug Discov 10:507–519
2. Ljosa V, Caie PD, Ter Horst R, Sokolnicki KL, Jenkins EL, Daya S et al (2013) Comparison of methods for image-based profiling of cellular morphological responses to small-molecule treatment. J Biomol Screen 18:1321–1329
3. Singh S, Carpenter AE, Genovesio A (2014) Increasing the content of high-content screening: an overview. J Biomol Screen 19:640–650
4. Reisen F, Sauty de Chalon A, Pfeifer M, Zhang X, Gabriel D, Selzer P (2015) Linking phenotypes and modes of action through high-content screen fingerprints. Assay Drug Dev Technol 13:150810081821009
5. Kümmel A, Selzer P, Siebert D, Schmidt I, Reinhardt J, Götte M et al (2012) Differentiation and visualization of diverse cellular phenotypic responses in primary high-content screening. J Biomol Screen 17:843–849
6. Garnett MJ, Edelman EJ, Heidorn SJ, Greenman CD, Dastur A, Lau KW et al (2012) Systematic identification of genomic markers of drug sensitivity in cancer cells. Nature 483:570–575
7. Barretina J, Caponigro G, Stransky N, Venkatesan K, Margolin AA, Kim S et al (2012) The cancer cell line encyclopedia enables predictive modelling of anticancer drug sensitivity. Nature 483:603–607
8. Perlman Z, Slack M, Feng Y, Mitchison TJ, Wu LF, Altschuler SJ (2004) Multidimensional drug profiling by automated microscopy. Science 306:1194–1199
9. Vincent F, Loria P, Pregel M, Stanton R, Kitching L, Nocka K et al (2015) Developing predictive assays: the phenotypic screening "rule of 3". Sci Transl Med 7:293ps15
10. Tanaka M, Bateman R, Rauh D, Vaisberg E, Ramachandani S, Zhang C et al (2005) An unbiased cell morphology-based screen for

new, biologically active small molecules. PLoS Biol 3:0764–0776

11. Caie PD, Walls RE, Ingleston-Orme A, Daya S, Houslay T, Eagle R et al (2010) High-content phenotypic profiling of drug response signatures across distinct cancer cells. Mol Canc Ther 9:1913–1926
12. Gustafsdottir SM, Ljosa V, Sokolnicki KL, Wilson JA, Walpita D, Kemp MM et al (2013) Multiplex cytological profiling assay to measure diverse cellular states. PLoS One 8:e80999
13. Warchal SJ, Dawson JC, Carragher NO (2016) Development of the theta comparative cell scoring method to quantify diverse phenotypic responses between distinct cell types. Assay Drug Dev Technol 14:395–406
14. Carpenter AE, Jones TR, Lamprecht MR, Clarke C, Kang IH, Friman O et al (2006) CellProfiler: image analysis software for identifying and quantifying cell phenotypes. Genome Biol 7:R100
15. Bray M-A, Fraser AN, Hasaka TP, Carpenter AE (2012) Workflow and metrics for image quality control in large-scale high-content screens. J Biomol Screen 17:266–274

Chapter 14

Accessing the Open PHACTS Discovery Platform with Workflow Tools

Daniela Digles, Andrei Caracoti, and Edgar Jacoby

Abstract

The Open PHACTS Discovery Platform integrates several public databases, which can be of interest when annotating the results of a phenotypic screening campaign. Workflow tools provide easy-to-customize possibilities to access the platform. Here, we describe how to create such workflows for two different workflow tools (KNIME and Pipeline Pilot), including a protocol to annotate compounds (e.g., phenotypic screening hits) with compound classification, known protein targets, and classifications of the targets.

Key words Open PHACTS, Workflow tool, KNIME, Pipeline Pilot, Phenotypic screening

1 Introduction

The aim of the Open PHACTS project [1] was to integrate several public databases, allowing scientists to answer selected questions relevant for research in the drug discovery process [2]. One possibility to access the data from the Open PHACTS Discovery Platform is the use of workflow tools such as KNIME [3] and Pipeline Pilot [4]. Workflow tools help to automatize tasks without the necessity for programming skills. Predefined tools for single tasks (such as reading a file, performing filtering steps, or showing a graph) are available and can be visually connected to each other. These building blocks are called "nodes" in KNIME and "components" in Pipeline Pilot. Nodes in KNIME are executed subsequently by default, while in Pipeline Pilot the data flows through "pipes" to each component at execution time.

These workflow tools were previously employed for selected use cases within the Open PHACTS project [5, 6]. Recently, workflows where data from Open PHACTS can support the annotation and interpretation of phenotypic screening results were presented [7]. One of the protocols introduced there is included in a step-by-step description in this chapter.

Bridget Wagner (ed.), *Phenotypic Screening: Methods and Protocols*, Methods in Molecular Biology, vol. 1787,
https://doi.org/10.1007/978-1-4939-7847-2_14,

To retrieve data from the Open PHACTS Discovery Platform, two main steps are necessary. The first step is accessing the web interface of Open PHACTS (the "API") to return the data in one of the available formats (e.g., JSON or XML). The second step is to convert the data into a table-based format native to the workflow tool. Here, we provide detailed instructions to perform these steps in either of the two workflow tools.

2 Materials

2.1 Open PHACTS KNIME Nodes

1. Download the Open PHACTS KNIME nodes from Github and place the folder in the plugins directory of the KNIME installation (*see* **Note 1**).
2. Restart KNIME. If installed correctly, the OPS_Swagger node should now be available in the Node Repository in KNIME.
3. The configuration of the API version is carried out directly in each OPS_Swagger node (*see* **Note 2**) as described in Subheading 3.

2.2 Open PHACTS Pipeline Pilot Components

1. BIOVIA implemented the API calls in their Open PHACTS Component Collection [8]. The Open PHACTS Component Collection enables protocol developers and users to access Open PHACTS data through easy-to-use components and provides full coverage of the API (*see* **Note 2**). The latest version connects to the Open PHACTS API version 2.1 (*see* **Note 3**).
2. Installation and configuration: Install the BIOVIA Open PHACTS Component Collection package on the Pipeline Pilot server. Configure the following properties in the administration portal (*see* **Note 4**) to enable the components to connect to the Open PHACTS API: the OPS Server URL parameter (e.g., https://beta.openphacts.org/2.1) and the Open PHACTS Application ID and Key (*see* **Note 5**).

3 Methods

Here we describe how to perform single queries to the Open PHACTS API both in KNIME and Pipeline Pilot. Settings for a specific example of retrieving bioactivity data for a list of compounds using the "Compound Pharmacology: List" API call (*see* **Note 6**) are given in the Notes.

3.1 Implementing Open PHACTS Workflows in KNIME

1. Use the "OPS_Swagger" node to create the API query. For this, first provide the API call definitions as link to a Swagger file (*see* **Note 7**) in the field "Swagger URL" and click "Fetch". Select the wanted query from the drop-down menu. Once the

API call is selected, the parameter page is updated with the available parameters for the query (*see* **Notes 8** and **9**). Parameters can also be provided via the input table (column names must match the parameter name and only the first row is taken into account). All additional columns need to be removed with a Column Filter node before the OPS_Swagger node, to prevent the addition of wrong parameters. If several input URIs are to be used, a Chunk Loop is necessary to provide each URI separately to the OPS_Swagger node.

2. Fetch the data from the web service with the "GET Request" node (*see* **Note 10**). In the configuration window select the output from the "OPS_Swagger" node (url) as the URL column. Set the timeout to 900 s to reflect the maximum execution time of the web service.
3. Use the "JSON Path" node to transform the data of interest from json format into a tabular structure (for XML usage *see* **Note 11**). To create JSONPath queries, click on the data value of interest and click on "Add single query" for single values or on "Add collection query" if multiple values should be returned (if necessary, modify the created query manually, *see* **Note 12**). For queries where the data is returned from the API as list of items (e.g., from pharmacology queries), a multistep procedure is useful: first return the items as a list of JSON objects, then use the "Ungroup" node to extract the list into single items, and finally extract the data of interest with a second "JSON Path" node (Fig. 1).
4. All necessary steps to retrieve the data can be wrapped into a metanode to increase the clarity of the resulting workflow and for easier reuse in the same or other workflows.

3.2 Implementing Open PHACTS Workflows in Pipeline Pilot

1. Select the appropriate Open PHACTS component for the task and add it to the protocol (*see* **Note 13**).
2. Set the parameters for the component. The parameter panel for a component becomes visible in Pipeline Pilot when clicking on the component (Fig. 2). Required parameters are displayed in red in the parameter panel and the component cannot run until a value has been supplied (*see* **Note 8**). In addition, many components support the setting of additional parameters (*see* **Note 9**). The value of a parameter can be a static string, but often it is better to use information on incoming records to supply the value for each record. By right-clicking on the parameter name and selecting Initialize Using>Data Property..., followed by entering the data record property name with the input of interest in the pop-up box and making sure that the "Reinitialize for each data" checkbox is checked, the value of the property from each data record will be used in the API

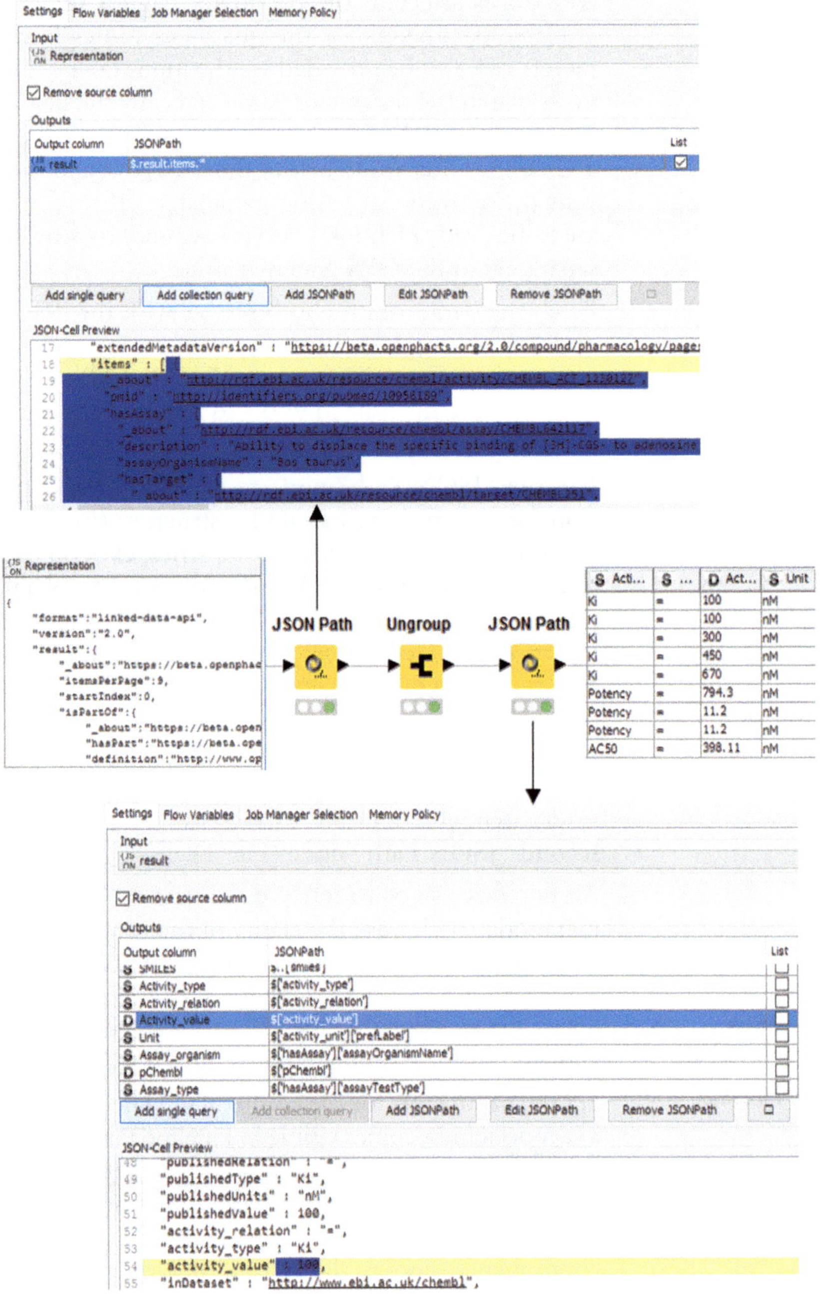

Fig. 1 Multi-step procedure in KNIME to transform a list of JSON items into a table. The first JSON Path node returns a list of JSON objects, which is extracted with the ungroup node. The second JSON Path node transforms the data of interest into columns

call for each record. Leaving a query parameter blank typically results in the default API behavior being used for that parameter. In most cases, that means that no filtering will be done using that parameter.

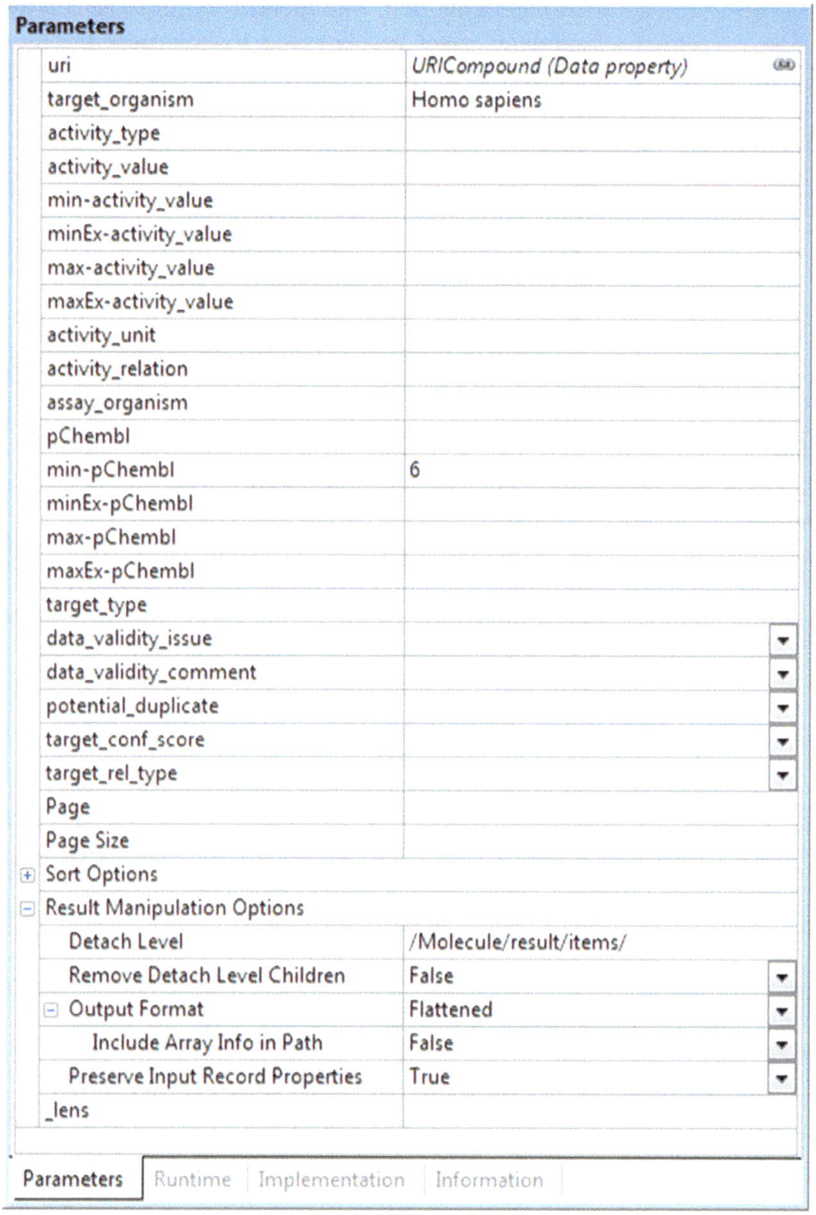

Fig. 2 Parameter panel for the Pipeline Pilot Compound Pharmacology. List (OpenPHACTS) component showing the data record property URICompound being used as the value of the uri parameter

3. Define how the results should be extracted from the JSON format in the Result Manipulation Option parameter group (*see* **Note 14**). Set the Detach Level parameter to the level in the JSON tree from where the data should be extracted. The full data tree can be visualized using the Data Record Tree Viewer component. The parameter Remove Detach Level Children is to be set to False to preserve the data deeper in

the hierarchy than the detach node. The Output Format parameter is set to Flattened to compress the hierarchy into one level of properties that are added to the data record. The parameter Include Array Info in Path is to be set to false. The parameter Preserve Input Record Properties is to be set to true to keep the full list of properties of the input record.

4. In general, an API call provides more output properties than really needed to address a specific question. Therefore the API call component is in general followed by a Remove Properties component which allows to remove any unneeded properties and simplify the output records for downstream processing or other API calls.

3.3 Workflow to Annotate Results from a Phenotypic Screen

These are the necessary steps for Protocol 1 (ChEBI/ChEMBL annotation and classification) as published previously [7]. The protocol contains two parts: retrieving ChEBI annotations for the molecules (**steps 1–3**) and retrieving ChEMBL classification of targets where the molecules show activity (**steps 1** and **4–9**). The original workflows are available at http://www.myexperiment.org/workflows/4866.html for KNIME and https://exchange.sciencecloud.com/exchange/browse#details:266 for Pipeline Pilot (Fig. 3).

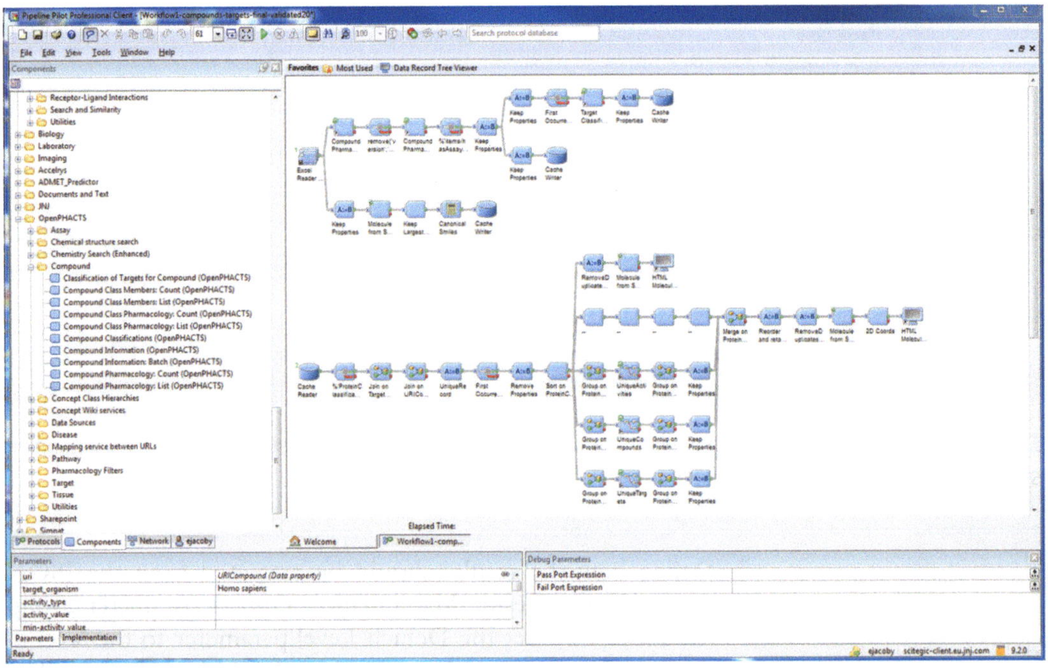

Fig. 3 BIOVIA Pipeline Pilot Professional Client user interface showing the compound target protocol. The central window with the blue boxes describes the arrangement of individual components. The parameters for each component are visible in the lower left pane

1. Start with a list of compound URIs for active molecules from a phenotypic screen (*see* **Note 8**).
2. To retrieve the ChEBI annotations, use the Compound Classifications API call. Set the tree parameter to "chebi". Keep the ChEBI classification URI ("_about") and the label ("prefLabel") (*see* **Note 15**).
3. Remove rows with missing values before generating any counts, and then visualize or export the resulting tables.
4. Use the Compound Pharmacology: List API call to retrieve bioactivity data for each compound URI from **step 1** (*see* Subheading 3.1 or 3.2). Set the activity filter in the API call to the wanted grade of activity (e.g., set min-pChembl = 6 for IC50, XC50, EC50, AC50, Ki, Kd, or Potency values equal or smaller than 1 μM) and filter for the target organism of interest.
5. Additional filters that cannot be set in the API call directly can be applied subsequently (e.g., keep only target types of single protein or protein complex, or targets with a confidence score of 9, i.e., a direct single-protein target was assigned to the assay).
6. Group the results according to the target URI to keep a list of unique targets and minimize the number of subsequent API calls.
7. For each target, retrieve the ChEMBL classifications with the Target Classifications API call. Keep the protein classification URI ("_about") and label ("prefLabel") in the "hasProteinClassification" part of the JSON hierarchy (*see* **Note 16**).
8. Join the classification results with the data from **step 5** (with the target URI as identifier) to generate the full list of results.
9. Visualize or export the resulting tables.

4 Notes

1. The original Open PHACTS KNIME nodes as created by Ronald Siebes (VU Amsterdam) are available at https://github.com/openphacts/OPS-Knime. New nodes are currently under development and will be available as part of the Erl Wood Community Nodes.
2. One major difference between the implementations of Open PHACTS in KNIME and Pipeline Pilot is how the individual API calls are accessed. The KNIME implementation follows a generic approach, where all API calls which are defined in the Swagger documentation file (*see* **Note 7**) can be selected from a drop-down menu in a single node (the OPS_Swagger node).

The Pipeline Pilot Component Collection provides single components for each API call.

3. Users are referred to the Open PHACTS discussion group on the BIOVIA Community Forum (https://community.3dsbiovia.com/Communities_Home) for the latest information.
4. The configuration steps are best completed by the system administrator. Once these global properties are configured, they are used by any protocol that uses components from the Open PHACTS collection. The global settings can be overridden by using the Open PHACTS Configuration component at the top of a protocol. In this case, the settings are only changed for the current protocol and do not affect the global server settings in any way.
5. The Application ID and Key can be obtained by registering at https://dev.openphacts.org/.
6. The "Compound Pharmacology: List" API call retrieves bioactivity data collected in the ChEMBL database [9]. The version 2.1 of the API includes ChEMBL version 20.
7. Swagger [10] is a definition format for REST-API services. The OPS_Swagger node was created to interpret any API definition using Swagger version 1. For the Open PHACTS API, an example of such a file is available at https://raw.githubusercontent.com/openphacts/OPS_LinkedDataApi/1.5.0/api-config-files/swagger.json. However, newer files (using Swagger version 2) cannot be interpreted. To create API requests for the latest Open PHACTS API version, use the version 1.5 swagger file link, and then a subsequent String Replacer node to replace https://beta.openphacts.org/1.5/ with https://beta.openphacts.org/2.1/ (set option replace all occurrences).
8. The only required input for the Compound Pharmacology: List API call is a compound URI. Such a compound URI can be any link to a molecule from a source that is supported by the Open PHACTS Discovery Platform (e.g., ChEMBL or Chemspider). If no identifier is available to the user, a compound URI can be retrieved via the text search or structure search API calls. Another important parameter for API calls returning lists of items is the _pageSize, which defaults to 10 from the API side (100 in Pipeline Pilot). To retrieve all result items the value can be set to "all". This works fine for results with less than 10,000 items. If more items are expected, a loop through all result pages is necessary (by changing the _page parameter for each loop step). The Pipeline Pilot components will retrieve all pages automatically if the Page parameter is left blank. A value is needed only when the output should be restricted to a

specific page of results (for instance when developing a protocol to be deployed through a web browser where the user has the ability to page through results by clicking Next and Previous buttons).

9. Additional parameters to specify the query contain the target organism (e.g., setting this parameter to *Homo sapiens* will return only data for assays from human targets) and activity filters (e.g., setting the min-pChembl parameter to 6 will only return activity values larger than 6 for activity types included in the pChembl parameter). All available parameters with options are specified in the API documentation.
10. Older available workflows use either the OPS_JSON node from the Open PHACTS KNIME nodes or the GET Resource node from the KNIME Community nodes. As both are no longer actively maintained, we recommend to use the GET Request node for new workflows.
11. If the use of XML is preferred over the use of JSON, the _format parameter in the OPS_Swagger node can be set to XML. The corresponding XML KNIME nodes (e.g., XPath) can then be used to interpret the data.
12. The created JSON path query is often including the position of the data of interest in the result set. As this position might change when using several different input URIs (e.g., compound IDs), it is best to keep the query as generic as possible. For this remove all unnecessary path definitions, e.g., use $..compoundPharmacologyTotalResults instead of the full path of $['result']['primaryTopic']['compoundPharmacologyTotalResults']. If the name of the result parameter of interest is not unique (e.g., "_about" is used for URIs from different datasets), you can specify the data source: $..[?(@.inDataset=='name_of_the_dataset')]._about. This type of query can also be used to specify other filters. For example, to retrieve the labels of the compound classification from the Compound Classification API call, but only when the classification is of the type "has role", the following query can be used:

    $..hasChebiClassification[?(@.classificationType.prefLabel=='has role')].prefLabel

13. The individual components are grouped in subfolders based on the data category they call (e.g., Assay, Compound, Disease, Target). This hierarchy matches the one in the API. In Pipeline Pilot protocols, components are put together by linking the output port of one component to the input port of another component. The data flows from left to right, and each component processes the record before passing it on to the next component downstream. In the default operation mode, Pipeline Pilot processes individual data records sequentially.

14. The Result Manipulation Option parameter group defines the manner the results are extracted from the default JSON hierarchical data tree and transformed in properties to be added to the output data record. In the given example, the Detach Level parameter needs to be set to /Molecule/results/item.
15. In KNIME the JSON Path node will retrieve either only a list of labels or one label, as the response from the API will not include singletons in a list, but provide only one entry. To retrieve all results, both types of queries need to be specified. For the list case, these are $..['hasChebiClassification'][*]['_about'] and $..['hasChebiClassification'][*]['prefLabel'] (checking the List checkbox). For the singleton cases, the JSON Path queries are the equivalent queries but without [*] and not checking List. To combine the results, first use the Ungroup node, followed by two Column Merger nodes. In the first Column Merger node specify the list result of the URI as Primary Column and the singleton result as Secondary Column, with the option to replace the primary column. Repeat this for the labels in the second Column Merger node. Filter out the columns that are no longer needed with a Column Filter.
16. In KNIME the JSON Path queries to achieve this are $..['hasProteinClassification'][*]['_about'] and $..['hasProteinClassification'][*]['prefLabel']. Check the List checkbox to retrieve all classifications if there are multiple classifications per protein and use the Ungroup node subsequently to show one classification per row.

Acknowledgment

We wish to acknowledge the Open PHACTS Foundation (http://www.openphactsfoundation.org/), the charitable organization responsible for the Open PHACTS Discovery Platform, without which this work would not have been possible.

References

1. Williams AJ, Harland L, Groth P, Pettifer S, Chichester C, Willighagen EL et al (2012) Open PHACTS: semantic interoperability for drug discovery. Drug Discov Today 17:1188–1198
2. Azzaoui K, Jacoby E, Senger S, Rodriguez EC, Loza M, Zdrazil B et al (2013) Scientific competency questions as the basis for semantically enriched open pharmacological space development. Drug Discov Today 18:843–852
3. Berthold MR et al (2007) KNIME: the Konstanz information miner. In: Studies in Classification, Data Analysis, and Knowledge Organization (GfKL 2007). Springer, New York, pp 319–326
4. BIOVIA Pipeline Pilot. http://accelrys.com/products/collaborative-science/biovia-pipeline-pilot/
5. Ratnam J, Zdrazil B, Digles D, Cuadrado-Rodriguez E, Neefs JM, Tipney H et al

(2014) The application of the open pharmacological concepts triple store (Open PHACTS) to support drug discovery research. PLoS One 9(12):e115460

6. Chichester C, Digles D, Siebes R, Loizou A, Groth P, Harland L et al (2015) Drug discovery FAQs: workflows for answering multidomain drug discovery questions. Drug Discov Today 20(4):399–405
7. Digles D, Zdrazil B, Neefs J-M, Van Vlijmen H, Herhaus C, Caracoti A et al (2016) Open PHACTS computational protocols for *in silico* target validation of cellular phenotypic screens: knowing the knowns. Med Chem Commun 7:1237–1244
8. Caracoti A (2015) BIOVIA Open PHACTS Component Collection. https://exchange.sciencecloud.com/exchange/browse#details:216. Accessed 24 May 2017
9. Bento AP, Gaulton A, Hersey A, Bellis LJ, Chambers J, Davies M et al (2014) The ChEMBL bioactivity database: an update. Nucleic Acids Res 42:1083–1090
10. OpenAPI Specification. http://swagger.io/specification/. Accessed 24 May 2017

Chapter 15

Quantitative Prioritization of Tool Compounds for Phenotypic Screening

Yuan Wang and Jeremy L. Jenkins

Abstract

Phenotypic screens are increasingly utilized in drug discovery for multiple purposes such as lead and/or tool compound finding, and target discovery. Using potent and selective chemical tool compounds against well-defined targets in phenotypic screens can help elucidate biological processes modulating assay phenotypes. Unfortunately the identification of such tools from large heterogeneous bioactivity databases is nontrivial and there is repeated use of published unselective compounds as phenotypic tools. Here we describe a computational model, the compound-target tool score (TS), which is an evidence-based quantitative confidence metric that can be used to systematically rank tool compounds for targets. The identified selective and nonselective tool compounds have applications in phenotypic assays for target hypothesis validation as well as assay development.

Key words Chemical probe, Tool compound, Phenotypic screen, Selectivity, Target hypothesis validation, Bioactivity data integration

1 Introduction

Phenotypic screening is widely used for target and lead discovery, and for connecting chemistry to pathway modulation [1–9]. It has proven its value in the discovery of over half of small-molecule first-in-class new molecular entities [1, 2], but presents a challenge of following up hits with target identification. To facilitate phenotypic drug discovery and systematic target validation, individual small molecules or designed sets of molecules having well-understood MOAs can be used as chemical probes to observe the phenotypic effect of target modulation [10–15]. A high-quality tool compound can also play a critical role as a positive control molecule for assay development, such as signal-to-noise optimization, or to help in preclinical in vivo target validation in a drug discovery project. As an example, the Wnt signaling pathway antagonist XAV939 typifies the workflow of phenotypic screening discovery, target identification, and tool compound application (Fig. 1).

Bridget Wagner (ed.), *Phenotypic Screening: Methods and Protocols*, Methods in Molecular Biology, vol. 1787, https://doi.org/10.1007/978-1-4939-7847-2_15,

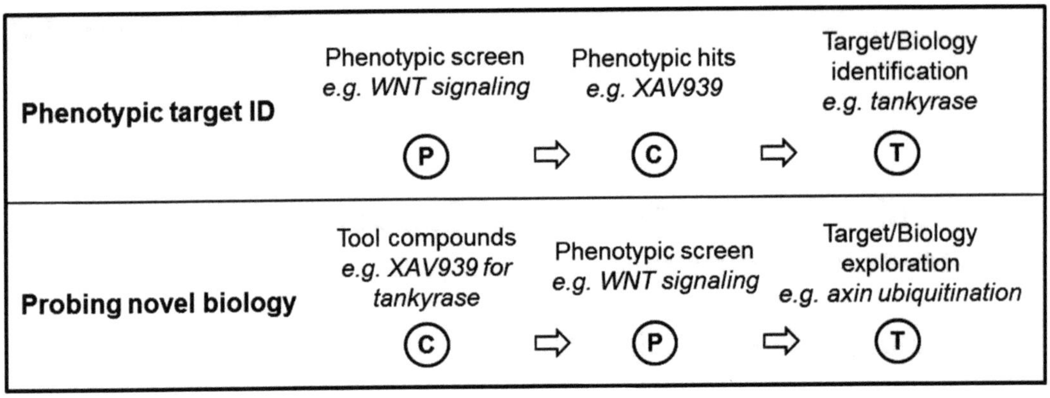

Fig. 1 Workflow of validating target hypotheses in phenotypic screens using tool compounds. Huang et al. (2009) found XAV939 in a WNT signaling phenotypic screen, identified its target tankyrase, and then used it as a tool compound to probe tankyrase biology, such as axin ubiquitination inhibition [16]. Phenotypic hits ("C" in top panel) and tool compounds ("C" in bottom panel) may be the same but not necessarily—the latter have more strict selection criteria

XAV939 was first identified as a phenotypic hit in a Wnt pathway phenotypic screen, after which its target (tankyrase) was elucidated by affinity chromatography (i.e., "chemical proteomics"). Having confirmed the compound to be selective for this mechanism, XAV939 can be put back into the phenotypic assays as a tool to probe downstream biology. Axin stabilization was shown to be the effect of tankyrase inhibition [16].

A good phenotypic tool compound needs to satisfy a set of criteria: (1) potency and selectivity on target both in cell-free and cell-based assays; (2) exposure at the site of action, or cell permeability; (3) proven utility as a probe, i.e., phenotypic relevance via a demonstrated proximal biomarker; and (4) identity of the active species and availability [10–15]. However, it is difficult to obtain the requisite data to select the ideal tool compound from large databases. As a result, researchers often consult vendor catalogs, although many well-known purchasable compounds may not be as selective as the catalog description. A recent crowd-sourcing initiative, the Chemical Probe Portal, has been started to collect qualifying tool compounds from researchers in the chemical biology community across the globe [11], yet the practical aspect of prioritizing existing tool compounds from large, diverse bioactivity databases still needs to be addressed.

One major difficulty for tool compound selection is compound polypharmacology (i.e., binding to more than one molecular target), which has been well established with extensive bioactivity profiling and data integration in bioactivity databases [17–19]. Polypharmacology can reduce selectivity and limit a compound's utility as a targeted tool, yet the knowledge of polypharmacology is unfortunately limited by the extent to which a

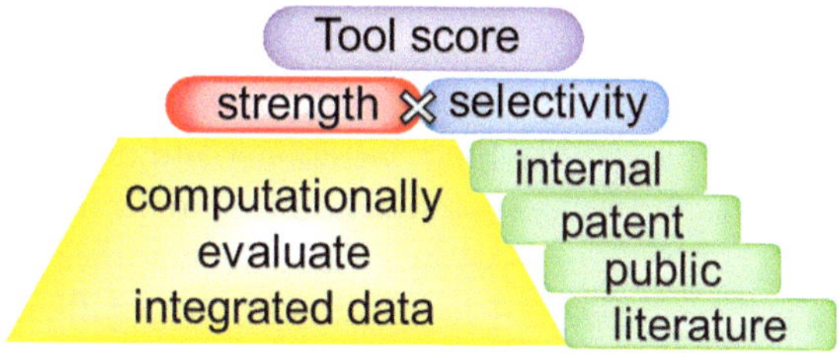

Fig. 2 Schematic of the workflow for TS calculation for prioritization of tool compounds. Integrated data from multiple sources (internal, patent, public, and literature) are consolidated to derive heuristics for both compound-target potency ("strength") and selectivity. The tool score (TS) is the combined product of these two

compound has been profiled. Many methods have been published on the selectivity of small molecules against a panel of targets and the quantification thereof [15, 20–22]. However, published profiling panels are so far all limited to a biased, sub-genomic set of targets which vary within different labs and institutes. While researchers have tried to integrate heterogeneous potency measurements [23–25], the potential errors in the data source [26] might still be propagated to the final quantification, leading to suboptimal selectivity evaluation on candidate tool compounds.

To address the issue of phenotypic tool compound selection, we have integrated large-scale heterogeneous data from many data sources and investigated critical features that drive the tool utility of a compound for a target. We quantify this utility as the tool score (TS), which is a computational summary of diverse potency and selectivity information independent of any predefined target panel: Potency information is captured by an aggregated assertion stack model, incremented by the amount of supporting evidence; selectivity is described by three heuristics related to on- and off-target bioactivity measurements; and TS is the combined product of these two pillars (Fig. 2). We demonstrate that using TS it is possible to identify both selective and nonselective compounds and apply them in phenotypic screens: high-TS compounds tend to be phenotypically specific and they can be used to probe relevant downstream pathway responses, whereas low-TS compounds having specific binding pocket preferences can be used for assay development.

2 Materials

We use R (www.r-project.org) as the programming language and code for computing the TS is available online (www.github.com/novartis/toolscore). An example input data file is also included with the code.

3 Methods

3.1 Integrating Bioactivity Data

1. First, we collect and normalize two branches of data from both internal and external sources. The first branch is bioinformatics knowledge, which comes from sources like NCBI Entrez Gene [27], GeneOntology [28, 29], and UniProt [30]. The other is cheminformatics knowledge, which is collected from sources like Drugbank [31], ChEMBL [32], and BindingDB [33, 34].
2. Target-related data are standardized and connected using NCBI Entrez Gene IDs. If there are multiple genes in the same homolog gene group, we merge them using HomoloGene (http://www.ncbi.nlm.nih.gov/homologene).
3. Compound-related data are standardized (CACTVS toolkit, Xemistry) and connected using the IUPAC InChIKey [35, 36] representation of the structures.
4. Compound bioactivity data in dose response are normalized to micromolar as the standard unit.

3.2 Defining Primary Target to Evaluate Compound-Target Potency and Selectivity

1. We define target allocation as the proportion of measurements on each target for a given compound.
2. All targets of the compound are ranked with their allocation and the target with the highest allocation is assigned the primary target (on-target).
3. In case of a tie, i.e., two targets have the same allocation, the one with the lowest micromolar activity measured is assigned the primary target of the compound (*see* **Note 1**).

3.3 Defining Assertions and Rules of Inference to Derive Compound-Target Activity Strength

1. We adapt the concept of computational inferences to build a stack of assertions and rules. Conventionally, computational inferences are done using binary assertions (true-false predicates): "if $\vdash X$ and $\vdash Y$ then $\vdash Z$," i.e., if assertions of the form X and Y have been proven true, then Z is also proven [37]. For example, "if 'a activates b' can be proven, and 'b inhibits c' can be proven, then 'activating a will activate b and cause inhibition of c' is proven." Yet we cannot simply binarize all bioactivities of compounds in real world. Therefore we extend this key logical foundation to make new assertions with quantitative attributes to describe knowledge of compound-target network.
2. As shown in Fig. 3, we define objects/entities as compounds and their targets. Their relationships are described as assertions with attributes. For example, one relationship, "ChEMBL reported IC50 activity of 0.4 μM between compound 1 and target X," can be described as assertion "CHEMBL_ACTIVE" with strength 1. And if there are multiple ChEMBL reports between compound 1 and target X, we define rules to increment the strength accordingly (Table 1, *see* **Note 2**).

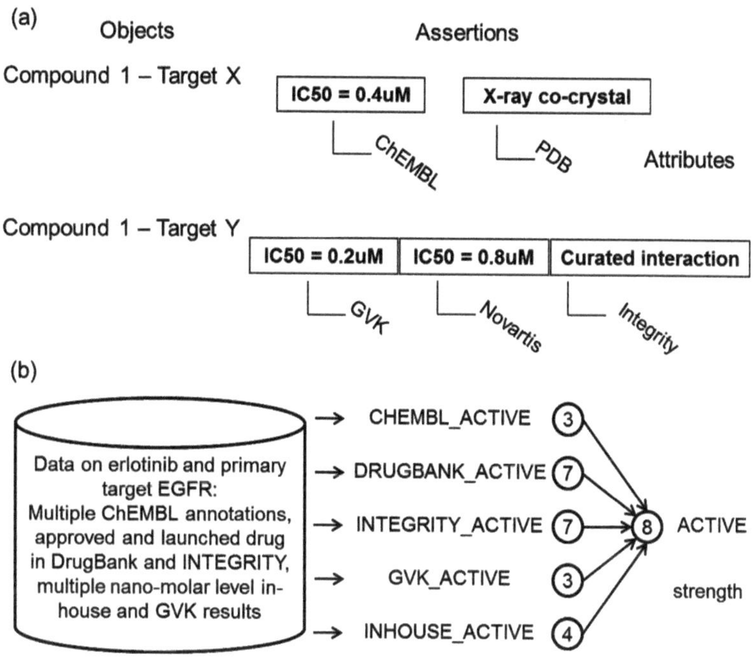

Fig. 3 Calculation of compound-target potency strength using stacks of assertions. (**a**) Modeling objects (compounds, targets) and their relationships as assertions with attributes. (**b**) Calculation example of erlotinib-EGFR potency strength. Each individual lower level assertion is assigned a strength value based on the supporting evidence, and the derived assertion has an aggregated strength according to the rules (Table 1)

Table 1
Assertions and rules

Assertion	Rules for assigning strength
DRUGBANK_ACTIVE (Qualitative, curated)	7 for approved drugs 2 for approved nutraceuticals, else 1
INTEGRITY_ACTIVE (Qualitative, curated)	7 for approved drugs 5 if Phase III or higher 3 if IND filed or higher 2 if preclinical, else 1
CHEMBL_ACTIVE (Quantitative assay data)	3 if more than one low nanomolar value for target ($\leq$50 nM) 2 if more than four activities (<1 micromolar) for target 1 if it has a micromolar value $\leq$1 for target
GVK_ACTIVE (Quantitative assay data)	3 if more than one low nanomolar value ($\leq$50 nM) for target 2 if more than four activities (<1 micromolar) for target 1 if it has a micromolar value $\leq$1 for target
INHOUSE_ACTIVE (Internally validated quantitative assay data)	4 if more than one low nanomolar value ($\leq$50 nM) for target 3 if more than four activities (<1 micromolar) for target 1 if it has a micromolar value $\leq$1 for target
ACTIVE (Derived high-level assertion)	Maximum of evidence strengths from above sources Plus 1 if evidence count >2

3. Some assertions (such as "CHEMBL_ACTIVE" described earlier) are derived directly from data sources, whereas others are derived from other assertions. The latter high-level assertions require input from more than one source of evidence, and we define rules to aggregate the strength values from the supporting low-level assertions. The final strength between a compound-target pair is the strength of the top-level assertion "ACTIVE" (Fig. 3 and Table 1).
4. These assertion logic rules are programmed so that all assertions derived from all data we collected from in-house and external sources are precalculated and stored in our database tables. Both the integrated data and the dependent assertions are maintained and updated regularly.

3.4 Describing Selectivity Using Derived Heuristics

1. We use three features to describe compound-target selectivity: on/off-target Q1 (25-percentile) potency difference, high-potency (Q1) target allocation, and on/off-target measurement distribution difference (Fig. 4, *see* **Note 3**).

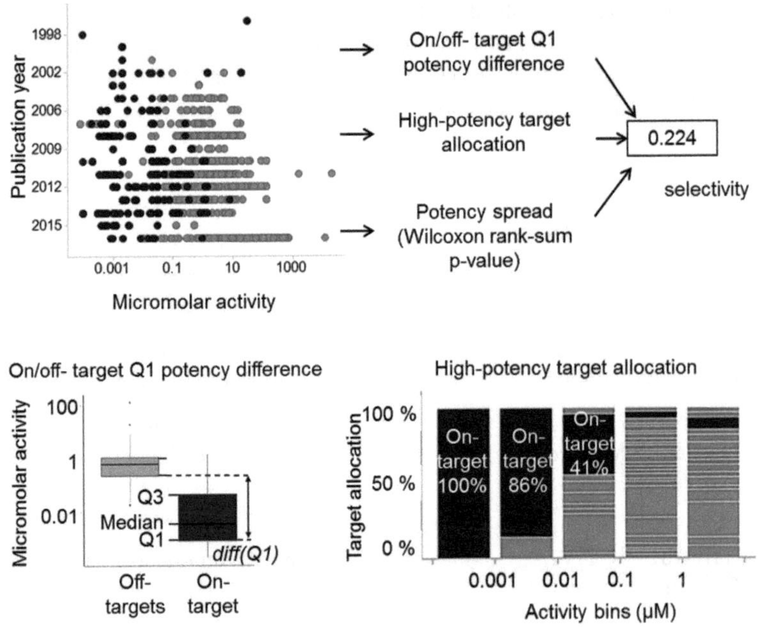

Fig. 4 Calculation of selectivity heuristics. From integrated bioactivity data, three selectivity heuristics were derived: difference of Q1(IC50) between on- and off-targets (lower left), high-potency target allocation (lower right), and Wilcoxon rank-sum *p*-value of on- and off-target activity distribution. Selective compounds have larger difference in the Q1 statistics of activities against on- and off-targets, high allocation for on-target at high potency, and wider spread of on- and off-target activity distribution (example: erlotinib with on-target EGFR in dark gray and other off-targets in light gray)

2. On/off-target Q1 potency difference (*Q1_IC50_diff*) is calculated by separating all on-target and off-target bioactivity measurements of a compound, and comparing the difference in their lower Q1 (25-percentile) statistics in log scale.
3. High-potency target allocation (*TA_Q1_R*) is calculated by taking all bioactivity measurements, keeping the lowest 25%, and computing on-target allocation (as described in Subheading 3.2) on this high-potency subset.
4. On/off-target measurement distribution difference is represented by Wilcoxon rank-sum *p*-value (*Wilcoxon_RS_Pvalue*), calculated by separating all on-target and off-target bioactivity measurements of a compound, and conducting Wilcoxon rank-sum test (*see* **Note 4**).
5. We define selectivity as

$$\text{Selectivity} = (Q1\_IC50\_diff/10 + TA\_Q1\_R - \log 10(Wilcoxon\_RS\_Pvalue)/32)/3$$

3.5 Calculating Tool Score (TS)

1. For each compound-target pair, tool score is defined as

$$\text{TS} = \text{Strength} \times \text{Selectivity}$$

For each target, compounds can be ranked by their TS and the top-ranking compounds can be prioritized in a phenotypic assay as potential good tools. For example, in Fig. 5 we show the phenotypic profiles of compounds having different TS against three targets: ADRB2, TGFBR1, and CDK9, in a reporter gene assay (RGA) panel consisting of 41 reporters against multiple pathways. Compounds having high TS tend to be more potent and selective in the corresponding phenotypic assay (*see* **Note 5**).

2. As new data become available, updating TS is important for the best prioritization of tool compounds. Unlike target panel-based Gini coefficients [21], entropy [22], or KIBA score [25], the TS is computable independent of the number of targets profiled, and the score computed with unequal number of measurements can still be compared with each other (*see* **Note 6**).

3.6 Systematic Identification of Selective and Nonselective Tools

1. Using TS and corresponding phenotypic profiles, a collection of good tool compounds for phenotypic screens can be maintained dynamically (Fig. 6, compounds in upper-right quadrant), as well as a "blacklist" of nonselective compounds (Fig. 6, compounds in lower-left quadrant).
2. Many members of the latter have been frequently used in literature as tools and we advise the readers to be more cautious of the reported phenotype of such compounds (*see* **Note 7**).

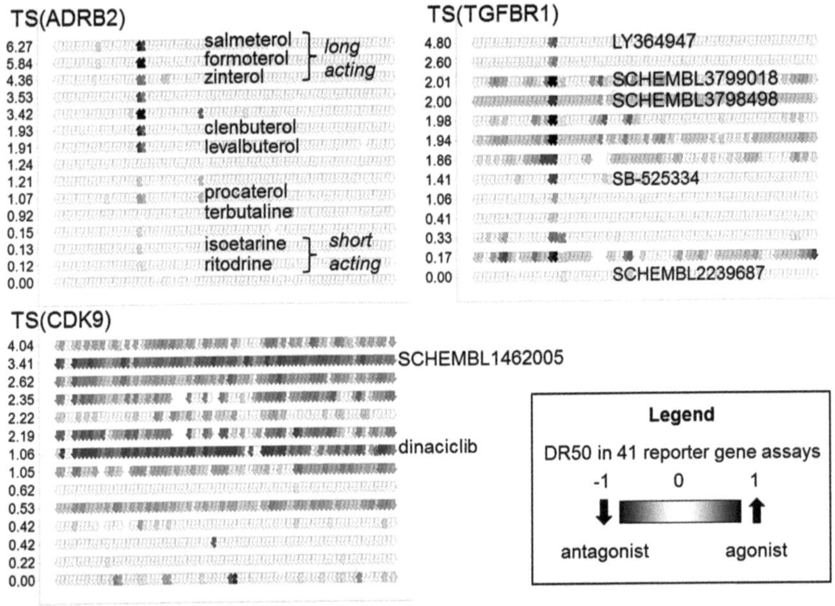

Fig. 5 Compound target tool scores and corresponding phenotypic reporter gene assay (RGA) panel profiles. For each compound, 12- and 24-h DR50s of the 41 RGAs are plotted next to each other (DR50 = 1.0: strong agonist represented as black up-arrow; DR50 = 0: inactive represented as white bar; DR50 = −1.0: strong antagonist represented as black down-arrow). Compounds having high TS on ADRB2 and TGFBR1 were more likely to be potent and selective against the corresponding pathway reporter assays (cyclic AMP signaling and TGFβ signaling, respectively), while compounds having higher TS on CDK9 were more potent on almost all tested assays because the target itself is upstream of multiple processes including apoptosis. Among ADRB2 agonists, long-acting compounds such as formoterol had higher TS and stronger phenotypic selectivity than short-acting ones such as ritodrine. Due to incomplete profiling data on which the TS depends, some TGFBR1 inhibitors having low TS also had high phenotypic potency/specificity (e.g., SB-525334, 33 measurements in ChEMBL vs. LY364947, 600 measurements in ChEMBL)

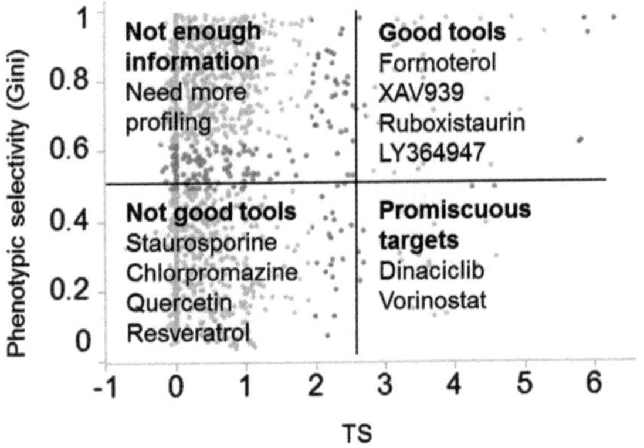

Fig. 6 Identification of selective and nonselective tool compounds. For each compound we have calculated its TS and phenotypic Gini activity in the RGA panel. Overall, compounds having high TS are more likely to be phenotypically specific and therefore TS can be used to rank good tool compounds (upper right) against undesirable ones (lower left). In case not enough information is available, more profiling is needed to update the TS of a compound that is potentially a good tool (upper left). For promiscuous targets such as CDKs or HDACs, our RGA panel could not distinguish the pathway specificity of their tools and such compounds end up in the lower-right quadrant

3.7 Calculating Pocket Specificity for Nonselective Tools

1. PFAM allocation is defined to be the proportion of a PFAM domain [38] represented by all targets of a compound. For example, if a compound is active against two kinase targets with each *m* and *n* measurements, the PFAM allocation of the kinase domain is *m+n/all measurements on all domains.*
2. The PFAM domain that has the highest allocation is assigned the primary PFAM domain of the compound, and the maximum PFAM allocation can be used to evaluate the compound's pocket specificity.
3. Compounds having high pocket specificity are specific to one type of pocket such as ATP-binding site (local polypharmacology), while compounds having low pocket specificity are generally promiscuous.

3.8 Applying Tool Compounds in Phenotypic Assays

1. As shown in Fig. 1, good tools can be applied in a properly designed phenotypic assay to probe the on-target biology. However, it is important to understand the role of the tool compound in the context of assay setup. We propose the following checklist before applying selective tool compounds:
2. Compounds cannot act on targets that are not expressed in the tissue. A potent and selective tool will appear inactive if its target is absent in the phenotypic assay.
3. At high concentrations even selective tool compounds tend to be active against multiple off-targets. For example, erlotinib hits EGFR at nM potency level, but multiple other kinases at 10 μM or above. The experimental design should take into consideration what the effective concentration within the phenotypic system would be.
4. Targets affecting multiple downstream signaling pathways in a phenotypic assay setup, or inducing apoptosis/toxicity, are less likely to be specifically probed, even with highly potent and selective tool compounds.
5. Additionally, nonselective compounds are not completely unavailing. Given the proper context, they can also provide useful insight into a phenotypic assay in the following applications:
6. Compounds with polypharmacology can still help deconvolute the biology when the tissue expression of their individual targets is taken into consideration [39, 40] (*see* **Note 8**).
7. Nonselective compounds with local polypharmacology such as pan-kinase inhibitors can be used to stress-test a phenotypic system and determine whether the readout is specific enough to desired biology. The readout under multiple perturbations can be evaluated and help experimenters design better counter-assay or control experiments.

4 Notes

1. One limitation of the data normalization process is the identification of subunits of complexes and target isoforms. When targets were converted to Entrez Gene IDs and HomoloGene IDs, it is possible that one subunit or isoform is assigned as the on-target, and the remaining subunits/isoforms off-targets even though they are all required for the binding affinity. In such cases comparable potency values on both on- and off-targets led to apparent lack of selectivity.
2. The authors develop the particular numeric values for lower level assertions on the basis of personal expertise and experience working with the named data sources. For example, we expect that a drug that progresses to a launched status is better validated in vivo for its primary target than a tool from preclinical studies, and therefore gives the former higher strength. Although we provide here our definition of assertions and rules, we advise the reader to implement personal and organizational domain expertise to tune strength on their available data sources.
3. The issue of multiple inconsistent measurements has been a substantial challenge to selectivity and the common belief that it is possible to use mean, geometric mean, or minimum to aggregate multiple measurements and get good understanding of the potency and selectivity of a compound is an oversimplification of the problem. We launched a systematic evaluation of candidate features that describe compound selectivity using supervised machine learning. These three features have been shown to be better at distinguishing selective compounds than other features, including those comparing fold IC50s [41].
4. Wilcoxon rank-sum *p*-values which are not computable were filled with 1, and which exceed the significance value of 1E-32 were replaced with 1E-32 for ease of computation.
5. Overall, compounds having high TS have better phenotypic potency and specificity. However, there are two exceptions: First, when the on-target of a good tool compound is promiscuous or toxic, upstream of multiple pathways, we cannot observe a clean phenotypic response of this high-TS compound. Second, since TS depends on collected data, poorly tested compounds for which too little are known could not have high TS even if it could potentially be a good tool.
6. There is no hard cutoff on the number of targets profiled, yet more profiling would clearly enlarge the knowledge space of the compound and facilitate better decision-making: With all other conditions equal on potency and selectivity, more

measurements against both on- and off-target will improve the Wilcoxon rank-sum test significance and result in higher TS.

7. One example of popular tool compound is quercetin [42–45], for which over 300 quantitative investigations have been conducted among a diverse set of targets. It also appears as a bound ligand in 20 different published structures in RSCB PDB bound to diverse proteins and even G-quadruplex DNA. However, we do not see high TS between quercetin and any of its targets. In other words, quercetin is not a potent or selective tool compound to the best of our knowledge.
8. For example, Gujral et al. have applied regularized regression to kinase expression levels on multi-kinase inhibitors and validated cell type-specific kinases that regulate cancer cell migration [40].

References

1. Swinney DC (2013) Phenotypic vs. target-based drug discovery for first-in-class medicines. Clin Pharmacol Ther 93:299–301
2. Moffat JG, Rudolph J, Bailey D (2014) Phenotypic screening in cancer drug discovery: past, present and future. Nat Rev Drug Discov 13:588–602
3. Hopkins AL (2008) Network pharmacology: the next paradigm in drug discovery. Nat Chem Biol 4:682–690
4. Lee JA, Uhlik MT, Moxham CM, Tomandl D, Sall DJ (2012) Modern phenotypic drug discovery is a viable, neoclassic pharma strategy. J Med Chem 55:4527–4538
5. Jones JO, Diamond MI (2007) Design and implementation of cell-based assays to model human disease. ACS Chem Biol 2:718–724
6. Hart CP (2005) Finding the target after screening the phenotype. Drug Discov Today 10:513–519
7. Schirle M, Jenkins JL (2016) Identifying compound efficacy targets in phenotypic drug discovery. Drug Discov Today 21:82–89
8. Feng Y, Mitchison TJ, Bender A, Young DW, Tallarico JA (2009) Multi-parameter phenotypic profiling: using cellular effects to characterize small-molecule compounds. Nat Rev Drug Discov 8:567–578
9. King FJ, Selinger DW, Mapa FA, Janes J, Wu H, Smith TR et al (2009) Pathway reporter assays reveal small molecule mechanisms of action. J Assoc Lab Autom 14:374–382
10. Frye SV (2010) The art of the chemical probe. Nat Chem Biol 6:159–161
11. Arrowsmith CH, Audia JE, Austin C, Baell J, Bennett J, Blagg J et al (2015) The promise and peril of chemical probes. Nat Chem Biol 11:536–541
12. Eggert US (2013) The why and how of phenotypic small-molecule screens. Nat Chem Biol 9:206–209
13. Bunnage ME, Chekler EL, Jones LH (2013) Target validation using chemical probes. Nat Chem Biol 9:195–199
14. Workman P, Collins I (2010) Probing the probes: fitness factors for small molecule tools. Chem Biol 17:561–577
15. Uitdehaag JC, Verkaar F, Alwan H, De Man J, Buijsman RC, Zaman GJ (2012) A guide to picking the most selective kinase inhibitor tool compounds for pharmacological validation of drug targets. Br J Pharmacol 166:858–876
16. Huang SM, Mishina YM, Liu S, Cheung A, Stegmeier F, Michaud GA et al (2009) Tankyrase inhibition stabilizes axin and antagonizes Wnt signalling. Nature 461:614–620
17. Gregori-Puigjane E, Setola V, Hert J, Crews BA, Irwin JJ, Lounkine E et al (2012) Identifying mechanism-of-action targets for drugs and probes. Proc Natl Acad Sci U S A 109:11178–11183
18. Roth BL, Sheffler DJ, Kroeze WK (2004) Magic shotguns versus magic bullets: selectively non-selective drugs for mood disorders and schizophrenia. Nat Rev Drug Discov 3:353–359
19. Hopkins AL (2009) Drug discovery: predicting promiscuity. Nature 462:167–168

20. Jester BW, Gaj A, Shomin CD, Cox KJ, Ghosh I (2012) Testing the promiscuity of commercial kinase inhibitors against the AGC kinase group using a split-luciferase screen. J Med Chem 55:1526–1537
21. Graczyk PP (2007) Gini coefficient: a new way to express selectivity of kinase inhibitors against a family of kinases. J Med Chem 50:5773–5779
22. Uitdehaag JC, Zaman GJ (2011) A theoretical entropy score as a single value to express inhibitor selectivity. BMC Bioinformatics 12:94
23. Kalliokoski T, Kramer C, Vulpetti A, Gedeck P (2013) Comparability of mixed IC(5)(0) data: a statistical analysis. PLoS One 8:e61007
24. Kramer C, Kalliokoski T, Gedeck P, Vulpetti A (2012) The experimental uncertainty of heterogeneous public K(i) data. J Med Chem 55:5165–5173
25. Tang J, Szwajda A, Shakyawar S, Xu T, Hintsanen P, Wennerberg K et al (2014) Making sense of large-scale kinase inhibitor bioactivity data sets: a comparative and integrative analysis. J Chem Inf Model 54:735–743
26. Tiikkainen P, Bellis L, Light Y, Franke L (2013) Estimating error rates in bioactivity databases. J Chem Inf Model 53:2499–2505
27. Maglott D, Ostell J, Pruitt KD, Tatusova T (2011) Entrez gene: gene-centered information at NCBI. Nucleic Acids Res 39:D52–D57
28. Ashburner M, Ball CA, Blake JA, Botstein D, Butler H, Cherry JM et al (2000) Gene ontology: tool for the unification of biology. The Gene Ontology Consortium. Nat Genet 25:25–29
29. Gene Ontology C (2015) Gene Ontology Consortium: going forward. Nucleic Acids Res 43:D1049–D1056
30. Uniprot C (2015) UniProt: a hub for protein information. Nucleic Acids Res 43: D204–D212
31. Law V, Knox C, Djoumbou Y, Jewison T, Guo AC, Liu Y et al (2014) DrugBank 4.0: shedding new light on drug metabolism. Nucleic Acids Res 42:D1091–D1097
32. Bento AP, Gaulton A, Hersey A, Bellis LJ, Chambers J, Davies M et al (2014) The ChEMBL bioactivity database: an update. Nucleic Acids Res 42:D1083–D1090
33. Liu T, Lin Y, Wen X, Jorissen RN, Gilson MK (2007) BindingDB: a web-accessible database of experimentally determined protein-ligand binding affinities. Nucleic Acids Res 35: D198–D201
34. Gilson MK, Liu T, Baitaluk M, Nicola G, Hwang L, Chong J (2015) BindingDB in 2015: a public database for medicinal chemistry, computational chemistry and systems pharmacology. Nucleic Acids Res 44: D1045–D1053
35. Heller S, Mcnaught A, Stein S, Tchekhovskoi D, Pletnev I (2013) InChI: the worldwide chemical structure identifier standard. J Chem 5:7
36. Heller SR, Mcnaught A, Pletnev I, Stein S, Tchekhovskoi D (2015) InChI, the IUPAC international chemical identifier. J Chem 7:23
37. Hoare CAR (1969) An axiomatic basis for computer programming. Commun Acm 12:576–583
38. Finn RD, Bateman A, Clements J, Coggill P, Eberhardt RY, Eddy SR et al (2014) Pfam: the protein families database. Nucleic Acids Res 42:D222–D230
39. Emig D, Kacprowski T, Albrecht M (2011) Measuring and analyzing tissue specificity of human genes and protein complexes. EURASIP J Bioinform Syst Biol 2011:5
40. Gujral TS, Peshkin L, Kirschner MW (2014) Exploiting polypharmacology for drug target deconvolution. Proc Natl Acad Sci U S A 111:5048–5053
41. Wang Y, Cornett A, King FJ, Mao Y, Nigsch F, Paris CG et al (2016) Evidence-based and quantitative prioritization of tool compounds in phenotypic drug discovery. Cell Chem Biol 23:862–874
42. Shin YJ, Kim JJ, Kim YJ, Kim WH, Park EY, Kim IY et al (2015) Protective effects of quercetin against HgCl(2)-induced nephrotoxicity in Sprague-Dawley rats. J Med Food 18:524–534
43. Gryglewski RJ, Korbut R, Robak J, Swies J (1987) On the mechanism of antithrombotic action of flavonoids. Biochem Pharmacol 36:317–322
44. Vita JA (2005) Polyphenols and cardiovascular disease: effects on endothelial and platelet function. Am J Clin Nutr 81:292S–297S
45. Sankari SL, Babu NA, Rani V, Priyadharsini C, Masthan KM (2014) Flavonoids—clinical effects and applications in dentistry: a review. J Pharm Bioallied Sci 6:S26–S29

Chapter 16

A Quantitative Systems Pharmacology Approach to Infer Pathways Involved in Complex Disease Phenotypes

Mark E. Schurdak, Fen Pei, Timothy R. Lezon, Diane Carlisle, Robert Friedlander, D. Lansing Taylor, and Andrew M. Stern

Abstract

Designing effective therapeutic strategies for complex diseases such as cancer and neurodegeneration that involve tissue context-specific interactions among multiple gene products presents a major challenge for precision medicine. Safe and selective pharmacological modulation of individual molecular entities associated with a disease often fails to provide efficacy in the clinic. Thus, development of optimized therapeutic strategies for individual patients with complex diseases requires a more comprehensive, systems-level understanding of disease progression. Quantitative systems pharmacology (QSP) is an approach to drug discovery that integrates computational and experimental methods to understand the molecular pathogenesis of a disease at the systems level more completely. Described here is the chemogenomic component of QSP for the inference of biological pathways involved in the modulation of the disease phenotype. The approach involves testing sets of compounds of diverse mechanisms of action in a disease-relevant phenotypic assay, and using the mechanistic information known for the active compounds, to infer pathways and networks associated with the phenotype. The example used here is for monogenic Huntington's disease (HD), which due to the pleiotropic nature of the mutant phenotype has a complex pathogenesis. The overall approach, however, is applicable to any complex disease.

Key words Quantitative systems pharmacology, QSP, Chemogenomics, Heterogeneity, Pittsburgh Heterogeneity Index, PHI, Huntington's disease, Precision medicine

1 Introduction

The focus of the traditional paradigm to drug discovery over the past 30–40 years has been on the target-centric approach. In this approach, molecular targets associated with a disease phenotype are identified by a variety of methods and then biochemical or cell-based assays reporting on the target activity are developed to screen compound libraries and identify hits that interact with and modulate the target activity. The advantages of this approach are that the molecular target associated with the disease state is known, relatively high-throughput screening assays can be set up to test large

Bridget Wagner (ed.), *Phenotypic Screening: Methods and Protocols*, Methods in Molecular Biology, vol. 1787,
https://doi.org/10.1007/978-1-4939-7847-2_16,

numbers of chemical structures, and structure–activity relationships (SAR) are facilitated by knowing how the compounds interact with the target. However, while this approach has been successful in development of many essential medicines, the overall approach is inefficient and expensive, often taking 10–15 years and costing over $1 billion, and having a high (>90%) attrition rate in clinical trials for particular indications such as oncology and neurodegenerative diseases [1, 2]. Many factors can contribute to the lack of efficacy in the clinic, but perhaps the most significant is an inadequate understanding and validation of the function of the target in the disease. Also, it is rarely one target alone that is involved in the disease process, especially in complex diseases such as cancer and neurodegeneration. In order to adequately design effective therapeutic strategies for such diseases, more comprehensive and unbiased approaches at the systems level need to be implemented to understand biological networks involved in the disease pathology and how they can be modulated to halt the progression or reverse the disease state.

Quantitative systems pharmacology (QSP) is an emerging approach to drug discovery that integrates computational and experimental methods to more completely understand the molecular pathogenesis of a disease at the systems level, thereby enabling a more optimal design of a therapeutic strategy [1–3]. A working definition of QSP is the determination of the mechanism(s) of disease progression and mechanism(s) of action of drugs on multi-scale systems through iterative and integrated computational and experimental methods to optimize the development of therapeutic strategies. In practice, there are several complementary components to QSP as outlined by Stern et al. [1] that have been applied in the traditional drug discovery paradigm. The integration and iterative analysis of observations made within these different modules has led to a more comprehensive understanding of the biological system. Through computational analysis of patient data, disease-relevant molecular pathways are inferred which define a starting point for understanding the mechanisms of the disease. In conjunction with these pathway inferences, machine learning can be implemented to predict drug-target interactions. The biological impact of these predicted interactions can then be studied using the corresponding small-molecule drugs as probes in disease-relevant phenotypic assays to test the hypotheses generated from the inferred pathways. At this point, other probes such as focused or annotated compound libraries and RNAi libraries can be tested to gain additional valuable information on the mechanism of disease modification. Critical to a more thorough understanding of the biology is the assessment of the heterogeneity in the cellular response to the probes [4, 5]. This iterative cycle of generating hypotheses, testing probes, developing experimental and computational models, and refining the hypotheses leads to disease-specific emergent properties. In turn, the knowledge gained from this process can be used to design optimal therapeutic strategies and identify pharmacodynamic, prognostic, and predictive biomarkers. An important component in this approach is the chemogenomics analysis where information about the probes reveals clues to the mechanisms at the systems level.

Chemical genomics or chemogenomics is the systematic analysis of chemical-biological interactions with the goal of identifying novel drugs and drug targets [6, 7]. Conventional chemogenomic analysis has focused on the compound interactions within protein families where information about the protein family is used to predict new chemical interactions within the family [6–8]. Alternatively, compounds active in a phenotypic assay can be used to identify molecular targets involved in the phenotype through profiling the compounds in genomics and proteomics assays [6, 9]. In QSP, the chemogenomic strategy uses information about phenotypically active compounds to learn about the pathways associated with phenotype modulation, focusing on identifying biological mechanisms based on compounds interacting on different targets and analyzing the convergence of pathways, establishing networks, and associating these with disease functions. Here relatively small screens are performed on libraries of compounds with diverse, well-annotated mechanisms. Linking the known targets of the active compounds to cellular pathways enables a systems-level analysis of the phenotype.

Described here is a chemogenomics method to elucidate potential targetable mechanisms in complex diseases. The example used is Huntington's disease (HD), which is a familial neurodegenerative disease that has eluded the identification of a successful therapy. HD is caused by a mutation in the huntingtin gene (HTT). Insufficient knowledge of the biological function of HTT and mutant HTT (mHTT) and the complexity of their pleiotropic effects has been attributed to the lack of progress in HD therapy. While the method described here has been applied to HD [10] as an example, the approach is generally applicable to any complex disease.

A phenotypic assay assessing a relevant clinical phenotype is required for the elucidation of disease-dependent pathways. The assay needs to enable the assessment of phenotype modulation at the single-cell level, and needs to be unbiased regarding molecular mechanisms that lead to the phenotype. A hallmark of neurodegenerative diseases is neuronal death and a relevant phenotypic assay would be to measure the breakdown of the cellular membrane because this is a late-stage, irreversible step in cytotoxicity that is common to all mechanisms of cell death. Induction of cytotoxicity should reflect clinical biology such as the accumulation of disease-specific protein aggregates in the case of tauopathies, or withdrawal of critical growth factors such as BDNF in the case of HD. The cell model should encompass key features relevant to the clinical biology such as known disease genetics. In the case of HD this would be the presence of the extended polyglutamine tract in the huntingtin gene which tracks with the onset and severity of the disease. The ST*Hdh*^{Q111} cell line is a mouse cell line with a stretch of 111 glutamines inserted into the huntingtin gene and is derived from animals that display many clinical characteristics of HD [11]. The clinical phenotype is cell death induced by serum deprivation (removal of growth factors). The ST*Hdh*^{Q7} isogenic cell line has the wild-type huntingtin gene with a polyglutamine tract of seven amino acids and serves as the non-disease control. After selection and development of the phenotypic assay, the chemogenomic process for inferring pathways is (1) assay optimization; (2) screening for modulators of the phenotype; (3) characterizing the active compounds; (4) evaluating effects of combinations of compounds; (5) computationally identifying targets associated with the active compounds; and (6) inferring pathways and networks associated with the identified targets.

2 Materials

1. Complete medium: DMEM, 10% FBS, 5% sodium pyruvate, 0.3% penicillin-streptomycin.
2. Serum-free medium: DMEM, 5% sodium pyruvate, 0.3% penicillin-streptomycin.
3. Phosphate-buffered saline (PBS): Dulbecco's PBS, no $CaCl_2$, no $MgCl_2$.
4. 0.25% Trypsin.
5. DMSO.
6. Solution of 1 mg/mL propidium iodide and 10 mg/mL Hoechst (*see* **Note 1**).
7. Black clear-bottom 384-well assay plates for high-content imaging.
8. Conditionally immortalized mutant huntingtin (mHTT) knock-in ST*Hdh*^{Q111} cells and isogenic wild-type ST*Hdh*^{Q7} cells [11]: To maintain consistency across experiments, a single batch of cells was grown, and cells are frozen down at 1.5 million/mL in complete medium with 10% DMSO (*see* **Note 2**).
9. Automated high-content imager.
10. Databases: Access to DrugBank, STITCH, PubChem, and KEGG databases.
11. Software: CompuSyn.

3 Methods

3.1 Selection and Development of a Disease-Relevant Phenotypic Assay

1. Culture cells under assay-specific conditions. For the HD neuroprotection assay example, 1 week prior to an assay, thaw a cryo-vial of ST*Hdh*^{Q111} and ST*Hdh*^{Q7} cells in 33 °C water bath.
2. Aspirate cells and place in a 15 mL conical tube containing 8 mL of complete medium pre-warmed to 33 °C. Centrifuge cells at 150 × *g* for 8–10 min.
3. Aspirate the supernatant, resuspend cells in 10 mL of complete medium pre-warmed to 33 °C, and then seed into a T-150 flask containing 15–20 mL of complete medium pre-warmed to 33 °C. Incubate at 33 °C, 5% CO_2. Check daily for confluence.
4. When confluence reaches 70–80% (about 3–4 days in culture) aspirate medium and add 2 mL of 0.25% trypsin pre-warmed to 33 °C. Incubate for 3–5 min at 33 °C (*see* **Note 3**). Gently shake the flask to make sure that all of the cells are floating.
5. Add 8 mL of pre-warmed complete medium and transfer cells to a 15 mL conical tube. Centrifuge at 150 × *g* for 8–10 min. Carefully aspirate the supernatant and resuspend the cells in 10 mL of complete medium, pipetting the cells up and down to remove clumps.
6. Count the cells using hemocytometer.
7. Dilute cells in complete medium to 60,000/mL (*see* **Note 4**). Plate cells at 3000 cells/well in clear-bottom, black-walled 384-well plates (50 μL/well) and allowed to attach in complete medium at 33 °C and 5% CO_2 for 24 h (Fig. 1; *see* **Note 5**).
8. After 24 h, remove the medium, wash the cells twice with 50 μL PBS, and then add 45 μL of serum-free medium. Incubate at 37 °C and 5% CO_2 for 24 h.
9. After 24 h add 25 μL of PBS containing 15 μg/mL Hoechst 33342 and 12 μg/mL propidium iodide (PI). Incubate for 30 min at room temperature on a rocker protected from light.
10. Image the cells on an automated high-content imager collecting two fluorescent channels: Hoechst (Ch1) with excitation at 405 nm and emission at 447 nm, and Texas Red (Ch2), with excitation at 561 nm and emission at 685 nm. Collect one field per well using a 10× Plan Fluor objective.
11. Analyze the images using image analysis software such as Image-J or MetaMorph. Identify the cell nuclei using the Hoechst stain and quantify the level of PI in each nucleus. Set a threshold for distinguishing positively PI-stained nuclei. Record the total number of nuclei and positive PI-stained nuclei, and calculate the percent positive PI cells (percent positive cells). The positive PI-stained cells are those whose membranes are compromised and are dying.

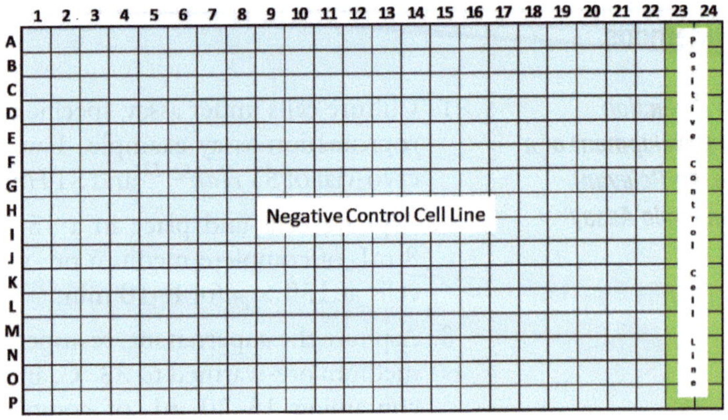

Fig. 1 Three hundred and eighty-four-well plate layout for plating cells for the HD propidium iodide assay. ST*Hdh*^{Q111} cells are plated in columns 1–22 and ST*Hdh*^{Q7} cells are plated in columns 23 and 24. Columns 1 and 2 are the negative control wells while columns 23 and 24 are the positive control wells. Compounds are tested in columns 3–22. For the HD PI assay the negative and positive controls are ST*Hdh*^{Q111} cells and ST*Hdh*^{Q7} cells, respectively, treated with DMSO. For assays where the negative controls are untreated cells and the positive controls are cells treated with a positive compound, the same cells are plated in all 384-wells

3.2 Assay Optimization and Characterization

1. Optimize the assay readout for maximal signal window and reproducibility. Titrate the cell density between 1000 and 12,000 cells/well in a 384-well plate, and assess the assay window at various times after serum withdrawal (2–72 h).
2. Determine the DMSO tolerance of the cell model. Under the above optimized conditions, assess the effect of 0, 0.1, 0.25, 0.5, 1, 2, 5, and 10% DMSO on cell viability and assay signal. The final concentration of DMSO should be low enough for cell tolerability but sufficient to facilitate compound solubility. A typical DMSO concentration used is 0.5% or less.
3. Using the optimized conditions above, calculate the Z-factor following the standard protocol [12] to ensure reasonable window and variance across the plate at the well level. The statistic is

$$1 - \left(\frac{3\left(\delta_{\mathrm{p}} + \delta_{\mathrm{n}}\right)}{|\mu_{\mathrm{p}} - \mu_{\mathrm{n}}|} \right) \tag{1}$$

where δ_{p} and δ_{n} are the standard deviation of the replicates of the positive and negative controls samples, respectively, and μ_{p} and μ_{n} are the mean of the replicates of the positive and negative control samples, respectively. For this analysis, half of the wells on a 384-well plate are negative control samples and half are positive control samples. For complex HCS phenotypic assays a Z-factor above zero is acceptable (*see* **Note 6**).

4. Use the Pittsburgh Heterogeneity Indices (PHI) [4, 5] to assess the heterogeneity of the phenotype response. The PHI profiles are used to identify biological heterogeneity in the samples, which will guide the selection of the appropriate approach for data analysis, and are used as a quality control assessment of the biology (*see* **Note 7**).
5. Calculate the quadratic entropy to assess the degree of diversity in the population:

$$\mathrm{QE} = \sum_{i>j=1}^{N} d_{ij} \times p_i \times p_j \tag{2}$$

where N is the number of data values, d is a linear matrix of intensity differences between data points i and j, and p is the probability distribution of data points. Here all elements of d are 1. For each well, the N cells are placed into 64 equally spaced bins based on nuclear Texas Red intensity in the HD example. The probability of a cell-occupying bin i is $p_i = n_i/N$, where n_i is the number of cells in bin. The quadratic entropy is a measure of diversity in the population.

6. Calculate the normKS to assess the deviation of the population distribution from a normally distributed population:

$$\mathrm{norm}KS = \max|\mathrm{CDF}_{\mathrm{dat}} - \mathrm{CDF}_{\mathrm{ref}}| \tag{3}$$

where CDFdat and CDFref are the cumulative distribution functions of the data and a reference distribution, respectively. The reference distribution is generated by calculating a normal distribution with the same mean, standard deviation, and total cell number as the data distribution. The normKS is thus a measure of the normality of the data distribution.

7. Calculate the percent outliers:

$$\%\mathrm{OL} = 1 - \frac{\mathrm{Count}(\mathrm{UIF} - \mathrm{LIF})}{\mathrm{Total\ Count}} \tag{4}$$

where UIF and LIF are the upper and lower inner fence, respectively, of the distribution of the phenotypic response at the cell level, and total count is the total number of cells in the distribution. The measurement of outliers enables the assessment of minor subpopulations of extreme signals.

8. After optimization and establishing the baseline heterogeneity in the assay response, assess the reproducibility of the assay across plates and across experiments. Run four plates (each consisting of half of the wells being positive samples and half being negative samples) in each of the three independent experiments. Assess the Z-factor across plates within each experiment and across experiments. For HCS assays a positive Z-factor is a reasonable assay.
9. For the plates above, calculate the PHI and verify that they are comparable to the baseline values determined during assay optimization. Deviation of the PHI measurements across plates or experiments is an indication of changes in the biological response. This needs to be controlled prior to and during screening.

3.3 Compound Screening

1. Select compound libraries whose biological activities have been well characterized and annotated. The LOPAC library, which contains 1200 compounds annotated for their pharmacologic activity, is our preferred first screening set. Other libraries such as those that contain FDA-approved drugs and mechanistically diverse compounds can also be used.
2. Screens can be performed at a single concentration (e.g., 10 or 20 μM), though screening at multiple concentrations is recommended (e.g., 3, 10, and 30 μM).
3. Screen compounds for activity in the HCS phenotypic assay. Follow the optimized protocol established above. For the HD assay example: Prior to compound treatment, remove medium, wash the cells once with PBS (pH = 7.4), and then add 45 μL of complete medium. Dilute 2 μL of test compounds (in DMSO) with 38 μL of complete medium and mix to create a 10× compound solution. Add 5 μL of the 10× solution to the cells and incubate at 33 °C and 5% CO_2 for 2 h (*see* **Note 8**). Wash the cells twice with PBS and then add 45 μL of serum-free medium. Dilute 2 μL of test compounds (in DMSO) with 38 μL of serum-free medium and mix to create a 10× compound solution. Add 5 μL of the 10× solution to the cells and incubate for 24 hr at 37 °C and 5% CO_2.
4. Label the cells with the Hoechst/PI stain as above and image the cells.
5. Calculate the compound activity as percent recovery:

$$\left(1-\frac{(\text{Percent PI-positive cells for Cpd})-(\text{Percent PI-positive cells for positive Ctrl})}{(\text{Percent PI-positive cells for negative Ctrl})-(\text{Percent PI-positive cells for positive Ctrl})}\right)*100 \quad (5)$$

where positive controls are the STHdh^{Q7} cells, and negative controls are the STHdh^{Q111} cells treated with DMSO.

6. Select hits based on criteria established during assay development. Common hit criteria are 1 or 3 standard deviations from the average signal of all of the library compounds on a given plate.
7. Prepare compounds for confirmation of the activity. Test the hit compounds in a concentration response assay. A typical concentration range to test consists of 10 points starting at 50 μM and decreases in either two- or threefold steps. Compounds are tested in triplicate titrations per plate.
8. Resuspend compounds in DMSO to 10 mM stocks (*see* **Note 9**).
9. Prepare compound titrations in DMSO. 50 μL of 10 mM stock is placed in columns 3 and 13 on a 384-well polypropylene plate. For a twofold dilution scheme, 25 μL is aspirated and dispensed into 25 μL of DMSO in the next column, and mixed three times. From this dilution, 25 μL is aspirated and dispensed into 25 μL of DMSO in the next column, and mixed three times. Continue the serial dilution until a total of ten concentrations are created. Spin plates at 50 × *g* for 1 min (Fig. 2; *see* **Note 10**).
10. Aspirate 2 μL of the DMSO dilutions and place into a 384-well polypropylene daughter plate. Spin plates at 50 × *g* for 1 min. Cover with a foil seal. Store master and daughter plates at −20 °C.

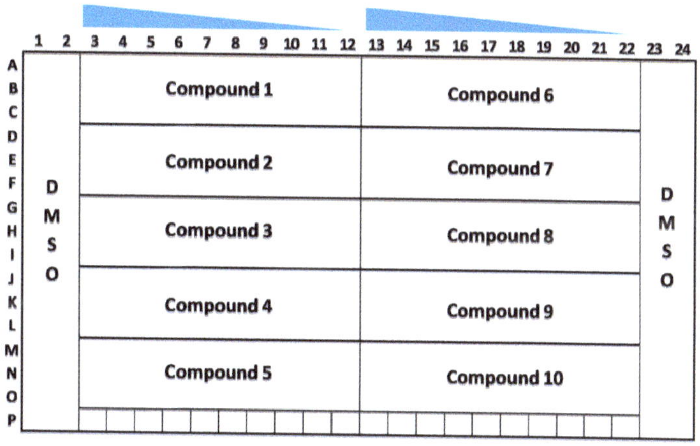

Fig. 2 Compound dilution plate map. Stock compounds are prepared in DMSO and 50 μL is dispensed into columns 3 and 13, rows A–O. DMSO (25 or 50 μL depending on the dilution scheme) is added to columns 4–12 and 14–22. Twenty-five microliter of the stock in columns 3 and 13 is transferred to columns 4 and 14, respectively, and mixed. Subsequently 25 μL of compound from 4 and 14 is transferred to columns 5 and 15 and mixed. This dilution continues to columns 12 and 22. DMSO is added to columns 1, 2, 23, and 24. After making the dilutions, 2 μL of each well is transferred to fresh plates, sealed, and stored at −20 °C. These are assay-ready daughter plates

11. Test the concentration response as above. Plot the percent recovery in Spotfire, GraphPad Prism, or other software package to generate a four-parameter logistic curve. Assess the potency and efficacy of the compounds.
12. Generate the PHI profile of the compound samples. Heterogeneity in compound response can be seen as a change in the PHI profile from the control samples as well as partial efficacy in modulating the phenotypic response. When heterogeneity is detected, test combinations of compounds to look for enhanced response. This will provide additional information for pathway inference (*see* below).

3.4 Drug Combination Analysis: Single Concentration Analysis

1. Mix compounds together in pairs using concentrations that were on the plateau of the respective concentration curves for the individual compounds (Fig. 3; *see* **Note 11**).
2. Run the assay as described above.
3. Calculate the Bliss combination index:

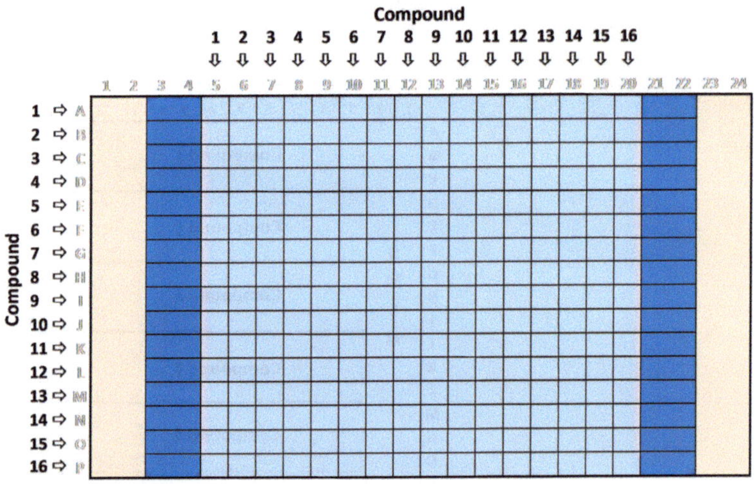

Fig. 3 Single-concentration compound combination layout. To test up to 16 compounds in combination with each other at a single concentration of each, 2 μL of a 400× final concentration of compounds is dispensed down columns 5–20 and another 2 μL is dispensed across rows in columns 3–22. DMSO (2 μL) is dispensed in columns 3, 4, 21, and 22. DMSO (4 μL) is dispensed in columns 1, 2, 23, and 24. When completed each well will have 4 μL volume with compound combinations in columns 5–20; single-compound controls in columns 3, 4, 21, and 22; and DMSO controls in columns 1, 2, 23, and 24. After making the combinations, 2 μL of each well is transferred to fresh plates, sealed, and stored at −20 °C. These are assay-ready daughter plates

$$\text{Bliss combination index} = \text{RF12}/((\text{RF1} + \text{RF2}) - (\text{RF1}^*\text{RF2})) \quad (6)$$

where RF12 is the activity of the combination of compounds 1 and 2, RF1 = the activity of compound 1, and RF2 is the activity of compound 2. When the Bliss combination index is >1 then the combination is considered to be acting synergistically, if the Bliss combination index is <1 then the combined activity is considered to be antagonistic, and if the Bliss combination index = 1 the activity of the compounds is considered to be independent [13].

4. Confirm positive combination effects in a multiple concentration analysis.

3.5 Drug Combination Analysis: Multiple Concentration Analysis

1. Prepare compound mixtures as described above using the plate map in Fig. 4. Test a minimum of four concentrations for each compound with two concentrations on the plateau and two concentrations on the slope for each respective compound activity curve.
2. Run the assay as described above.
3. Calculate the combination index (CI) using the method of Chou and Talalay [14, 15]:

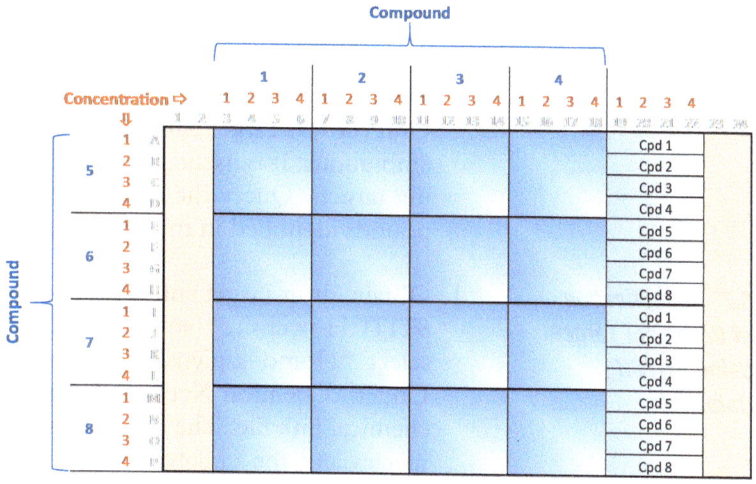

Fig. 4 Multiple-concentration compound combination layout. To test up to four compounds in combination with up to four other compounds at multiple concentrations, prepare DMSO stocks of each compound at 400× the final test concentrations. Dispense 2 μL of compound stock as shown with compounds 1–4 being dispensed down columns 3–18, and compounds 5–8 being dispensed across rows. DMSO (2 μL) is dispensed in columns 19–22. DMSO (4 μL) is dispensed in columns 1, 2, 23, and 24. When completed each well will have 4 μL volume with compound combinations in columns 3–18, single-compound controls in columns 19–22, and DMSO controls in columns 1, 2, 23, and 24. After making the combinations, 2 μL of each well is transferred to fresh plates, sealed, and stored at −20 °C. These are assay-ready daughter plates

$$\mathrm{CI} = \frac{D_1}{D_{x1}} + \frac{D_2}{D_{x2}} \quad (7)$$

where D1 and D2 denote the doses of compound 1 and compound 2 required to reach an effect of x% as single treatment, while Dx1 and Dx2 are the doses needed in combination to inhibit x%, respectively. Combinations were examined for induction of antagonism ($CI > 1.1$), additivity ($0.9 < CI < 1.1$), synergy ($CI < 0.9$), and strong synergy ($CI < 0.3$). The CompuSyn program (http://www.combosyn.com/index.html) can be used to calculate the CI.

3.6 Identification of Canonical Targets Using DrugBank

1. Obtain drug-target interaction data from DrugBank. On the DrugBank website (www.drugbank.ca), click Downloads→External Links. Under Target Drug-UniProt Links, download the CSV file for Approved drugs. If this is your first time using DrugBank, you will need to create a free account before downloading files. The file contains five columns, with headings for "DrugBank ID" (a unique identifier for the compound), "Name" (the compound name), "Type" (describes whether the compound is a small molecule, biotech antibody, natural product, etc.), UniProt ID (a unique identifier for the target), and UniProt Name (the target name). Each row represents a drug-target interaction.
2. Generate a dictionary from the DrugBank data, identifying each compound through its name and each target through its UniProt ID. Target names, and even UniProt Names, are often ambiguous. It is useful to use unique UniProt IDs for identifying targets. Query the dictionary using the names of hit compounds identified in the earlier screens (*see* **Note 12**).

3.7 Identification of Canonical Targets Using STITCH Database

1. Obtain drug-target interaction data from STITCH. On the STITCH website (stitch.embl.de), click "Download" and select "Homo sapiens" (or the organism of your choice). Under "Chemical Network," download the detailed protein-chemical link file. The exact filename depends on the selected species and the update version. For example, the file "9606.protein_chemical.links.detailed.v05.tsv.gz" is the fifth version (v05) of the drug-target interaction file for *Homo sapiens* (species ID 9606). Details of the file format and definitions of scores contained therein can be found in the README file on the STITCH download page. Briefly, each row represents one interaction between a drug and a target. The file contains seven columns: "chemical" (PubChem ID of the compound), "protein" (ENSMBL Protein ID of the target), and the five score columns ("experimental," "database," "channel," "text

mining," and "combined"). Retain only interactions that have experimental scores greater than 700, corresponding to a STITCH confidence value of 0.7.

2. Convert the names of hit compounds to PubChem IDs. This will allow each compound to be associated with targets from the STITCH database. Make sure that PubChem IDs are consistent with those that appear in the downloaded file, as explained in the STITCH README file.
3. Filter the STITCH data, keeping only interactions involving PubChem IDs from the hit compounds.
4. Convert all ENSEMBL protein IDs in the resulting data set to UniProt IDs. This conversion allows the STITCH and DrugBank results to be merged. Convert compounds IDs back to DrugBank-compatible names. The resulting list of drug-target interactions will match DrugBank-compatible compound names with UniProt IDs of targets.

3.8 Identify Pathways Affected by Canonical Targets

1. Generate a global list of all canonical targets for hit compounds by merging results of DrugBank and STITCH queries. Remove any duplicate entries.
2. Convert the UniProt ID of each target to an Entrez gene ID. This will allow pathways to be identified in KEGG.
3. Download the KEGG pathway-gene link file (http://rest.kegg.jp/link/pathway/hsa). This file is for academic research only. It has two columns. The first is the protein ID in KEGG format (hsa stands for "*Homo sapiens*," and the remainder is the Entrez gene ID). The second column is the KEGG pathway ID. Pathway names can be found for each pathway ID using the KEGG website (http://www.kegg.jp). A list of pathway names appears in http://rest.kegg.jp/list/pathway/hsa. The first column can be easily formatted to match Entrez gene IDs.

3.9 Identify Putative Pathways Involved in Synergistic Interactions

1. Tabulate all overlapping pathways for each synergistic pair of compounds. Count the number of times that each pathway is shared by a synergistic pair of compounds, and rank the pathways by the number of synergistic interactions in which they are involved. Select, based on biological insight, an interesting subset of frequently occurring pathways for analysis.
2. Identify the location of each target on the pathway of interest. When possible, identify the action of the compound (activation or inhibition) on the target. Drug activity information for some drugs can be found from DrugBank and STITCH. Manually follow the effect of each synergistic compound pair through the pathway. If multiple pairs of compounds potentially converge on a single downstream target, initiate a literature search for possible mechanisms.

4 Notes

1. Prepare a 3× stock of propidium iodide/Hoechst staining solution by mixing 19.5 μL of 10 mg/mL Hoechst, 156 μL of 1 mg/mL propidium iodide, and 13 mL of PBS. Store protected from light at 4 °C.
2. Prepare a single batch of cells by seeding monolayers in T-150 flasks in complete medium consisting of DMEM supplemented with 10% FBS, 5% sodium pyruvate, and 0.3% Pen-Strep, and allowed to reach 70–80% confluence before splitting. Aspirate medium and wash cells with 5 mL of Dulbecco's PBS (no $CaCl_2$ or $MgCl_2$) pre-warmed to 33 °C. Aspirate PBS. Add 2 mL of 0.25% trypsin pre-warmed to 33 °C, and incubate for 3–5 min at 33 °C. Gently shake the flask to make sure that all of the cells are detached and floating. Add 8 mL of complete medium, transfer cell suspension to a 15 mL conical tube, and centrifuge at 150 × *g* for 8–10 min. Aspirate medium being careful not to disturb the cell pellet. Add 10 mL of complete medium and resuspend the cells. Pipet the cells up and down a few times to ensure that there are no clumps. Remove 10 μL of cell suspension and add to 90 μL of trypan blue solution (1:1 mixture of 0.4% trypan blue and PBS). Count cells using a hemocytometer. Centrifuge cell suspension as above and aspirate the medium. Add an appropriate volume of 90% complete medium/10% DMSO to the cell pellet to obtain a cell concentration of 3×10^6 cells/mL. Dispense 1 mL of cell suspension into each cryo-vial and immediately place cryo-vials in a freezing chamber containing isopropanol, and place in a −80 °C freezer. Allow cells to freeze for 24–36 h and then move cryovials to vapor-phase liquid nitrogen for long-term storage. Leaving cells at −80 °C for more than a few days can lead to reduction in cell viability.
3. Q111 cells need longer time to detach than Q7 cells.
4. The final dilution of cells will be determined by the final density of cells used in the assay. This needs to be determined empirically during assay development and optimization. For the HD assay the optimal plating density was determined to be 3000 cells/well.
5. To help cells settle uniformly throughout the well, allow the plates to sit at room temperature on the bench for 15–30 min before putting them in the incubator.
6. The Z-factor provides information about the separation of the average positive and negative control signals and the ability to distinguish signals that belong to one or the other population. For high-throughput screening Z-factors >0.5 are considered good assays. For complex high-content phenotypic screens positive Z-factors <0.5 are often acceptable [16, 17].

7. The PHI profiles are used to determine if the response among cells is normally distributed. Distributions whose QE ≥ 0.03 (based on 64 equally sized bins and unit distance between all phenotypes), and normKS $\geq 1.031/\sqrt{N}$, where $N =$ sample size, are considered not to be a normal distribution and thus cell-level analysis (e.g., % cells responding or % cells above or below a defined threshold) should be used in analyzing the data rather than population average statistics (e.g., well-averaged signal intensity). For quality control analysis, the PHI profiles of the control samples on each plate are compared to each other and those generated during assay optimization. Plates whose PHI profiles deviate from the original profile should be discarded and the samples rerun.
8. For the HD PI assay, the optimal conditions included a pretreatment of compound under the non-stress conditions in complete medium for 2 h before the extended compound treatment under serum-free stress conditions.
9. 10 mM Stock will allow for a 50 μM final compound concentration with the cells with 0.5% DMSO. If a higher final concentration is desired then a higher DMSO stock needs to be prepared. The final concentration of DMSO should not exceed 0.5% for these cells. If other cell types are used, the maximum tolerated DMSO concentration will need to be determined empirically.
10. For a threefold dilution scheme, 25 μL of DMSO stock is aspirated and dispensed into 50 μL of DMSO in the next column, and mixed three times. From this dilution, 25 μL is aspirated and dispensed into 50 μL of DMSO in the next column, and mixed three times. Continue the serial dilution until a total of ten concentrations are created.
11. For testing combinations, individual compounds are prepared in 400× concentrations in DMSO. 2 μL of each 400× compound is mixed together to give 4 μL of 200× concentration of each compound. 76 μL of medium is added to the 4 μL of combined compounds to give 80 μL of a 10× compound concentration. 5 μL of this intermediate dilution is added to 45 μL cells in the assay.
12. Note on compound names: DrugBank identifies compounds with human-readable compound names, which can lead to ambiguities and hinder the process of finding compounds in the database. Some compounds have multiple names that differ only by one or two characters (e.g., isoetarine/isotharine). DrugBank will use either one name or the other. In many cases, DrugBank's search tool will be able to identify alternative names for compounds. Compounds that are shipped as salts will appear in DrugBank in their ionic forms only. For example, isotharine mesylate appears in DrugBank as isoetarine.

Acknowledgments

The authors wish to thank Laura Vollmer and Seia Comsa for their technical assistance in developing the HD propidium iodide assay, and Tongying Shun for her assistance with the compound combination analysis. This work was supported by funds from the University of Pittsburgh Brain Institute (Taylor/Stern), PA Commonwealth grants SAP#4100054875 (Taylor) and SAP#4100068731 (Stern), and NIH R01NS039324 (Friedlander).

References

1. Stern AM, Schurdak ME, Bahar I, Berg JM, Taylor DL (2016) A perspective on implementing a quantitative systems pharmacology platform for drug discovery and the advancement of personalized medicine. J Biomol Screen 21:521–534
2. Sorger PK, Allerheiligen SRB, Abernethy DR, Altman RB, Brouwer KLR, Califano A et al. (2011) Quantitative and systems pharmacology in the post-genomic era: new approaches to discovering drugs and understanding therapeutic mechanisms. NIH White Paper, QSP Workshop Group
3. Taylor DL (2012) A new vision of drug discovery and development. European Pharmaceutical Review 17(6)
4. Gough A, Shun TY, Taylor DL, Schurdak M (2016) A metric and workflow for quality control in the analysis of heterogeneity in phenotypic profiles and screens. Methods 96:12–26
5. Gough AH, Chen N, Shun TY, Lezon TR, Boltz RC, Reese CE et al (2014) Identifying and quantifying heterogeneity in high content analysis: application of heterogeneity indices to drug discovery. PLoS One 9:e102678
6. Bredel M, Jacoby E (2004) Chemogenomics: an emerging strategy for rapid target and drug discovery. Nat Rev Genet 5:262–275
7. Haggarty SJ, Clemons PA, Schreiber SL (2003) Chemical genomic profiling of biological networks using graph theory and combinations of small molecule perturbations. J Am Chem Soc 125:10543–10545
8. Harris CJ, Stevens AP (2006) Chemogenomics: structuring the drug discovery process to gene families. Drug Discov Today 11:880–888
9. Wagner BK, Schreiber SL (2016) The power of sophisticated phenotypic screening and modern mechanism-of-action methods. Cell Chem Biol 23:3–9
10. Pei F, Li H, Henderson MJ, Titus SA, Jadhv A, Simeonov A et al. (2017) Connecting Neuronal Cell Protective Pathways and Drug Combinations in a Huntington's Disease Model through the Application of Quantitative Systems Pharmacology. Scientific Reports 7:17803
11. Trettel F (2000) Dominant phenotypes produced by the HD mutation in STHdhQ111 striatal cells. Hum Mol Genet 9:2799–2809
12. Zhang JH, Chung TD, Oldenburg KR (1999) A simple statistical parameter for use in evaluation and validation of high throughput screening assays. J Biomol Screen 4:67–73
13. Zhao W, Sachsenmeier K, Zhang L, Sult E, Hollingsworth RE, Yang H (2014) A new bliss independence model to analyze drug combination data. J Biomol Screen 19:817–821
14. Chou TC, Talalay P (1984) Quantitative analysis of dose-effect relationships: the combined effects of multiple drugs or enzyme inhibitors. Adv Enzyme Regul 22:27–55
15. Chou TC, Talalay P (1981) Generalized equations for the analysis of inhibitions of Michaelis-Menten and higher-order kinetic systems with two or more mutually exclusive and nonexclusive inhibitors. FEBS J 115:207–216
16. Buchser W, Collins M, Garyantes T, Guha R, Haney S, Lemmon V et al (2004) Assay development guidelines for image-based high content screening, high content analysis and high content imaging. In: Sittampalam GS, Coussens NP, Brimacombe K et al (eds) Assay guidance manual. Eli Lilly & Company and the National Center for Advancing Translational Sciences, Bethesda, MD
17. Bray MA, Carpenter A (2004) Advanced assay development guidelines for image-based high content screening and analysis. In: Sittampalam GS, Coussens NP, Brimacombe K et al (eds) Assay guidance manual. Eli Lilly & Company and the National Center for Advancing Translational Sciences, Bethesda, MD

Chapter 17

Phenotype-Based High-Content Screening Using Fluorescent Chemical Bioprobes: Lipid Droplets and Glucose Uptake Quantification in Live Cells

Young-Hee Shin and Seung Bum Park

Abstract

Phenotypic screening in live cells has emerged as a promising strategy for drug discovery in pharmaceutical communities. For relevant phenotype-based screening setups, it is critical to develop adequate reporters in order to selectively visualize subcellular compartments or phenotypic changes that represent disease-related characteristics during compound screening. In this chapter, we introduce two phenotype-based high-content/high-throughput assays using fluorescent bioprobes that have been designed and refined to selectively stain cellular lipid droplets (LDs) and to show cellular glucose uptake. In conjunction with target identification process for the hit compounds from phenotypic screening, these fluorescent chemical probe-based screening techniques are expected to drive a great advancement for the discovery of novel first-in-class therapeutics.

Key words Phenotypic screening, Image-based screening, Fluorescent bioprobe, Seoul-Fluor 44 (SF44), Glucose bioprobe (GB2-Cy3)

1 Introduction

Discovery of novel therapeutics has been driven by conventional target-based screenings that utilize in vitro setups to monitor the perturbation of a single-gene or -protein targets [1]. However, numerous drug candidates derived from those screenings turned out to be a failure at the later stages of drug development due to their off-target effects [2]. Also, because it is not easy to understand the protein networks with the whole complexity in the cellular system, only limited targets are available for drug screening [2]. Alternatively, phenotype-based screening focuses on the changes of the characteristic disease-relevant phenotypes in living cells while the molecular mode of action and protein targets of hit compounds remain unknown [3–5]. As sophisticated target identification techniques became available for candidate compounds [6–8], phenotype-based high-throughput screening (HTS) has

Bridget Wagner (ed.), *Phenotypic Screening: Methods and Protocols*, Methods in Molecular Biology, vol. 1787, https://doi.org/10.1007/978-1-4939-7847-2_17,

been an efficient approach for the early stage of drug discovery, enabling researchers to avoid cytotoxic drug candidates caused by off-target effects and to explore novel therapeutic targets with new modes of action.

Automated imaging technologies have facilitated an efficient investigation of cellular phenotypes by providing multiple pieces of information, including biological changes from the compound activity, cell morphology, and cell population alteration with a single experiment, and thereby are called a high-content screening (HCS) technique. Usage of fluorescent reporters is considered the most reasonable way to visualize compound activities in live cells, due to its excellent signal-to-noise ratio [9, 10]. Fluorescent reporters have been used either by expressing proteins of interest fused with fluorescent proteins in cells [11] or by treating cells with small-molecule fluorescent probes [12–14]. Since fluorescent chemical probes are not limited in their application ranges such as cells, tissues, and organisms, unlike genetically modified fluorescent reporters, many research groups have been extensively developing novel fluorescent molecules to monitor cellular changes [12–14]. In this chapter, we introduce two phenotypic HTS/HCS screening systems utilizing rationally designed fluorescent probes to monitor cellular lipid droplets (LDs) and cellular glucose uptake in live cells.

1.1 Monitoring Cellular Lipid Droplets

Lipid droplets (LDs) are dynamic cytoplasmic organelles and nearly all cells store lipids as the form of LDs under the nutrient-sufficient conditions [15]. LDs are implicated in a number of cellular functions such as lipid storage for energy generation and membrane synthesis, and protein degradation [16]. These functional processes are associated with metabolic diseases, including cardiovascular diseases, type 2 diabetes, and cancer [17]. Recently, numerous reports have demonstrated the roles of macroautophagy in cellular lipid homeostasis by removing lipids, namely lipophagy [18–20]. Macroautophagy (for simplicity hereinafter referred to as autophagy), translated from Greek to mean "self-eating," is an evolutionally conserved catabolic process for the maintenance of cellular and tissue homeostasis [21]. Chemical modulators that can enhance autophagic process, while not inducing cellular cytotoxicity, can be developed as promising therapeutics for protein aggregation-related neurodegenerative diseases [22, 23]. The efficacy of autophagy modulators can be assessed based on the up- or downregulation of cellular LDs. Therefore, the HCS/HTS systems monitoring cellular LDs are advantageous to find effective small molecules for the regulation of lipid metabolism, lipid biogenesis, as well as autophagy.

The autophagic process starts with the formation of autophagosome to engulf cytoplasmic materials for the cellular recycling while recruiting microtubule-associated protein light chain (LC3)-

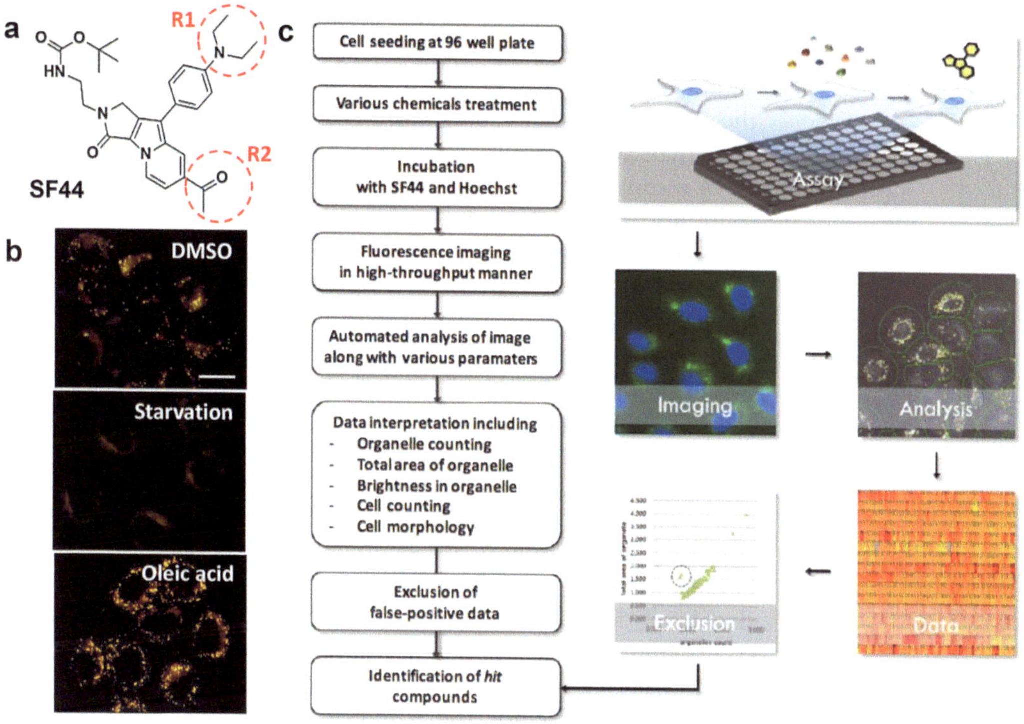

Fig. 1 Lipid droplet (LD) screening. (**a**) Structure of SF44. (**b**) Representative fluorescence images showing cellular LDs in HeLa cells stained with SF44. Cells were incubated with DMSO, serum-free medium, oleic acid (5 mM), scale bar: 20 mm (reproduced from Ref. [29] with permission from The Royal Society of Chemistry). (**c**) Flowchart depicting LD-based phenotypic HCS/HTS process (adapted from Ref. [5] with permission from The Royal Society of Chemistry)

I to convert to LC3-II. Autophagosomes then fuse with lysosomes to form autolysosomes, which degrade the contained materials [19]. To track and examine the autophagic process, widely used methods are LC3-green fluorescent protein (GFP) puncta and mCherry-EGFP-LC3 puncta assays [24, 25]. However, these LC3-based HCS systems are limited, in that LC3 cannot represent the entire flux of autophagy process and LC3 can be affected by both autophagy activators and inhibitors, resulting in misleading analyses. To overcome these aspects, fluorescent bioprobes can be applied to visualize autophagic process by taking advantage of the knowledge on the lipid droplet (LD) metabolism, or lipophagy [18–20].

Since LDs uniquely constitute a hydrophobic phase in the aqueous cytosols, we have developed a fluorescent bioprobe, namely Seoul-Fluor 44 (SF44), that has been optimized to selectively turn on its fluorescence in the hydrophobic environment (Fig. 1a) [26–29]. We have reported a series of Seoul-Fluor (SF) compounds containing a novel indolizine-based molecular framework that is tunable and predictable in its photophysical properties simply by changing electronic natures of R_1 and R_2 substituents (Fig. 1a) [30]. SF44 has an electron-donating diethylamine at the R_1 position and an electron-accepting acetyl group at the R_2 position that causes more charge separation at the excited state than at the ground state in SF44. This structural element of SF44 means that the light-mediated excited state of SF44 interacts more readily with the polar environment, while losing their energy via a non-radiative decay, than with the hydrophobic condition. Thus, SF44 can selectively turn on its fluorescence in the hydrophobic circumstances probably through photoinduced intramolecular charge transfer (ICT) process [26]. The lipophilicity and specificity of SF44 against LDs were additionally tuned by adjusting *c*Log*P* values [28]. Therefore, the cellular LD staining with SF44 does not require additional washing or fixation steps since SF44 loses its fluorescence in the polar circumstances such as medium and cytoplasm. When a fluorescent bioprobe is used for high-throughput screening, it is essential to exclude false-positive data because intense fluorescence signals could be induced from the shrinking dead cells, dividing cells, or non-soluble aggregated compounds. These misleading fluorescent signals can be avoided in an image-based HCS by checking cell morphologies. In other words, we can exclude the false-positive data via identifying the deviated data from the correlation plots between organelle (LDs) count per cell, total organelle area per cell, and total fluorescence intensity per cell (Fig. 1) [27]. Therefore, SF44-based HCS can be an excellent high-throughput screening platform to identify the novel modulators of cellular autophagic process as well as lipid biogenesis without washing and fixation steps.

1.2 Monitoring Cellular Glucose Uptakes

Alteration in cellular glucose uptake is one of the major biomarkers for cancers and metabolic diseases such as type 2 diabetes [31]. Fast-proliferating cancer cells exhibit high metabolic demands, resulting in an enhanced cellular glucose uptake compared to normal cells, known as the Warburg effect [32]. Therefore, the increase of glucose uptake is one of the key markers in tumor malignancy. Effective anticancer drug candidates can be initially found by investigating their effects on the reduction in glucose uptake by cancer cells. Type 2 diabetes mellitus is characterized by insulin resistance that causes the reduction in glucose uptake by metabolically active tissues such as muscle, fat, and liver, and brings about the glucose accumulation in blood, namely hyperglycemia [33]. The efficacy of antidiabetic agent candidates can be evaluated by measuring the increase of glucose uptake in muscle or fat cells. Therefore, an efficient HCS/HTS system for monitoring cellular glucose uptake could be a critical tool in discovering drug candidates for cancers and metabolic diseases.

To quantify the cellular glucose uptake, we have successfully developed a glucose bioprobe (GB2-Cy3) by conjugating a

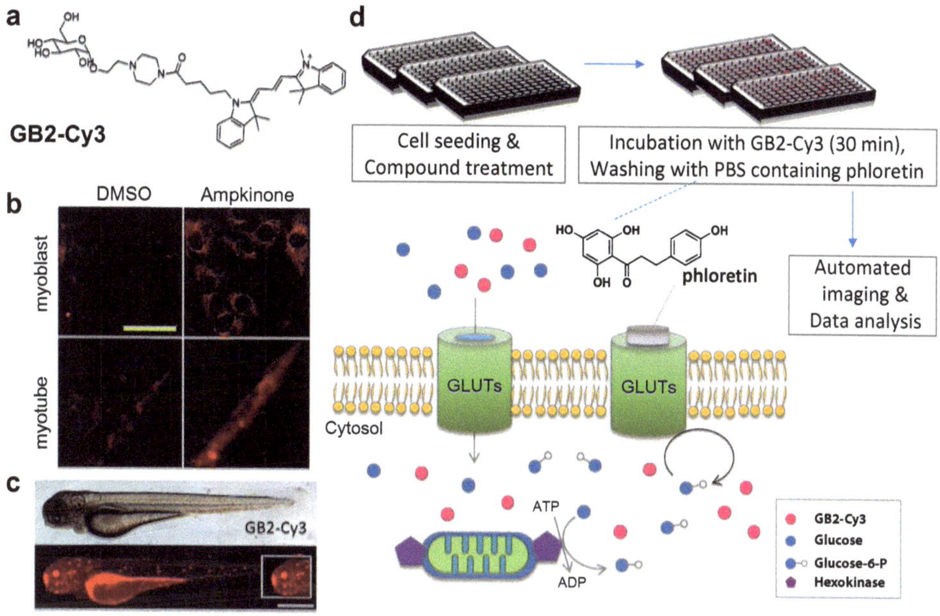

Fig. 2 Glucose uptake screening. (**a**) Structure of GB2-Cy3. (**b**) Representative fluorescence images showing GB2-Cy3 uptake by C2C12 myoblasts and myotubes. Scale bar: 10 μm (reproduced from Ref. [44] with permission from Wiley-VCH). (**c**) GB2-Cy3 uptake in zebrafish. White box highlights the GB2-Cy3 uptake into the zebrafish larval eyes. Scale bar: 500 μm [35]. (**d**) Depiction of phenotypic HCS/HTS process monitoring glucose uptake. The illustration describes that cellular uptake of GB2-Cy3 requires a GLUT blocking step using phloretin because GB2-Cy3 is not a substrate of hexokinase. Phosphorylated glucose does not leak out through GLUTs [5]. (Fig. 2c, d was reproduced from Refs. [5, 35], respectively, with permission from The Royal Society of Chemistry)

fluorophore to D-glucose using a linker that showed a great cellular uptake kinetics and an excellent signal-to-noise ratio (Fig. 2a) [34–37]. GB2-Cy3 contains a D-glucose with an α-glycosidic bond that outperformed in cellular uptake than a β-glycosidic bond analogue [34]. A linker between the glucose and a fluorophore has been optimized to control the water solubility, cytotoxicity, and GLUT-mediated cellular uptake of the glucose bioprobe. Among various linkers including polyethylene glycol (PEG), cyclohexane, etc., the piperazine linker showed the best kinetics for the GLUT-mediated cellular uptake of glucose bioprobes [34, 35]. In the case of organic fluorophores, Cy3 outperformed all other tested fluorescent dyes. Interestingly, the glucose bioprobe containing Cy3 with +1 net charge showed the best cellular uptake than other charged analogues, with the bright fluorescence and excellent photostability in the aqueous environment [37]. Our optimized glucose bioprobe, GB2-Cy3, was proved to be transported to the cells selectively through glucose transporters (GLUTs) in a competitive manner with D-glucose, but not with L-glucose [34]. GB2-Cy3 is also applicable to in vivo as demonstrated in zebrafish larva as well as various mammalian cells including

C2C12 myocytes (Fig. 2b, c) [37]. For the practical application to high-throughput screening, the cellular up-taken GB2-Cy3 needs to be blocked from its efflux by treatment of cells with phloretin, a transient and selective GLUT inhibitor [38]. This additional step is essential to sustain the fluorescence intensity within the timeframe for the fluorescence measurement in the high-throughput format since GB2-Cy3 is not phosphorylated by hexokinases; otherwise GB2-Cy3 will efflux out of cells via GLUT on the basis of concentration gradient after the washing step (Fig. 2d).

The phenotype-based screening system can pose a great impact on the development of novel therapeutics when extended to the target identification processes at the early stages of researches [6, 8, 39–41]. This is because without the knowledge on the target proteins of the hit compounds, it will be extremely difficult to efficiently improve their potency or reveal the mechanism of their actions. Once target proteins were identified along with their structures and functions, target-based rational drug design approaches can be applied to enhance the efficiency in lead optimization of initial hits from phenotypic screening [42, 43]. Using our target identification method, namely fluorescence difference in two-dimensional gel electrophoresis (FITGE), developed in our group [8], we were able to identify the target protein of initial hit compounds that enhance glucose uptake using GB2-Cy3-based phenotypic screening, and significantly improve the potency of hit compounds using structure-based rational design [44, 45]. Several effective compounds that reduce cellular LDs have been identified as autophagy inducers and investigated as the promising drug candidates for neurodegenerative diseases from the SF44-based phenotypic screening and their targets are currently under investigation [6, 27].

2 Materials

2.1 Cellular Lipid Droplet Quantification

1. HeLa human cervical cancer cells.
2. Cell culture medium: RPMI medium 1640 supplemented with 10% (v/v) fetal bovine serum (FBS) and 1% (v/v) antibiotic-antimycotic (AA) in 100 mm petri dishes. Cells were maintained at 37 °C in a humidified atmosphere of 5% CO_2.
3. Assay plate: Sterile 96-well black-wall clear-bottom plate with lid.
4. Pinning tool: Multi-blot replicator with 96 pins.
5. Oleic acid, prepared as 1 mM stock in isopropyl alcohol.
6. DMSO (bio-grade).
7. Hoechst 33342 (trihydrochloride, trihydrate, 10 mg/mL solution in water).

8. Seoul-Fluor 44 (SF44) was obtained from a commercial source (www.sparkbio.co.kr). SF44 was stored as 20 or 5 mM DMSO stocks and kept in a −20 °C freezer.
9. Compounds to be tested during screening were prepared as 5 mM DMSO stocks in 96-well plates.
10. High-content screening was performed by InCell Analyzer 2000.
11. Data were analyzed by InCell Analyzer 1000 workstation program according to the manufacturer's protocol. Alternatively, Developer Toolbox was also used for the image analysis.

2.2 Cellular Glucose Uptake Quantification

1. Assay plate, pinning tool, instrument, and software: same as in Subheading 2.1.
2. C2C12 (mouse myoblasts) cells were cultured in high glucose.
3. Cell culture medium: Dulbecco's modified Eagle medium (DMEM) supplemented with 10% (v/v) FBS and 1% (v/v) AA. Cells were maintained at 37 °C in a humidified atmosphere of 5% CO_2.
4. DPBS buffer (Dulbecco's phosphate-buffered saline without calcium chloride, magnesium chloride).
5. Glucose bioprobe 2 (GB2-Cy3) was obtained from a commercial source (www.sparkbio.co.kr): 10 mM GB2-Cy3 stock in DMSO was stored at −20 °C.
6. Phloretin was stored as 20 mM stock in DMSO in a −20 °C freezer.
7. Washing buffer: DPBS containing 20 μM phloretin.

3 Methods

3.1 Cellular Lipid Droplet Screening

1. Culture HeLa cells (*see* **Note 1**).
2. Seed HeLa cells on 96-well black clear-bottom plates at a density of 2500 cells/well in 100 μL of culture medium.
3. Incubate the plates in a 37 °C, 5% CO_2, incubator overnight (*see* **Note 2**).
4. Add compounds to the cells in 96-well plates using a 0.2 μL (or 0.1 μL) pinning tool or a pipette as a final concentration of the compounds to be 10 μM in the assay. When a pinning tool is used, compounds to be tested should be prepared in 96-well plates (*see* **Notes 3** and **4**). For the control experiments, four wells are added with just DMSO and two wells are treated with 5 μM oleic acid (positive control, *see* **Note 5**). As a negative control, cells in two wells are starved for 18 h by changing to serum-free medium 6 h after the compound treatment. Serum

starvation medium should be changed to culture medium just before adding SF44 for imaging.

5. Incubate the plates in a 37 °C incubator for 24 h.
6. Add medium-diluted SF44 (LD probe) to the cells as a final concentration of 10 μM in the assay.
7. Incubate the plates in a 37 °C, 5% CO_2, incubator for 10 min (*see* **Note 6**).
8. Add medium-diluted Hoechst 33342 (2 μg/mL) to the cells.
9. Incubate the plates in a 37 °C, 5% CO_2, incubator for further 20 min (*see* **Note 7**).
10. Scan the plates in an imaging instrument without any washing steps. Measure the fluorescence intensity of SF44 for lipid droplets (LDs) and Hoechst for nuclei using the filtering sets: ex/em 430/605 nm (CFP/Cy3) for LDs; ex/em 350/455 nm (DAPI) for nuclei. Bright-field images are also taken to check the cell morphologies (*see* **Note 8**).
11. Analyze the images to quantify LDs per cell (*see* **Note 9**). Exclude false positives from the data based on the bright-field images (cell morphology) and the correlation plots between various LD parameters (*see* **Note 10**).

3.2 Cellular Glucose Uptake Screening

1. Culture C2C12 (mouse myoblasts) cells (*see* **Note 1**).
2. Seed C2C12 cells on 96-well black wall with clear-bottom plates at a density of 2000 cells/well in culture medium.
3. Incubate the plates in a 37 °C, 5% CO_2, incubator overnight (*see* **Note 2**).
4. Treat compounds to the cells using a pinning tool or a pipette as a final concentration of the compounds to be 10 μM in the assay (*see* **Notes 3** and **4**).
5. Incubate the plates in a 37 °C, 5% CO_2, incubator for 24 h.
6. Aspirate the medium, and then add culture medium containing 10 μM GB2-Cy3 (*see* **Note 11**).
7. Incubate the plates in a 37 °C, 5% CO_2, incubator for 20 min.
8. Add culture medium-diluted Hoechst 33342 (2 μg/mL) to each well.
9. Incubate the plates in a 37 °C, 5% CO_2, incubator for further 10 min (*see* **Note 7**).
10. Wash the wells three times with PBS (warm or room temperature) containing 20 μM phloretin. Add 100 μL of PBS containing 20 μM phloretin to the wells before performing image screening (*see* **Note 12**).
11. Scan the plates in an imaging instrument in the presence of PBS buffer containing phloretin. Measure the fluorescence intensity

of GB2-Cy3 for cellular glucose uptake and Hoechst for nuclei using the filtering sets: ex/em 543/605 nm (Cy3) for glucose uptake; ex/em 350/455 nm (DAPI) for nuclei. Bright-field images are also taken to check the cell morphologies (*see* **Note 7**).

12. Analyze the images to quantify glucose uptake per cell by measuring fluorescence intensity of GB2-Cy3 per cell (*see* **Note 13**).

4 Notes

1. Other types of cells or tissues can also be used following the protocol.
2. Cover slides can also be used depending on the purpose of the experiment.
3. Make sure that all the DMSO stock compounds are fully thawed if the compound plates were stored in a freezer. Compounds were usually prepared as 5 mM DMSO stocks so that the final concentration of the compounds in the assay is 10 μM when 0.2 μL pinning tool is used.
4. For dose-response experiments, compounds are serial diluted in DMSO and then further diluted in culture medium. 10 μL of the compound in culture medium is added to the cells in 96-well plates. Make sure that the final DMSO concentration is not over 0.5% (v/v) since DMSO could be toxic to the cells.
5. 0.5 μL of 1 mM oleic acid stock prepared in isopropyl alcohol is added to the cells in 96-well plates. For further dilution of oleic acid, water can be used.
6. 5 mM SF44 in DMSO should be freshly diluted in the culture medium before adding to the cells. Recommend to add 10 μL of medium-diluted SF44 to the cells so that the desired SF44 concentration is achieved in the assay. SF44 requires 30 min to stain LDs.
7. Hoechst 33342 (10 mg/mL) is diluted by 1/500 in culture medium, and then 10 μL of it is added to the cells so that the final dilution becomes approximately 1/5000 (2 μg/mL).
8. Images are recommended to be taken from at least four different fields (randomly chosen) per well in a 96-well plate. Images can be taken by either 20× scale or 60× scale. Make sure that the image brightness contrast is not automatically adjusted while taking images since it will make all the image background different.
9. Cellular LD quantification is performed by measuring organelles (LDs) count per cell, total area of organelles per cell, and

total fluorescence intensity of organelles per cell. Individual cells are recognized by nuclei staining.

10. Be careful that you are not choosing hit compounds based on the false-positive data. You might observe intense bright fluorescent signals when cells are shrinking dead or dividing. If the solubility of compounds is low, the compounds might aggregate and show intense fluorescence. Check the deviated spots from the correlation plots between different parameters obtained from the screening. The deviated spots could be false-positive data.
11. 10 μM of GB2-Cy3 in culture medium should be prepared freshly before use.
12. 20 μM Phloretin in PBS should be freshly prepared before use.
13. Cellular glucose uptake quantification is performed by measuring total fluorescence intensity in the cytoplasm per cell. Individual cells are recognized by nuclei staining.

Acknowledgment

This work was supported by the National Creative Research Initiative Grant (NRF-2014R1A3A2030423) and the Bio & Medical Technology Development Program (2012M3A9C4048780) through the National Research Foundation of Korea (NRF) funded by the Korean Government (Ministry of Science and ICT). This research was also supported by Basic Science Research Program (NRF-2009-0093826, 2016R1A6A3A01011351) through NRF funded by the Korean Ministry of Education.

References

1. Fishman MC, Porter JA (2005) Pharmaceuticals: a new grammar for drug discovery. Nature 437:491–493
2. Sams-Dodd F (2005) Target-based drug discovery: is something wrong? Drug Discov Today 10:139–147
3. Swinney DC (2013) Phenotypic vs. target-based drug discovery for first-in-class medicines. Clin Pharmacol Ther 93:299–301
4. Zheng W, Thorne N, McKew JC (2013) Phenotypic screens as a renewed approach for drug discovery. Drug Discov Today 18:1067–1073
5. Jo A, Jung J, Kim E, Park SB (2016) A high-content screening platform with fluorescent chemical probes for the discovery of first-in-class therapeutics. Chem Commun (Camb) 52:7433–7445
6. Park H, Ha J, Koo JY, Park J, Park SB (2017) Label-free target identification using in-gel fluorescence difference via thermal stability shift. Chem Sci 8:1127–1133
7. Hart CP (2005) Finding the target after screening the phenotype. Drug Discov Today 10:513–519
8. Park J, Oh S, Park SB (2012) Discovery and target identification of an antiproliferative agent in live cells using fluorescence difference in two-dimensional gel electrophoresis. Angew Chem Int Ed Engl 51:5447–5451
9. Stoughton RB, Friend SH (2005) How molecular profiling could revolutionize drug discovery. Nat Rev Drug Discov 4:345–350
10. Zhang J, Campbell RE, Ting AY, Tsien RY (2002) Creating new fluorescent probes for cell biology. Nat Rev Mol Cell Biol 3:906–918
11. Bivona TG, Philips MR (2005) Analysis of Ras and Rap activation in living cells using

fluorescent Ras binding domains. Methods 37:138–145
12. Xu W, Zeng Z, Jiang JH, Chang YT, Yuan L (2016) Discerning the chemistry in individual organelles with small-molecule fluorescent probes. Angew Chem Int Ed Engl 55:13658–13699
13. Lee MH, Kim JS, Sessler JL (2015) Small molecule-based ratiometric fluorescence probes for cations, anions, and biomolecules. Chem Soc Rev 44:4185–4191
14. Yin J, Hu Y, Yoon J (2015) Fluorescent probes and bioimaging: alkali metals, alkaline earth metals and pH. Chem Soc Rev 44:4619–4644
15. Martin S, Parton RG (2006) Lipid droplets: a unified view of a dynamic organelle. Nat Rev Mol Cell Biol 7:373–378
16. Walther TC, Farese RV (2012) Lipid droplets and cellular lipid metabolism. Annu Rev Biochem 81:687–714
17. Greenberg AS, Coleman RA, Kraemer FB, McManaman JL, Obin MS, Puri V, Yan QW, Miyoshi H, Mashek DG (2011) The role of lipid droplets in metabolic disease in rodents and humans. J Clin Invest 121:2102–2110
18. Singh R, Cuervo AM (2011) Autophagy in the cellular energetic balance. Cell Metab 13:495–504
19. Singh R, Kaushik S, Wang Y, Xiang Y, Novak I, Komatsu M, Tanaka K, Cuervo AM, Czaja MJ (2009) Autophagy regulates lipid metabolism. Nature 458:1131–1135
20. Ward C, Martinez-Lopez N, Otten EG, Carroll B, Maetzel D, Singh R, Sarkar S, Korolchuk VI (2016) Autophagy, lipophagy and lysosomal lipid storage disorders. Biochim Biophys Acta 1861:269–284
21. Kondo Y, Kanzawa T, Sawaya R, Kondo S (2005) The role of autophagy in cancer development and response to therapy. Nat Rev Cancer 5:726–734
22. Rubinsztein DC, Codogno P, Levine B (2012) Autophagy modulation as a potential therapeutic target for diverse diseases. Nat Rev Drug Discov 11:709–U784
23. Cheng Y, Ren XC, Hait WN, Yang JM (2013) Therapeutic targeting of autophagy in disease: biology and pharmacology. Pharmacol Rev 65:1162–1197
24. Hansen TE, Johansen T (2011) Following autophagy step by step. BMC Biol 9:39
25. Mizushima N, Yoshimori T, Levine B (2010) Methods in mammalian autophagy research. Cell 140:313–326
26. Kim E, Lee S, Park SB (2012) A Seoul-Fluor-based bioprobe for lipid droplets and its application in image-based high throughput screening. Chem Commun 48:2331–2333
27. Lee S, Kim E, Park SB (2013) Discovery of autophagy modulators through the construction of a high-content screening platform via monitoring of lipid droplets. Chem Sci 4:3282–3287
28. Lee Y, Na S, Lee S, Jeon NL, Park SB (2013) Optimization of Seoul-Fluor-based lipid droplet bioprobes and their application in microalgae for bio-fuel study. Mol Biosyst 9:952–956
29. Choi Y, Kim H, Shin YH, Park SB (2015) Diverse display of non-covalent interacting elements using pyrimidine-embedded polyheterocycles. Chem Commun 51:13040–13043
30. Kim E, Koh M, Lim BJ, Park SB (2011) Emission wavelength prediction of a full-color-tunable fluorescent core skeleton, 9-aryl-1,2-dihydropyrrolo[3,4-b]indolizin-3-one. J Am Chem Soc 133:6642–6649
31. Hsu PP, Sabatini DM (2008) Cancer cell metabolism: Warburg and beyond. Cell 134:703–707
32. Heiden MGV, Cantley LC, Thompson CB (2009) Understanding the Warburg effect: the metabolic requirements of cell proliferation. Science 324:1029–1033
33. Lee CH, Olson P, Evans RM (2003) Minireview: lipid metabolism, metabolic diseases, and peroxisome proliferator-activated receptors. Endocrinology 144:2201–2207
34. Park J, Lee HY, Cho MH, Park SB (2007) Development of a Cy3-labeled glucose bioprobe and its application in bioimaging and screening for anticancer agents. Angew Chem Int Ed 46:2018–2022
35. Lee HY, Lee JJ, Park J, Park SB (2011) Development of fluorescent glucose bioprobes and their application on real-time and quantitative monitoring of glucose uptake in living cells. Chemistry 17:143–150
36. Jo A, Park J, Park SB (2013) Exploiting the mechanism of cellular glucose uptake to develop an image-based high-throughput screening system in living cells. Chem Commun 49:5138–5140
37. Park J, Um JI, Jo A, Lee J, Jung DW, Williams DR, Park SB (2014) Impact of molecular charge on GLUT-specific cellular uptake of glucose bioprobes and in vivo application of the glucose bioprobe, GB2-Cy3. Chem Commun 50:9251–9254
38. Fuhrmann GF, Dernedde S, Frenking G (1992) Phloretin keto-enol-tautomerism and inhibition of glucose-transport in human erythrocytes (including effects of phloretin on anion transporter). Biochim Biophys Acta 1110:105–111
39. Schenone M, Dancik V, Wagner BK, Clemons PA (2013) Target identification and

mechanism of action in chemical biology and drug discovery. Nat Chem Biol 9:232–240

40. Park J, Koh M, Koo JY, Lee S, Park SB (2016) Investigation of specific binding proteins to photoaffinity linkers for efficient deconvolution of target protein. ACS Chem Biol 11:44–52

41. Lee S, Nam Y, Koo JY, Lim D, Park J, Ock J, Kim J, Suk K, Park SB (2014) A small molecule binding HMGB1 and HMGB2 inhibits microglia-mediated neuroinflammation. Nat Chem Biol 10:1055–1060

42. Levoin N, Calmels T, Poupardin-Olivier O, Labeeuw O, Danvy D, Robert P, Berrebi-Bertrand I, Ganellin CR, Schunack W, Stark H, Capet M (2008) Refined docking as a valuable tool for lead optimization: application to histamine H-3 receptor antagonists. Arch Pharm 341:610–623

43. Bollini M, Domaoal RA, Thakur VV, Gallardo-Macias R, Spasov KA, Anderson KS, Jorgensen WL (2011) Computationally-guided optimization of a docking hit to yield catechol diethers as potent anti-HIV agents. J Med Chem 54:8582–8591

44. Koh M, Park J, Koo JY, Lim D, Cha MY, Jo A, Choi JH, Park SB (2014) Phenotypic screening to identify small-molecule enhancers for glucose uptake: target identification and rational optimization of their efficacy. Angew Chem Int Ed 53:5102–5106

45. Bae H, Jang JY, Choi SS, Lee JJ, Kim H, Jo A, Lee KJ, Choi JH, Suh SW, Park SB (2016) Mechanistic elucidation guided by covalent inhibitors for the development of anti-diabetic PPAR gamma ligands. Chem Sci 7:5523–5529

Chapter 18

Developing High-Throughput Assays to Analyze and Screen Electrophysiological Phenotypes

Jen Q. Pan, David Baez-Nieto, Andrew Allen, Hao-Ran Wang, and Jeffrey R. Cottrell

Abstract

Ion channels represent nearly a quarter of all targets that currently available medications modulate, and their dysfunction underlies increasing number of human diseases. Functional analysis of ion channels have traditionally been a bottleneck in large-scale analyses. Recent technological breakthroughs in automated planar electrophysiology have democratized the technique to enable high-throughput patch clamping at scale. In this chapter, we describe the methodology to perform a phenotypic screen on voltage-gated calcium channels across many different genetic coding variations and against small-molecule modulators. We first describe the procedures to establish inducible heterologous ion channel expression in HEK293 cells, where each cell incorporates one copy of a target protein cDNA—a step that is critical for producing stable and consistent expression of ion channels. We then describe the experimental and analytical methods for analyzing the function of ion channels using high-throughput planar electrophysiology.

Key words Patch clamp, Planar electrophysiology, Ion channels, Voltage-gated calcium channels, High throughput, Electrophysiology, Phenotypic screen

1 Introduction

Ion channels are pivotal to every aspect of physiology. Dysfunction of ion channels is implicated in complex common disorders including diabetes, heart disease, and psychiatric illnesses. Their central role in signaling, sensing, and cellular homeostasis is further underscored by a sizable number of "channelopathies" including cystic fibrosis [1], long QT syndrome [2], cancer [3], epilepsy, and many more. More recently, advances in DNA-sequencing technology helped identify a large number of genetic variations of ion channels in association with disease conditions. As such, ion channels represent one of the largest target classes for current medicines, and the future prospect of ion channel as a target class is tied to the ability to phenotype genetic variation-associated disorders and the ability to identify compounds that modulate channel properties at scale.

Bridget Wagner (ed.), *Phenotypic Screening: Methods and Protocols*, Methods in Molecular Biology, vol. 1787, https://doi.org/10.1007/978-1-4939-7847-2_18,

While surrogate tools to measure cytoplasmic ionic influx exist, such as thallium flux and calcium flux [4–6], these assays routinely identify false positives and nonspecific interactions. On the other hand, patch clamp electrophysiology monitors ion channel biophysical activities directly and provides highly precise readouts of channel function. While the patch clamp technique is the gold standard to study cellular electrophysiology, it is traditionally laborious and of low throughput. Recent technological breakthroughs in automated planar electrophysiology have advanced the technique sufficiently to enable patch clamp at scale [7, 8]. In this chapter, we describe a protocol for establishing an inducible heterologous expression system that is critical for consistent, reproducible, and reliable high-throughput channel characterization. We then describe the recording protocols and assays that monitor $Ca_V3.3$ voltage-gated calcium channel activities using high-throughput electrophysiology with sensitivity and precision, illustrated mainly by the SyncroPatch 384PE, an assay system that enables electrophysiological recording in 384-well plates with giga-ohm seal resistance [9]. Many of the protocols and notes given in this chapter can be applied to other ion channels, similarly.

2 Materials

2.1 Cell Culture

1. FlpIn-TREx HEK293 parental cells.
2. Advanced DMEM/F12 medium, fetal bovine serum (FBS), 0.25% trypsin.
3. Accutase and trypsin.
4. Selection antibiotics: Hygromycin B, zeomycin, blasticidin S.

2.2 Molecular Reagents

1. pFRT-TO cDNA expression vector and pOG44 vector expressing FLP recombinase.
2. Doxycycline hyclate (prepare 1 mg/mL stock solution and store at −20 °C).
3. Opti-MEM media.
4. Lipofectamine 2000.

2.3 Recording Solutions

All solutions are prepared with ultrapure MilliQ water (18 MΩ-cm). Hygroscopic reagents must be kept in a salt desiccator container, as moist salts may lead to inadequate osmolarity, a key parameter in electrophysiology experiments. All solutions are vacuum filtrated with polyethersulfone (PES) 0.22 μm membrane and stored at 4 °C until use. The following solutions are used to analyze voltage-gated T-type calcium channels (*see* **Notes 1** and **2**).

1. Physiological extracellular solution (pECS): 10 mM HEPES, 140 mM NaCl, 5 mM glucose, 4 mM KCl, 2 mM $CaCl_2$, 1 mM $MgCl_2$, 295–305 mOsm, adjust to pH 7.4 with 1 N NaOH.
2. Internal recording solution: 20 mM EGTA, 10 mM HEPES, 50 mM CsCl, 10 mM NaCl, 60 mM CsF, 285 mOsm, adjust to pH 7.2 with 1 N CsOH.
3. External recording solution: 10 mM HEPES, 80 mM NaCl, 5 mM glucose, 60 mM NMDG, 4 mM KCl, 6 mM $CaCl_2$, 1 mM $MgCl_2$, 300 mOsm, adjust to pH 7.4 with 1 N HCl.
4. Seal enhancer: 10 mM HEPES, 80 mM NaCl, 5 mM glucose, 60 mM NMDG, 4 mM KCl, 10 mM $CaCl_2$, 1 mM $MgCl_2$, 310 mOsm, adjust to pH 7.4 with 1 N HCl.

2.4 SyncroPatch 384PE System

1. SyncroPatch 384PE automatic patching system (Nanion Technologies®) is able to form giga-ohm seals in every well of 384-well plates [8]. The high-resistance seal in each well enables precise voltage control in all 384 wells simultaneously, thereby providing high-quality biophysical analyses with throughput.
2. The patch system is further integrated with robotics for compound handling and solution exchanges, as well as a cell hotel to store cells prior to be transferred to the patch plate. The technical details of the SyncroPatch 384PE system have been described previously [8].

3 Methods

3.1 Establish and Validate Cell Lines

1. Maintain FlpIn-TREx 293 parental cells in DMEM/F12 medium supplemented with 10% FBS. For maintenance, seed 1–1.5 × 10^6 cells in one 10 cm culture dish with 9 mL medium in the presence of 100 μg/mL Zeomycin and 15 μg/mL Blasticidin. Passage cells with 0.25% trypsin (*see* **Note 3**).
2. To establish inducible cell lines, seed 3 × 10^5 FlpIn TREx 293 parental cells in each well of a 6-well culture dish with 2 mL cell culture medium in the absence of drug selection.
3. Two days after seeding, transfect cells (which should be approximately 75% confluent) with pFRT-TO vector containing the cDNA of the gene of interest and the pOG44 vector encoding FLP recombinase. Perform three separate transfections for establishing one cell line, each differing in the amount of the pFRT-TO vector (0.5, 1, or 2 μg).
4. For each transfection, mix 10 μL Lipofectamine into 150 μL of warm Opti-MEM serum-free medium and quickly vortex in one tube.

5. In a separate tube, add the DNA mix (pFRT-TO and 4.5 μg pOG44) to 150 μL of Opti-MEM, pulse-vortex five times, and incubate at room temperature for 5–10 min.
6. Combine the content of the two tubes by adding the DNA mix to the Lipofectamine mix, pulse-vortex five times, and incubate at room temperature for another 5–10 min.
7. Add the DNA complexes dropwise into each well and gently shake the plate to distribute.
8. Two days after transfection, replate the cells from each well to a 6 cm dish in the presence of 200 μg/mL hygromycin for selection. Change the medium every 3–4 days in the presence of the selection.
9. In a week, a few to 50 healthy colonies should be growing from one transfected well. The number of colonies grown largely depends on the size and potential cytotoxicity in the presence of the specific cDNA of the gene of interest.
10. A week after the colonies become visible, trypsinize, wash, and replate in a 6-well culture dish with hygromycin selection to establish a polyclonal cell line. Alternatively, you can pick and maintain clonal lines at this point if desired. As each cell is likely to incorporate a single copy of the cDNA in the FlpIn-TREx cellular system, polyclonal cells perform quite uniformly.
11. Grow and expand the polyclonal cells to a 10 cm dish with blasticidin and hygromycin selection. At this point, cells can be seeded for long-term cryo-storage in drug-free medium or tested for function.
12. To validate the established cell line, seed ~2×10^5 cells per well in a 6-well plate, and leave two wells un-induced for qPCR-negative control. Use genomic DNA analyses to validate the presence of single-copy cDNA insertion, qPCR to confirm the expression of cDNA, Western blotting for protein expression, and electrophysiology for functional validation. All inductions are done in the absence of antibiotics.

3.2 Induce Channel Expression and Harvest for Recordings

1. After establishing the cell line, proceed to induce expression and to harvest cells for recording.
2. In order to perform one recording of a 384-well plate on the SyncroPatch 384PE, 8–10 million cells are required at a 250,000 cells/mL density. One T175 flask of cells seeded at ~1.5×10^6 in the presence of 1 μg/mL doxycycline is typically used 3 days after induction at 50–75% confluence (*see* **Note 4**).
3. To harvest the cells from the T175 flask, rinse cells once with 5 mL of PBS, treat with 3 mL of Accutase (*see* **Note 5**) for 5 min at 37 °C, resuspend in 10 mL of serum-free medium, and pellet at 233 × *g* for 3 min at room temperature.

4. Aspirate the supernatant and resuspend cells in either 50%/50% serum-free medium:pECS (v/v) or pECS. The cell viability should be above 90% for best performance.
5. Add the cell suspension to the cell hotel of the SyncroPatch 384PE (pre-chilled to 10 °C when appropriate for specific channels) with gentle shaking at 100 rpm (*see* **Note 6**).
6. Keep the cells in the cell hotel prior to experiments in a 50 mL single-well reservoir or in a 4–6-well reservoir with 15 mL of solution in each well.
7. Cells may be used for electrophysiological analyses upon thawing from frozen stocks [9] directly. For this application, the thawing of frozen cells should be done quickly to ensure maximum survival rate.
   (a) Thaw the cells by incubating the frozen vial in a 37 °C water bath for 1–2 min. When the ice starts to dislocate from the wall, transfer the cells (with the ice) to a tube containing serum-free medium or external buffer (10 mL) at room temperature.
   (b) Mix well and centrifuge cells at low speed (~200 × *g*, 3–5 min), remove the supernatant, and resuspend gently with serum-free medium. With gentle rotation/shaking, incubate the cell mixture for 0.5–1 h to recover (at room temperature or 37 °C). Count cells and proceed to the assay.
8. Different automated patch platforms require different amount of cells. For example, the SyncroPatch 384PE requires >5 million cells for each run (>0.2 million/mL, >25 mL); Barracuda/Quattro system requires at least 10 million cells for each run (>2 million/mL, 5–7 mL); and Q-patch requires around 4 million cells for each run (2 million/mL, 2 mL). The cells used should always be 50–80% confluent at harvest regardless of the platform used.

3.3 Stimulus Protocols for Phenotypic Characterization of Ion Channels

The principle for designing stimulus (voltage or ligand) protocols on automated patch clamp instruments is to balance the clarity and amount of information collected with the assay performance (*see* **Note 7**). Protocols for voltage-gated channels are discussed in this chapter, with a few notes on the ligand-gated channels that can be studied efficiently using the SyncroPatch 384PE system (*see* **Notes 8–11**). We discuss several considerations for stimulus protocols using the voltage-gated calcium channel $Ca_V3.3$ as an example.

3.3.1 Define the Holding Potential

We have defined the holding potential for $Ca_V3.3$ at −100 mV, to balance the recording stability and channel availability. $Ca_V3.3$ belongs to the family of low-voltage-activated Ca^{2+} channels, whose activation and inactivation processes happen at more hyperpolarized potential compared with other voltage-gated Ca^{2+} channels. Thus, it is important to define a holding potential negative enough to remove most of the channels from the inactivated state, prior to the application of the test voltage to open the channel. On the other hand, holding cells at very negatives potentials for prolonged periods of time might reduce the stability of the seal during the length of the experiment.

3.3.2 Determine Steady-State Parameters

One efficient way to assay steady-state parameters is to use a dual-pulse protocol to recover voltage dependence of both activation and inactivation parameters. An example dual-pulse protocol and its corresponding $Ca_V3.3$ channel current traces obtained from a SyncroPatch 384PE recording plate are shown in Fig. 1a. From this protocol, we can generate both voltage dependence of activation (G/V) plots (not shown) and steady-state voltage dependence of inactivation plots (Fig. 1b).

3.3.3 Generate the Current Voltage (I–V) Relationship Plot for the Voltage Dependence of Activation

1. Define the peak current amplitude from the episodic voltage stimulus for each test pulse.
2. Divide the peak current amplitude at each test pulse by the cell's capacitance to produce ionic current density.
3. Plot each current density versus each episodic voltage applied (V).
4. Establish the reversal potential (V_r) by finding the intersection in the V-axis.
5. Obtain the conductance (G) by dividing the peak current density by the driving force ($V - V_r$).
6. Normalize all the conductance (G) values by the maximum value (G_{max}).
7. G/G_{max} plot should be well described by single Boltzmann function, where the fitted parameters correspond to the slope ($z\delta$) and the midpoint or V_{half} (V_h; *see* **Note 12**).

3.3.4 Generate the Current Voltage (I–V) Relationship Plot for the Voltage Dependence of Inactivation

1. Measure the inward current amplitude induced by a short test pulse, typically elicited at a voltage that is 5–10 mV more depolarized to the voltage that produces the maximum current amplitude. For $Ca_V3.3$, −20 mV was used in this system.
2. Divide the current amplitude by the cell's capacitance to obtain current density.
3. Plot each current density versus the voltage applied (V) during the long pre-pulse to induce inactivation.
4. Normalize each current density (I) by the maximum current density (I_{max}).

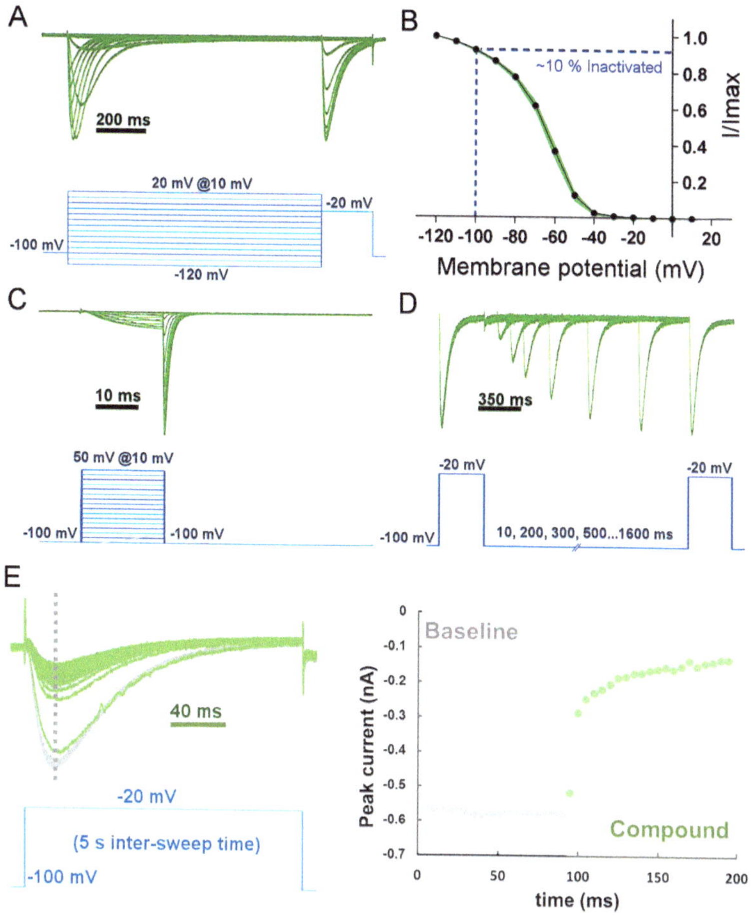

Fig. 1 Sample experimental protocols: (**a**) $Ca_V3.3$ current traces (top) and voltage protocol (bottom) to obtain steady-state activation and inactivation readouts. The voltage protocol includes (1) 200 ms at −100 mV holding potential, (2) 1000 ms voltage steps (Δ10 mV), from −120 mV up to +40 mV (pre-pulse); (3) 200-ms test depolarization to −20 mV, inter-sweep interval is 5 s at −100 mV. (**b**) Voltage-dependent steady-state inactivation for $Ca_V3.3$. (**c**) $Ca_V3.3$ current traces (top) and voltage protocol (bottom) for tail currents. The voltage protocol includes (1) 100 ms at −100 mV holding potential, (2) 25-ms voltage steps (Δ10 mV), from −120 mV up to +40 mV (pre-pulse), (3) 200 ms at −100 mV (test pulse), and (4) interval of 1 s between each sweep, at −100 mV. For analysis, normalize the tail peak current by the maximum current elicited (G/G_{max}), and plot the normalized values (G/G_{max}) against the voltage from the pre-pulse. This relationship should be well described by a Boltzmann function. (**d**) $Ca_V3.3$ current traces (top) and voltage protocol (bottom) to study the recovery from inactivation. The voltage protocol includes (1) 100 ms at −100 mV holding potential, (2) 350 ms at −20 mV (test pulse 1), (3) variable interval 10 ms (Δ ms), and (4) 350 ms test depolarization at −20 mV. For analysis, normalize peak current from test pulse 2 with the peak current from test pulse 1 and plot the fraction (P2/P1) versus the time between the two pulses. This relationship should be well described by an exponential function. (**e**) Voltage protocol for pharmacological screening. Left: $Ca_V3.3$ current traces (top) in response to the voltage protocol in blue (bottom). Gray and green traces were obtained in the presence of DMSO first and then at single concentration of a $Ca_V3.3$ blocker. The voltage protocol consists of 200 ms depolarization at −20 mV, from a holding potential of −100 mV. The inter-sweep time was set to be 10 or 20 s. Right: The peak current density (amplitude at dashed line) was monitored. Baseline (grey dots) is obtained in DMSO treatment. After the addition of the compound (green dots, 10 μM), the peak current is reduced from −600 pA to −100 pA

5. I/I_{max} should be described by a single Boltzmann function, where the fitted parameters correspond to the slope ($z\delta$) and the midpoint or V_{half} (V_h; *see* Fig. 1 and **Note 12**).

3.3.5 Tail Current Protocol

While the steady-state activation parameters can be extracted from the *I–V* protocol described above, the most appropriate way to obtain the activation dependence on voltage is to measure the tail current amplitudes at a constant driving force, using a "tail-current protocol." This protocol opens the channel briefly and repolarizes the cell before the inactivation process starts. This is a way to measure the proportion of channels in the open state at the end of the pre-pulse (G/G_{max}). A tail current protocol and its corresponding $Ca_V3.3$ current traces obtained from a SyncroPatch384PE recording are shown in Fig. 1c.

3.3.6 Kinetic Parameter Determination

Some of the classic parameters for channel kinetics can be recovered from current traces described in Fig. 1a–c. For instance, the protocol described in Fig. 1a provides the time constant associated with channel opening ($\tau_{activation}$) and inactivation ($\tau_{inactivation}$) by fitting a bi-exponential function to the current trace during the test pulse. In a similar way, the time constant associated with the closing of the channel can be described by an exponential decay function fitted to the tail protocol (Fig. 1c). A protocol specific to describe how the channel recovers from inactivation consists of a pair of test pulses to −20 mV with an interval of variable time length between the two pulses. The voltage stimulus protocol to obtain the recovery from inactivation and the corresponding $Ca_V3.3$ current traces are shown in Fig. 1d.

3.3.7 Pharmacology and EC_{50}/IC_{50}

During high-throughput screening and/or secondary validation assays, a simple and robust voltage protocol is preferred to measure aspect(s) of channel function that is monitored before and after chemical compound treatment. Most phenotypic screening of ion channels by chemical modulators is geared to identify either inhibitors or enhancers. To accomplish this goal, most protocols are designed to capture the change in the peak current density or area under the curve at a given voltage. An example pharmacology protocol for $Ca_V3.3$ is presented below (Fig. 1e).

1. Hold cells at −100 mV and depolarize to −20 mV briefly (100 ms). Repeat this depolarization every 20 s.
2. After the baseline stabilizes, add compounds to each well and continue the protocol for another 5–10 min.
3. The fold change in the peak current density before and after compound addition is calculated and compared to the change in wells treated with DMSO controls. Those compounds that increase or decrease the current density or the total amount of calcium influx are selected for further analyses.

4. In addition to directly blocking the ionic conductance, compounds may alter how channels respond to depolarization and therefore demonstrate an increase or a decrease in the current density preferentially at a given voltage or channel state. In order to differentiate these possibilities and address the mechanism of action of any particular chemical modulator, the steady-state and kinetic parameters can be extracted (Fig. 1a–d).

3.4 Recording Procedure for the SyncroPatch 384PE

1. Inspect the cells to be harvested. The confluence should be between 50 and 80%.
2. Turn on the SyncroPatch 384PE, the vacuum line, and the water pump to prime the washing station and remove air bubbles in the tubing.
3. Turn on the temperature control and the shaker for the cell hotel.
4. Flush and prime the internal solution perfusion system with deionized water. The internal solution should be at room temperature; otherwise, air bubbles may form during the experiment.
5. Prepare and set up solution reservoirs for the sealing enhancer (50 mL), pECS (100 mL), and reference/external recording solution (150 mL). Fill the solvent boat with 50% ethanol and 2 mM EDTA (pH 8) for tip washing (for ligand-gated channels, omit ethanol to avoid problems with stacked liquid pipetting). Load a new tip box.
6. Prepare compound plates. One to six compounds plates can be accommodated on the Biomek robotic platform.
7. Establish a SyncroPatch 384PE experimental tree that details each experiment with sequential steps in pressure control, robotics control, patch control, recording quality control, time control, voltage control, and setup and cleanup logistic sequences.
8. For pressure control, set pressures during the cell catch to approximately −300 mbar, −100 mbar for catching the cell in each well, −250 mbar for break-in whole-cell configuration, and ambient or slightly more negative pressure during the experiment.
9. Inspect all stimulus protocols used for the experiment run.
10. Inspect liquid-handling protocols including solution-exchange procedures.
11. Inspect the system flow for quality check, tip control, and solution-exchange sequences.

12. Home all axes for the robotics without pipette tips on liquid-handling manifolds.
13. Harvest the cells (*see* Subheading 3.2), resuspend and transfer the cells to the precooled cell hotel, and let cells recover for 10–15 min. Cells can be used for recording in the subsequent 1–3 h with shaking.
14. Load a new patch plate and remove lids of all plates.
15. Click RUN in the control panel to execute the experimental tree. Each run consumes about 100 mL of internal solution, 20 mL of pECS, 20 mL of seal enhancer solution, 50–150 mL of external recording buffer (depending on the number of solution exchanges), and 20 mL of washing solution.
16. At the end of each day, cleanup includes emptying and flushing the internal perfusion system with 70% ethanol. Remove all buffer boats/reservoirs; rinse with water, deionized water, and ethanol sequentially; and permit them to air-dry.
17. Unclamp the pump to drain the water from tubing. Turn off the HV1 water pump, turn off the vacuum-line valve, and turn off the cooler and shaker for the cell hotel before turning off the instrument.

3.5 Automated Electrophysiology Workflow

3.5.1 Solution Exchanges

It is important to exchange solutions during the experiment, optimally before every new voltage protocol to reset the ionic conditions for each measurement, if there is no continuous solution flow during the experiment. Ion accumulation/depletion through ion influx in the immediate vicinity of the plasma membrane may change the ionic driving force, thereby altering channel behavior. This effect can be worse in automated platforms, where the extracellular volume is less than 100 μL and the currents recorded can be as large as several nanoamperes. Figure 2 illustrates the experimental design for the most common applications on the SyncroPatch 384PE.

3.5.2 Establish QC Parameters During Setup Phase (Fig. 2, Top Row)

A key step is to establish quality control (QC) parameters at the end of each step, which permits tractability and control during the experiment. For instance, if seal resistance increases after a solution exchange, it will be discovered immediately by the altered QC parameters, facilitating the assay development. The other crucial role for defining the QC parameters is during the analysis process, as using these parameters to filter artifacts and control the quality of the recorded data is crucial for quality results (Fig. 3).

1. Seal resistance indicates the quality of the patch seal. Typically, remove recordings with less than 500 MΩ from analysis.
2. Capacitance can be a useful parameter to discriminate empty wells and deficient cells. Disregard anything less than 5–7 pF or over 35 pF for recordings of HEK293 cells.

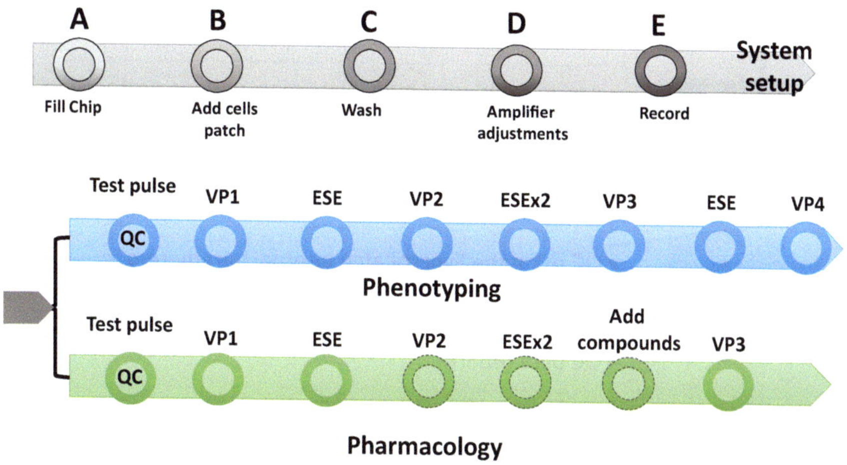

Fig. 2 Procedures for setting up the SyncroPatch384 PE system for an experiment. Experimental steps in this phase include the following: (**a**) (1) create a new experiment file, (2) add external physiological solution, (3) perfuse internal solution chamber, and (4) measure QC resistance (contact success rate); (**b**) (1) junction potential/Cfast correction, (2) cell addition (250,000 cells/mL), (3) add seal enhancer, (4) stacked seal enhancer, (5) QC resistance (seal success rate); (**c**) (1) add external recording solution (three times), (2) measure QC parameters, (3) go to whole-cell configuration, (4) report on seal resistance (establishing whole-cell patch success rate); D set holding potential, (2) Cslow compensation, (3) series resistance compensation. (**d**) Check the experimental tree and run the experiment. The two bottom rows indicate the typical procedures for two types of experiments after the initial setup. *VP* Voltage protocol, *ESE* external solution exchange. In the blue experiments, after the setup, a test pulse is delivered to measure the seal and capacitance QC, followed by four different voltage protocols (VP1, VP2, VP3, and VP4) to sequentially probe for channel properties, including steady-state activation, inactivation, tail protocol, and recovery from inactivation, and to systematically extract the maximum amount of information within a SyncroPatch 384PE run (20–30 min). The green row illustrates the procedure for a pharmacology experiment that starts with a test pulse to measure the seal and capacitance QC, followed by a current voltage relationship protocol, and a simple test pulse every 20 s to evaluate the current density before and after the compound treatment. The dotted circle indicates the voltage protocol lasted through compound addition

3. Access resistance is used to define wells with optimal voltage clamp properties, which can be very useful when evaluating voltage control (clamp) quality and analyzing *I–V* and *G–V* relationships.
4. Utilize online marking tools to define parameters to be extracted during recordings. The most common parameter to follow compound action on ion channels is the peak current amplitude, and this parameter can be followed online while experiments are ongoing and can be used as a simple way for baseline control during analyses.

3.5.3 Experimental Phase

As shown in Fig. 2, the blue and green processes represent two typical types of experiments after the initial setup phase. The blue experiment is biased toward recording through many different voltage protocols to maximize the amount of the information in

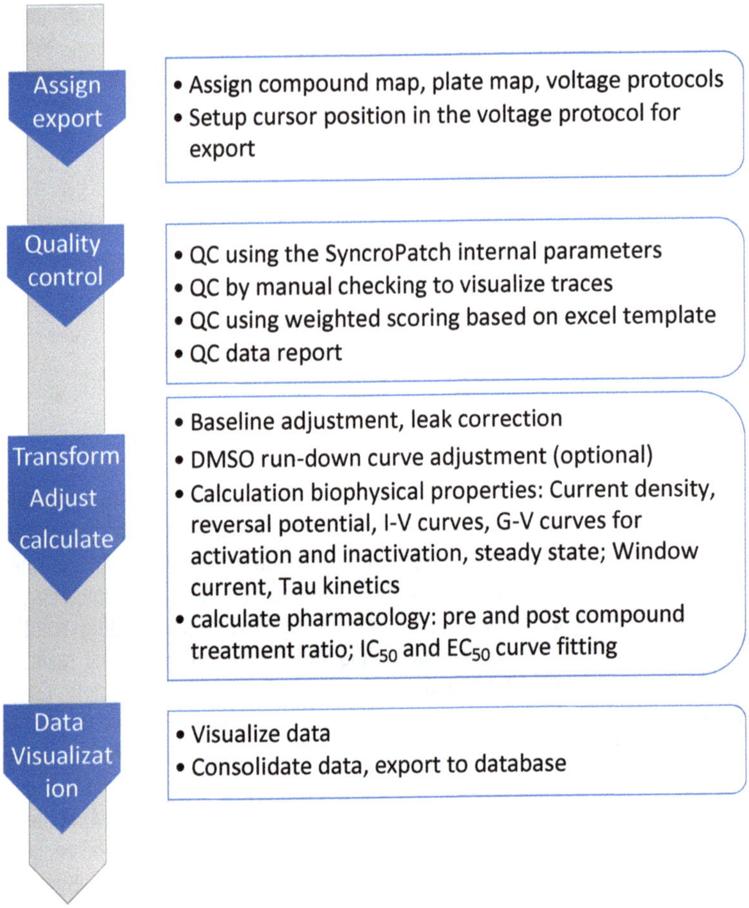

Fig. 3 Principles to filter, normalize, filter, and visualize the data obtained from the SyncroPatch 384PE system

probing channel physiology and is useful for investigating the functional impact of many genetic coding variations simultaneously. In comparison, the green experiment takes advantage of a robust readout and uses a simple voltage protocol continuously to measure the effect of compound addition, with adequate QC.

1. When studying multiple variants, we typically include 5–20 channel variants (16–64 wells for each variant) on one 384-well plate. It is critical to harvest the cells at similar 50–80% confluence to ensure consistency, as not all cell lines grow at the same rate.
2. Establishing a consistent baseline is critical to study the effect of compound treatment.
3. Ensure the same volume of solvent per well during the experiment.
4. Include solution exchanges in the baseline, to avoid artifacts in the recording as a result of the solution-exchange process.

5. Design compound plates that contain both negative and positive control treatments, located at different positions within the plate, to avoid plate location-associated artifacts.
6. For calcium channels, rundown is a particular problem in controlling the baseline (*see* **Note 13**).

3.6 Data Analysis

Depending on the purpose of the assay, the amount of data that can be extracted from a SyncroPatch 384PE experiment varies greatly. A direct pharmacology study of channel inhibitors may only need to extract two data points per well (average peak current amplitude before and after the compound treatment), while a systematic study of the channel biophysical properties, such as current-voltage relationship and channel decay kinetics, and the effects of pharmacological reagents on those properties, could easily generate 200 K data points/patch plate. Handling such large datasets requires specialized strategy for QC, data normalization, feature extraction, curve fitting, and data visualization. Although high-throughput patch-clamp systems, including the SyncroPatch384PE, provide software support for achieving some of the above goals, we still find that stringent data QC and more sophisticated analysis, such as curve fitting, EC_{50} calculation, or channel kinetics analysis (e.g., $\tau_{activation}$ and $\tau_{inactivation}$), are necessary for complex phenotyping purposes.

Establishing appropriate QC parameters *prior* to performing experiments is critical for effective data analysis (*see* **Notes 14** and **15**). Figure 3 explains the principles to filter, normalize, and visualize data obtained from the SyncroPatch3 84PE.

1. Use the established QC parameters during the experiment as the first filter to discard the data with suboptimal quality.
2. Quickly review the data manually as the second filter.
3. Export the data in a consistent manner, for use with Microsoft Excel templates, custom-built software, or other analysis solutions.
4. For the third filtering step, we often use a Microsoft Excel template that scores the quality of recording based on the weighted criteria including (1) seal resistance (>500 MΩ); (2) minimum acceptable current density and maximum acceptable leak currents that often depend on the channel types and the expression levels; (3) cell capacitance; (4) voltage control; and (5) stable baseline. A perfect recording scores 10.
5. In addition to QC control, the same Excel template also contains information including (1) the compound plate map and the recording plate map; (2) calculation of the baseline readjustment from the leak current; (3) calculation of the reversal potential and plotting *I–V* and *G–V* curves; and (4) calculation of and plotting IC_{50} and EC_{50} upon compound treatment.

6. Additional parameters may be directly extracted from the SyncroPatch 384PE that includes current amplitude, rise/decay time, and area under the curve. These parameters can all be incorporated in this summary Excel template or other analytic tools for analyses. A software solution and/or tool that integrates such a template with the automated SyncroPatch data acquisition may be developed in the future to streamline analyses and reporting.

4 Notes

1. Choice of buffer: Optimal buffer compositions maximize the signal and reduce the background. For example, the use of cesium to replace potassium blocks background potassium conductance in HEK293 and CHO cells. Certain buffers may help form a tight seal between the glass and the cell membrane, e.g., solutions containing NMDG or gluconate. An intracellular fluoride-containing buffer, together with a calcium-containing extracellular buffer or high calcium seal-enhancing solution, will help seal formation and stability. Such buffer combinations are often used in high-throughput patch clamping, wherein they are replaced by a regular external solution after seal formation. Typically, osmolarity for the intracellular solution is 290 mOsm and extracellular buffer is 300 mOsm.
2. Freshly prepared extracellular buffer is often preferred for planar patch clamp. For the SyncroPatch 384PE, cold (4 °C) buffer often works better. For other platforms, buffer may be warmed up and degassed (with filtering, negative pressure, or stirring) before the experiment. Do not reuse the buffer. Spot-check the buffer for correct pH and osmolarity. Adjust the pH if necessary. Supplement glucose if the osmolarity is lower than optimal.
3. The passage number of parental FlpIn TREx 293 cells should be kept below 10 before establishing new lines. Frozen cells should never be thawed directly into drug selection, and cells plated for the purpose of freezing should not be plated into drug selection. Freeze cells in serum-supplemented medium with 10% DMSO. Split cells every 3–4 days, and do not change the medium between passages.
4. Doxycycline can be added at the same time when plating cells or 24 h after cells have attached. Medium can be changed every 48 h if desired but is not necessary for robust induction of $Ca_V3.3$ and several other channels that we have tested. Three days post-induction is a robust time point for most experiments in our experience, but evaluating channel expression and function over a time course from 1 to 5 days upon induction is

recommended for best results. Do not overgrow the cells, as their membranes may become fragile and less suitable for planar patch. Frozen cells can also be used (*see* **Note 2**).

5. Alternatively, other gentle detachment reagents such as Detachin™ or TrypLE™ can be used to detach the cell. For calcium-sensitive channels like NMDA receptors, it is advisable to use Ca^{2+}/Mg^{2+}-free solutions.
6. Use cells immediately for IonWorks Barracuda/Quattro. For the SyncroPatch384PE, it is suggested to equilibrate the cells for 15–20 min in serum-free medium before patch clamping.
7. For mechanism of action studies, larger number of cells (N) and multiple protocols are utilized for exploring channel physiology. When we phenotype a large number of ion channel coding variants simultaneously, the experiment is designed to cover as many parameters as possible, as it is difficult to know a priori which may be affected by any specific amino acid substitution. Therefore, the stimulus protocols for such studies are designed to obtain as rich information as possible in a short period of time within an automated patching experiment.
8. To record ligand-gated channel, fluidics is a key consideration. To estimate the flow rate in the fluidic control, cell-expressing Kir channels can be used in the presence of high K^+ to trigger inward Kir currents, which reflect the fluid exchange speed, as a way to determine the fluidic flow speed [5].
9. Applying additional negative pressure to hold the cell tightly to the patch plate can resist the mechanical liquid flow and increase the rate of success in recording. For example, in order to study cells expressing NMDA receptors, cells are held at negative voltage, e.g., −60 mV, with zero Mg^{2+} in the recording solution, with a high holding pressure, typically the maximum permitted by the system.
10. Co-agonists often increase the robustness of the assay. For example, glycine or D-serine may be preincubated with cells expressing NMDA receptors, prior to the application of glutamate, to activate the channel and maximize the signal-to-noise ratio.
11. The SyncroPatch 384PE utilizes a liquid stacking approach to incorporate ligand (3 μL) at the tip of the pipette tip followed by a washing solution (37 μL). The kinetics of ligand-off by washing is feasible.
12. The V_h of activation parameter is the amount of energy necessary to open the channel half of the time. Leftward shifts in this parameter imply that the mutant channel requires less energy to open. On the other hand, rightward shifts imply greater voltage to reach the same amount of activation:

$$\frac{G}{G_{\max}} = \frac{1}{1 + e^{-\frac{z\delta F\left(V - V_{h_{\text{inact}}}\right)}{RT}}}$$

The opposite is true when it comes to the inactivation process: leftward shifts in the V_h of inactivation mean less energy is necessary to drive the channels into the inactivated state, while rightward shifts in this parameter mean more energy is required to inactivate the channel:

$$\frac{I}{I_{\max}} = \frac{1}{1 + e^{\frac{z\delta F\left(V - V_{h_{\text{inact}}}\right)}{RT}}}$$

The "window current" is a useful visualization of the interplay between these two curves and can be used to identify phenotypes caused by genetic variation. Loss-of-function (LoF) mutants will lead to narrower window currents (rightward shift in activation and/or leftward shift in inactivation), while gain-of-function (GoF) mutants will entail larger window currents (leftward shift in activation and/or rightward shift in inactivation).

13. To prevent run down, perforated patch (using amphotericin or escin) is often used and easily configured in the SyncroPatch384PE. 5 mg/mL Escin stock solution is prepared fresh in sterile water. Do not filter, vortex vigorously, and heat in 37 °C water bath for 10 min to aid solubility if necessary. The desired final concentration in the internal solution is 5–15 ug/mL. In the perforated configuration, escin is applied during the chip priming (Fig. 2) step. One advantage of using escin is that the seal is stable throughout the experiment, with no need to wash out the detergent after the seal is formed. The second, and most important, advantage is that planar patch under perforated conditions permits the same access resistance (Ra) as the whole-cell configuration (<10 MΩ), demonstrating excellent electrical access during the experiment. In addition to the perforated patch, the following optimization can be utilized to control rundown of Ca^{2+} channels:

    (a) Utilize Ba^{2+} as the charge carrier instead of Ca^{2+}

    (b) Utilize BAPTA instead of EGTA to chelate Ca^{2+} intracellularly

    (c) Apply 2–4 mM ATP in the intracellular buffer, internal solution exchange protocol may be used to reduce the ATP consumption. It usually takes 25ml ATP containing solution for internal solution exchange

14. Without the necessary automation, data analysis could take significant time compared to the experiment procedure. Appropriate quality control (QC) criteria help improve signal over noise, and are critical for data analysis. The SyncroPatch 384PE software (Data control) provides some analysis functionalities and exports various parameters including QC. Customized Excel-based templates and XLfit (Excel based data analysis package) could process data with certain automation and manual interaction. R code or MATLAB-based customized scripts may provide more comprehensive analysis for fixed data formats. Spotfire can be used to visualize data generated at the individual plate level. For data analyses for multiple plates at the screen level, Genedata APC may be useful to calculate, normalize, and store large datasets with compound information.
15. The following QC log from a reasonable quality run consists of recordings of cells expressing $Ca_V3.3$ in a 384-well plate where each well contains a single hole. Cells were kept in the cell hotel for about 1 h before the experiment began.

(% success rate)	Time (min)	Procedure and QC criteria
98%	1	Priming (R (resistance) = 4–5 MΩ)
91%	2	Cell contact ($R > 10$ MΩ)
91%	4	Seal enhancer (SE) addition ($R > 100$ MΩ)
80%	5	SE stacking ($R > 500$ MΩ)
77%	6	SE wash 1 ($R > 500$ MΩ)
79%	8	SE wash 2 ($R > 500$ MΩ)
81%	10	Cell perforation
83%	13	Establish whole cell, apply first pulse train ($R > 500$ MΩ)
85%	17	DMSO addition followed by a second pulse train ($R > 500$ MΩ)
65%	21	First compound addition and incubation (5 min) followed by a third pulse train ($R > 500$ MΩ)
59%	25	Second compound addition (with same or higher concentration) followed by a forth pulse train ($R > 500$ MΩ)
	30	Cleanup

References

1. Gadsby DC, Vergani P, Csanady L (2006) The ABC protein turned chloride channel whose failure causes cystic fibrosis. Nature 440:477–483
2. Szeliga MA, Hedley PL, Green CP, Moller DV, Christiansen M (2010) Long QT syndrome: a genetic cardiac channelopathy. Kardiol Pol 68:575–583
3. Litan A, Langhans SA (2015) Cancer as a channelopathy: ion channels and pumps in tumor development and progression. Front Cell Neurosci 9:86
4. Cai B, Chen X, Liu F, Li J, Gu L, Liu JR et al (2014) A cell-based functional assay using a green fluorescent protein-based calcium indicator dCys-GCaMP. Assay Drug Dev Technol 12:342–351
5. Kim JG, Park SW, Byun D, Choi WS, Sung DJ, Shin KC et al (2016) Fluid flow facilitates inward rectifier K+ current by convectively restoring [K+] at the cell membrane surface. Sci Rep 6:39585
6. Beacham DW, Blackmer T, OG M, Hanson GT (2010) Cell-based potassium ion channel screening using the FluxOR assay. J Biomol Screen 15:441–446
7. Milligan CJ, Moller C (2013) Automated planar patch-clamp. Methods Mol Biol 998:171–187
8. Obergrussberger A, Bruggemann A, Goetze TA, Rapedius M, Haarmann C, Rinke I et al (2016) Automated patch clamp meets high-throughput screening: 384 Cells recorded in parallel on a planar patch clamp module. J Lab Autom 21:779–793
9. Polonchuk L (2014) Industrializing electrophysiology: HT automated patch clamp on SyncroPatch(R) 96 using instant frozen cells. Methods Mol Biol 1183:81–92

Chapter 19

Testing Susceptibility of Patient-Derived Organoid Cultures to Therapies: Pharmacotyping

Richard A. Burkhart, Lindsey A. Baker, and Hervé Tiriac

Abstract

Increasingly, patient models of disease are being utilized to facilitate precision medicine approaches through molecular characterization or direct chemotherapeutic testing. Organoids, 3-dimensional (3D) cultures of neoplastic cells derived from primary tumor specimens, represent an ideal platform for these types of studies because benchtop protocols previously developed for 2-dimensional cell lines can be adapted for use. These protocols include directly testing the survival of these organoid cultures when exposed to clinically relevant chemotherapeutic agents, a process we have called pharmacotyping. In this protocol, established tumor-derived organoid cultures are dissociated into single cells, plated in a 3D gel matrix, and exposed to pharmacologic agents. While our protocol has been developed for use with patient-derived pancreatic ductal adenocarcinoma organoids, with minor modifications to the dissociation and medium conditions, this protocol could be adapted for use with a wide range of organoid cultures. We further describe our standard ATP-based assay to determine cellular survival. This protocol can be scaled for use in high-throughput assays.

Key words 3D culture, Organoids, Pharmacotyping, Drug testing, Tumor models

1 Introduction

The introduction of precision medicine approaches in oncology has revolutionized the treatment for patients with certain tumor types. For over 250,000 women diagnosed with breast cancer annually in the United States, for example, identification and targeted treatment based on hormone receptor status is a cornerstone of modern management [1]. Similarly, in gastrointestinal stromal cell tumors, the discovery and targeting of an activating mutation in the C-kit gene has led to dramatic improvements in prognosis [2]. Despite these successes, there are several difficult-to-treat tumors for which we have failed to make progress toward a precision approach to therapeutics.

Pancreatic ductal adenocarcinoma (PDAC) is one example of a difficult-to-treat tumor. Affecting nearly 50,000 individuals in the

Bridget Wagner (ed.), *Phenotypic Screening: Methods and Protocols*, Methods in Molecular Biology, vol. 1787,
https://doi.org/10.1007/978-1-4939-7847-2_19,

United States annually, a diagnosis of PDAC is almost uniformly fatal [1]. Much like gastrointestinal stromal cell tumors, PDAC tumors harbor a relatively bland mutational profile, with a single mutated gene (*KRAS*) serving as the hallmark for almost all tumors [3]. Despite decades of research on *KRAS*, and other genes frequently mutated in PDAC, we remain unable to utilize genetic information in the majority of patients to guide therapy. In fact, contemporary work from a large patient cohort undergoing routine prospective genetic sequencing identified clinically actionable mutations in fewer than 10% of patients with this disease [4]. As a result, clinical scientists are considering alternative approaches to guide precision medicine initiatives in difficult-to-treat diseases such as PDAC.

One alternative to genetic analysis includes rapid patient-derived tumor modeling and comprehensive molecular analysis. Tumor modeling has been proposed both via patient-derived xenograft development and via the ex vivo culturing of a patient's tumor cells [5]. Recently, a new cell culture technique has made the latter option more attractive [6]. Organoid cultures, consisting of primary tumor cells established and propagated in a 3-dimensional (3D) matrix, have shown to be a robust tool for phenotypic screening. The propagation of each patient's tumor as a unique organoid culture facilitates molecular analyses using a wide variety of methods developed for traditional molecular biology research. These include sequencing efforts, gene expression analysis, and more novel "omics" analyses. Perhaps even more notable, these patient-derived tumor lines can be tested directly for susceptibility against a variety of therapeutics in a molecular characterization assay that we have termed "pharmacotyping."

Pharmacotyping involves the process of testing an established organoid line against one or more potential therapies, such as commonly used cytotoxic chemotherapeutics, approved targeted agents, or novel small-molecule inhibitors. This testing can be done against single agents, or scaled up for use in high-throughput assays in which panels of agents are tested. The first steps in this methodology include the dissociation of cells into a single-cell solution and culture of equal numbers of cells into each well of a multi-well dish. Cells are allowed to recover from passaging before being exposed to therapeutics. After an incubation period of several days, cell viability can be assessed using a variety of methods.

2 Materials

Prepare all materials prior to the initiation of pharmacotyping. Organoids should be cultured in the appropriate conditions so they maintain optimal proliferation and viability (*see* **Note 1**). Established organoid cultures are ready to be dissociated for pharmacotyping when the organoids have become confluent inside the 50 μL domes of Matrigel in which they are propagated (*see* **Notes 2** and **3**). The average number of human organoid cells that can be isolated from one well of a 24-well plate varies, but is typically around 50,000–100,000 cells. We recommend harvesting at least half of a 24-well plate of organoids to maximize the number of viable cells available for therapeutics.

2.1 Generation of Single Cells from Established Organoid Cultures

1. Organoid splitting medium: Advanced DMEM/F12, 10 mM HEPES, Glutamax 1×, 100 μg/mL Primocin.
2. Human complete feeding medium: Advanced DMEM/F12, 10 mM HEPES, Glutamax 1×, 100 μg/mL Primocin, 500 nM A83-01, 50 ng/mL hEGF, 100 ng/mL mNoggin, 100 ng/mL hFGF10, 0.01 μM hGastrin I, 1.25 mM *N*-acetylcysteine, 10 mM nicotinamide, B27 supplement (1× final), R-spondin1 conditioned medium (10% final), Wnt3A conditioned medium (50% final).
3. TrypLE Express.
4. Dispase.
5. DNase solution (10 mg/mL stock in sterile DPBS).
6. LoBind protein tubes (1.5 and 5 mL).
7. 15 or 50 mL conical tube.
8. Pipetman (P2, P20, P200, P1000) and sterile filtered tips.
9. Digital scale.
10. Refrigerated 15/50 mL tube centrifuge and microfuge.
11. 37 °C Rocking/rotating incubator.
12. Tissue culture hood with aspirator.
13. Microscope.
14. Cell counting chambers and solution.

2.2 Cell Counting, Plating, and Assaying Cells in a Multi-Well Format

1. Organoid splitting medium.
2. Human complete feeding medium.
3. Rho kinase inhibitor (Y-27632: 10.5 mM stock in ddH_2O).
4. Growth factor-reduced Matrigel, phenol red free (on ice).
5. Sterile phosphate-buffered saline.
6. 15 mL Conical tube.
7. Sterile basin for multichannel pipette.
8. Pipetman (P2, P20, P200, P1000, single- and multichannel) and sterile tips.
9. White 384-well plate (or a plate of your choosing for multi-well experiments; ultralow attachment is preferred).

10. Clear 384-well plate (for visual inspection and quality control, ultralow attachment is preferred).
11. 37 °C Rocking/rotating incubator.
12. Cell Titer-Glo or preferred cell viability assay.
13. Plate reader with analysis software.
14. Microscope.

3 Methods

3.1 Generation of Single Cells from Established Organoid Cultures

1. Measure out dispase in a 50 mL conical tube. For each well of organoids you plan to harvest for study, you will require approximately 2 mg of dispase. We typically measure at least 20 mg to ease creation of the solution.
2. Transport tube of dispase to tissue culture hood. Sterilize tube prior to introducing into the hood.
3. Add room-temperature splitting medium to the dispase to bring it to a final concentration of 2 mg/mL. Invert the tube to ensure that dispase is suspended into solution.
4. Obtain organoids from the incubator, transfer to cell culture hood, wipe down culture dish with alcohol-moistened towelette, and introduce culture dish into the incubator.
5. Remove medium overlaying the organoid wells to be harvested for experimentation using an aspirator glass pipette. Leave the Matrigel domes adherent to the plastic during medium aspiration.
6. Add 500 μL of dispase solution to each well, and pipette up and down with a P1000 pipetman and a sterile tip to break up each Matrigel dome (*see* **Note 4**).
7. Incubate plate in the tissue culture incubator (37 °C, 5% CO_2, 20% O_2) for 10 min.
8. Use P1000 pipette to transfer dispase/organoid mixture to one or more LoBind 1.5 or 5 mL tube(s).
9. Use 0.5–1 mL of dispase to wash the culture wells and free any remaining cells and Matrigel from the plastic. Transfer this wash into the tubes with the dispase/organoid mixture.
10. Add 1–5 μL of DNase I stock solution (stock concentration 10 mg/mL) to the dispase/organoid mixture, to reach a final concentration of 10 μL/mL (*see* **Note 5**).
11. Incubate at 37 °C with rocking or rotation for 5 min. Remove and inspect solution quickly for single cells. If cells are still in groups of three or more, place back onto incubating rocker for five additional minutes.

12. Centrifuge tube at 200 × *g* for 5 min at 4 °C.
13. Remove the supernatant leaving the cell pellet at the base of the tube.
14. Wash the cells once in 1 mL TrypLE Express, inspect under the microscope for dissociation into single cells, and centrifuge tube at 200 × *g* for 5 min at 4 °C.
15. Remove the supernatant leaving the cell pellet at the base of the tube.
16. If clumps of cells were present during **step 14**, add 1–4 mL of TrypLE Express and 2–8 μL of DNase stock solution (to a final concentration of 10 μL/mL) to the cell pellet. The size of the cell pellet will dictate the volume of TrypLE Express to add. Pipette up and down gently approximately 20 times to suspend the cells in solution, and incubate at 37 °C with rocking or rotation for 5 min (*see* **Note 6**).
17. Inspect the solution under the microscope to evaluate for cell dissociation into a single-cell solution. If cells are not dissociated, repeat **steps 12–16** until a single-cell solution is obtained (*see* **Note 7**).
18. Centrifuge tube at 200 × *g* for 5 min at 4 °C.
19. Remove the supernatant leaving the cell pellet at the base of the tube.
20. Wash cells with splitting medium (+++), resuspending in 1–4 mL of medium, pipetting gently up and down to suspend cells in solution, and repeating a centrifuge at 200 × *g* for 5 min at 4 °C.
21. Resuspend the single-cell solution in human complete medium, supplemented with Rho kinase inhibitor.
22. Pipette this solution well dispersing the cells evenly throughout solution, take a small representative aliquot, and count the cells using a hemocytometer or automated cell counter.

3.2 Plating in Multi-Well Format and Pharmacotyping

3.2.1 Preparations for Plating

1. We typically pharmacotype lines in a 384-well format. The volumes and concentrations discussed are for this 384-well format but can be scaled up or down as necessary for other applications/techniques.
2. In this format, each well will be loaded with 20 μL of total volume. The volume will be composed of 10% Matrigel mixed with 90% human complete medium supplemented with Rho kinase inhibitor and containing a defined number of dissociated cells.
3. It is imperative for pharmacotyping that each well across the plate contains the same number of cells. We have tested cell numbers as high as 5000 cells/well and as low as 25 cells/well.

Table 1
Typical volumes required to generate a single-cell-Matrigel suspension for a 384-well plate

Single-cell concentration (cells/mL)	Volume of cell solution	Volume of human complete medium	Volume of 10.5 mM Rho kinase inhibitor stock	Volume of Matrigel	Total volume of mixture	Number of cells per well (i.e., cells per 20 μL)
1×10^6	0.2 mL	7 mL	8 μL	0.8 mL	8 mL	500

As a general guideline, we have found approximately 500 cells/well to reliably work, and the protocol herein describes plating this number.

4. If each well contains 20 μL of volume, an entire 384-well plate will require 8 mL single-cell-Matrigel suspension. Table 1 illustrates a typical calculation of the volumes required to generate a single-cell-Matrigel suspension for a 384-well plate in which each well will receive a total of 500 cells in 20 μL. In this example, the dissociated organoid cells were counted and determined to have a concentration of 1×10^6 cells/mL.

3.3 Plating and Pharmacotyping

1. Make your cell solution in a 15 mL conical tube on ice, in preparation for plating into multi-well culture dishes, by assembling the contents as follows: add ice-cold complete human medium, supplement with Rho kinase inhibitor, add Matrigel and mix thoroughly on ice using a pipette, add cell solution, mix thoroughly using a pipette, and maintain on ice.
2. Place sterile basin and white 384-well plate on ice.
3. Transfer the cell suspension to the sterile basin on ice and pipette up and down to mix well.
4. Use a 12-channel pipette to add 20 μL of the single-cell-Matrigel suspension to each well of the culture plate. Note that the outer rows/columns are susceptible to evaporation and should not be filled with cell suspension (*see* **Note 8**).
5. After filling the plate, pipette 20 μL of the remaining solution into the wells of a clear 384-well plate to allow for observation under the microscope during the experiment.
6. Centrifuge the plates briefly at 200 × *g*, 4 °C: allow the centrifuge to reach full speed and cancel the spin. The goal of this step is to push the cell-Matrigel suspension to the bottom of each well, not to pellet the cells down on the plastic.
7. In the wells along the outer edges of the plates (first and last row, first and last column) add 80 μL of sterile PBS to reduce evaporation from the center of the plate.

8. Transfer plates to a tissue culture incubator until ready for therapeutic testing. We suggest allowing 24–48 h for the cells to recover from harvesting/single-cell dissolution prior to adding therapeutics.
9. When preparing to add therapeutics, inspect the clear plate under the microscope to ensure that (1) the cells have recovered from single-cell dissociation and appear healthy and (2) plating appears even across all wells.
10. Prepare therapeutic stock(s) according to the manufacturer's recommendations. Commonly, suspension of a powdered drug in DMSO can be utilized to facilitate the use of automated dispensing instruments. We prefer the HP D300e digital dispenser (Tecan).
11. Add therapeutic stock(s) to the plate.
12. For each therapeutic, a rational, clinically relevant concentration range should be selected. Commonly, this range is derived from prior in vivo work and, for long-studied compounds, can be elucidated by a thorough review of the literature (*see* **Note 9**).
13. Often, if the range selected allows, log-fold increases in concentration can aid in data interpretation downstream.
14. The negative control, a series of triplicate wells in triplicate treated with vehicle only, should be included. A positive control, a treatment expected to potently cause cell death (e.g., a high concentration of gemcitabine for PDAC organoids), should be included as well.
15. If using an automated dispensing instrument with the capability to normalize vehicle concentrations across wells (i.e., ensure that each well receives the same total volume of DMSO), this option should be used.
16. Following addition of the therapeutic, the plate should be returned to the incubator for a period of 3–7 days. In our hands, 5 days has demonstrated to be ideal for organoids, allowing for adequate levels of cell death in the maximally treated wells without allowing overgrowth in those wells treated with vehicle.
17. At the conclusion of this period of growth, test for viability using your preferred assay and the manufacturer's recommendations. We prefer to use an ATP-based luminescence assay such as Cell-Titer Glo coupled with a plate reader capable of reading luminescence in 384-well format.

4 Notes

1. The ideal medium components can vary dramatically across tumor types. At the core, organoids rely on growth factor support and differentiation of the tumor's "stem cells." As such, the growth factors required for each tissue type vary. The medium presented here has been optimized for organoids derived from human pancreatic ductal adenocarcinoma. Medium formulations for organoids derived from other types of tissues can be found by reviewing the currently available literature.
2. There are a variety of methods that can be used to establish 3D cultures from primary human tumor tissue. For routine passaging of organoids, we prefer to culture organoids embedded in a dome of Matrigel centered in a well of a tissue culture-treated plastic culture dish, with medium overlaid. For high-throughput multi-well formats (96 wells per plate and upwards), it is impractical to culture in a dome and rather a cell-Matrigel suspension is layered along the bottom of each well.
3. Matrigel is a commercially available protein-rich substance that is secreted by EHS mouse sarcoma cells. Matrigel is frozen at −20 °C, liquid and workable at 4 °C, and begins to harden into gel as it warms from 4 °C. In practice, we have found that growth factor-reduced Matrigel with protein contents between 9.5 and 10 mg/mL is ideal for pancreatic organoid work.
4. As an alternative to single-cell preparation using dispase and TrypLE Express to harvest organoids, cell recovery solution (CRS) can be used instead. In collecting cells with CRS, aspirate the medium from each well leaving the Matrigel dome undisturbed and replace this volume with 0.5–1 mL of ice-cold CRS per well. Set the plate on ice for 5 min and collect solution and Matrigel with a P1000 pipetman into a 15 mL conical tube. Place tube on ice for 15–45 min to ensure complete Matrigel dissolution, freeing the organoids into solution. Centrifuge the slurry as above to pellet the cells, remove the CRS overlying the pellet, and then carry on with **step 16** of the single-cell protocol as described previously.
5. The purpose of the DNase I solution during single-cell preparation is to prevent DNA released from dying cells to cause other cells to clump together. If you find significant clumping of cells or difficulty with reaching a single-cell solution, check to make sure that the DNase I is not expired and is prepared at the correct concentration.
6. If moderate levels of cell clumping are preventing the generation of a single-cell solution, consider passing the cell solution through a 35 μM filter to eliminate larger clumps. Typically, we

use 5 mL polystyrene test tubes with cell strainer caps for this step.

7. If low cell number is an issue, low-attachment plasticware may be used to avoid losing precious cellular material during sample processing. Alternatively, coating the plasticware with a 0.1% BSA solution may help reduce loss of cells. Single-cell dissociation provides a reliable method for normalizing well-to-well variation. Therefore, it is the preferred cell preparation method for pancreatic organoids. However other types of organoid cultures, such as colorectal organoids, do not recover well from single cells. For cultures that cannot handle single-cell dissociation, organoids should be mechanically broken down to small, uniform clumps of cells. The concentration of clumps should be determined using a microscope and a hemocytometer and an equal number of clumps should be plated per well. It is imperative that treatments be performed in multiple replicates (at least triplicates) with these cultures as well-to-well variation will be more dramatic.
8. After plating for pharmacotyping, the white plate is used for testing when luminescence assays are used as the readout. The white plate enhances the luminescence signal and helps to eliminate signal noise from one well to the next when using a plate reader. The clear plate can be used as a visual control, ensuring that the cells recover from single-cell dissolution appropriately and are evenly plated from well to well when using a multichannel pipette.
9. When testing two or more drugs in combination, care should be taken not to overwhelm the cell with a lethal dose of a single agent as it will mask any synergistic effect of the combination. By first determining the lethal dose to kill 50% of the cells (LD_{50}) for each single compound we can ensure that the combinations are performed at the proper dose ratio.

References

1. Howlader N, Noone A, Krapcho M et al. (2016) SEER cancer statistics review, 1975–2013. National Cancer Institute. http://seer.cancer.gov/csr/1975_2013/. Accessed July 2016
2. DeMatteo RP, Ballman KV, Antonescu CR, Corless C, Kolesnikova V, von Mehren M et al (2013) Long-term results of adjuvant imatinib mesylate in localized, high-risk, primary gastrointestinal stromal tumor: ACOSOG Z9000 (alliance) intergroup phase 2 trial. Ann Surg 258:422–429
3. Alexandrov LB, Nik-Zainal S, Wedge DC, Aparicio SA, Behjati S, Biankin AV et al (2013) Signatures of mutational processes in human cancer. Nature 500:415–421
4. Lowery MA, Jordan EJ, Basturk O, Ptashkin RN, Zehir A, Berger MF et al (2017) Real time genomic profiling of pancreatic ductal adenocarcinoma: potential actionability and correlation with clinical phenotype. Clin Cancer Res 23:6094–6100. pii: clincanres.0899.2017
5. Witkiewicz AK, Balaji U, Eslinger C, McMillan E, Conway W, Posner B et al (2016) Integrated patient-derived models delineate individualized therapeutic vulnerabilities of pancreatic cancer. Cell Rep 16:2017–2031
6. Boj SF, Hwang CI, Baker LA, Chio II, Engle DD, Corbo V et al (2015) Organoid models of human and mouse ductal pancreatic cancer. Cell 160:324–338

Chapter 20

Techniques to Enable 1536-Well Phenotypic Screening

Sinéad Knight, Helen Plant, Lisa McWilliams, and Mark Wigglesworth

Abstract

Adaptation of phenotypic cell assays to 1536-well format brings major challenges in liquid handling for high-content assays requiring washing steps and coating of plates. In addition, problematic edge effects and reduced assay quality are frequently encountered. In this chapter, we describe the novel application of a centrifugal plate washer to facilitate miniaturization of 1536-well cell assays and a combination of techniques to reduce edge effects, all of which improved throughput and data quality. Cell assays currently limited in throughput because of cost and complex protocols may be enabled by the techniques presented in this chapter.

Key words Miniaturization, 1536-well, High-throughput, Cell assays, Phenotypic screening

1 Introduction

The identification of initial hits in a primary assay is of critical importance to the ultimate success of drug discovery programs and can significantly influence the molecules identified as hits, especially using phenotypic approaches. Recent studies suggest that the most predictive cell assays involve use of disease-relevant assay systems, including induced pluripotent stem cells (iPSCs) and donor-derived primary cells [1]. Major barriers to adopting these assays for high-throughput screening (HTS) are the complex techniques, along with the cost and availability of the relevant cells, with commercially sourced iPSC-based assays routinely costing more than $1000 per 384-well microtiter plate. Miniaturization of such assays would inevitably reduce costs, increase throughput, and ultimately improve the translation of hit identification screens.

Miniaturization of cell assays to 1536-well plates is estimated to reduce the cells required for a typical HTS by at least 50% and reduces the time taken to run a typical HTS campaign by up to fourfold. Despite the potential impact of this miniaturization, there are only limited examples of effective adaptation of complex cell-based screens to a 1536-well format. Common challenges

Bridget Wagner (ed.), *Phenotypic Screening: Methods and Protocols*, Methods in Molecular Biology, vol. 1787, https://doi.org/10.1007/978-1-4939-7847-2_20,

encountered include high variability, poor assay quality, and edge effects during longer incubations [2–4]. This chapter addresses some common issues observed for miniaturization to 1536-well format, such as liquid-handling challenges for immunohistochemistry assays requiring washing steps and coating of plates. Edge effects can also result in compromised assay quality and throughput. Our data demonstrate that novel applications of techniques such as centrifugal washing instead of tip-based aspirations can have significant impact on the quality and feasibility of miniaturization of cellular assays to 1536-well format [5].

2 Materials

2.1 Cell Reagents and Buffers

Unless otherwise stated all reagents were obtained from Sigma-Aldrich (St. Louis, MI) and cell culture medium was supplied by Gibco (ThermoFisher Scientific, Waltham, MA). Unless otherwise stated all test compounds used were solubilized in 100% v/v dimethyl sulfoxide (DMSO).

1. PathHunter eXpress GPR83 CHO-K1 β-Arrestin Orphan GPCR cells (DiscoverX, Fremont, CA) were screened in Ultra-CHO Serum Free CHO Medium and Supplement (Lonza, Basel, Switzerland) containing 100 U/mL penicillin and 100 μg/mL streptomycin. Fibronectin (Invitrogen, Carlsbad, CA) was solubilized at 50 μg/mL in sterile phosphate-buffered saline (PBS). EPIC® assay buffer contained HBSS with 2 mM HEPES pH 7.4.
2. HEK293 Low Density Lipoprotein Receptor-Green Fluorescence Protein (LDLR-GFP) reporter cell line was generated in-house and cultured in DMEM, containing 10% v/v fetal bovine serum (FBS) and selection antibiotics hygromycin and geneticin. Cells were screened in OptiMEM (Invitrogen) containing 1% v/v fetal calf serum (FCS), 100 U/mL penicillin, and 100 μg/mL streptomycin.
3. Pancreatic β cells were obtained from EndoCells and either cultured in low-glucose DMEM or screened in glucose-free DMEM supplemented with 2.8 mM glucose. Both culture and screening media contained 2% w/v albumin from bovine serum fraction V (Roche), 50 μM 2-mercaptoethanol, 10 mM nicotinamide, 5.5 μg/mL transferrin, 6.7 ng/mL sodium selenite, 100 U/mL penicillin, and 100 μg/mL streptomycin. Prior to culturing, microtitre plates and tissue culture flasks were coated with media (high glucose DMEM, 100 U/mL penicillin, 100 μg/mL streptomycin, 2 μg/mL fibronectin (Invitrogen), and 1% v/v Extracellular Matrix).

4. THP-1 cells were obtained from ATCC and cultured in RPMI containing 10% v/v FCS (PAA Labs, ThermoFisher Scientific, Waltham, MA) and 2 mM L-Glutamine (Invitrogen). THP-1 cells were screened in phenol-red free RPMI containing 10% v/v FCS, 2 mM L-Glutamine, 100 U/mL penicillin, and 100 μg/mL streptomycin. CellTiter-Blue Cell Viability Reagent was obtained from Promega (Madison, WI).

3 Methods

3.1 Centrifugal Plate Washing for Coating 1536-Well Plates

Many cells require fibronectin-, poly-D-lysine-, or collagen-coated plates for optimal adherence and growth. Although these plates are commercially available from many vendors, in-house coating protocols can significantly reduce costs for HTS.

1. Connect, initialize, and balance the BlueWasher™ centrifugal plate washer.
2. Check and update the BlueWasher™ configuration. Update dispenser manifold to 32-pin, and use offset to fine-tune dispense position to ensure that dispense jet hits well wall instead of bottom of the well.
3. Prime washer with H_2O followed by phosphate-buffered saline (PBS).
4. Ensure that the appropriate plate carrier for 1536-well plate is selected. Carriers come in three heights: "standard" for regular height (14 mm) 96- or 384-well plates, h10.4 for regular 1536-well plates, and h8 for thinner 8 mm plates; mismatched carrier may cause plates to dislodge and shatter during spinning, damaging the instrument. For Perkin Elmer (Waltham, MA) View Plate (6004460), h10.4 is recommended, for Greiner Bio-One (Kremsmünster, Austria) Screenstar (789866) and Perkin Elmer Cell Carrier (6004550) plates, h8 carrier is recommended, although other plate types may also be suitable if matched with the correct plate carrier.
5. Remove excess coating medium using predefined evacuation BlueWasher™ program "Gentle Spin: 1000 RPM, 5s Clockwise, 5s CounterClockwise." Removal of excess fibronectin post-coating was essential to maintain assay robustness and eliminated the requirement for a PBS wash step when using manual flicking of the plates (Fig. 1). This step improved dynamic mass redistribution (DMR), as measured by the Epic® reader responses and *Z'* values [6].
6. Immediately after removal of coating solution, dispense cells.

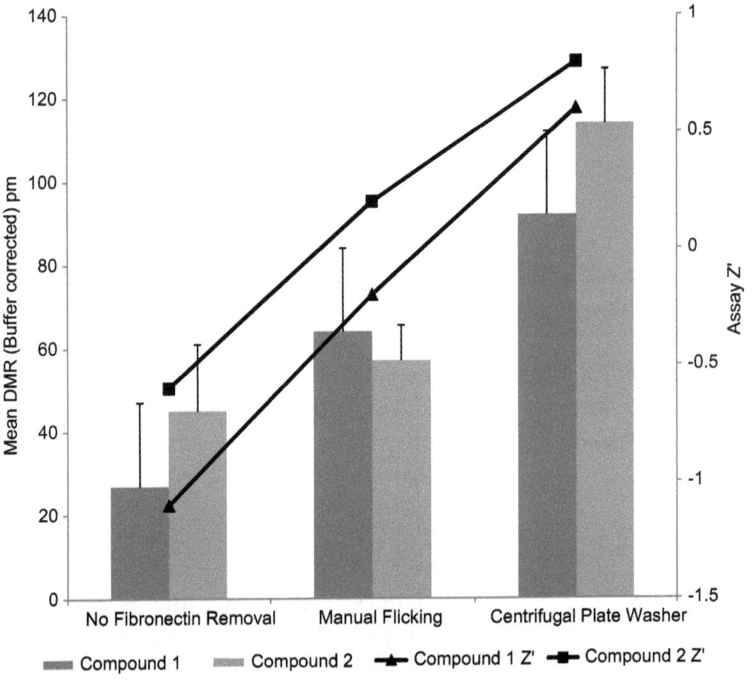

Fig. 1 Data from dynamic mass redistribution (DMR) assay to compare methods for removing excess fibronectin post-coating of microtiter plates. Epic® 1536-well plates were coated with fibronectin and excess solution removed by either (**a**) manually flicking out well contents, rinsing with PBS, and then flicking out again, or (**b**) by emptying centrifugally. CHO cells transfected with a GPCR were added to coated plates before being treated with tool compounds and DMR response measured 60 min post-compound addition. Assay *Z*′ and mean DMR responses are shown (corrected with buffer only controls)

3.2 Cell-Dispensing Techniques for 1536-Well Assays

3.2.1 Using the Multidrop Combi™

1. Prime a small tube plastic tip dispensing cassette sequentially with 10 mL of sterile H_2O, 10 mL of 70% (v/v) ethanol, 10 mL of sterile H_2O, and 10 mL of sterile PBS or cell culture medium.
2. If dispensing cells into large numbers of microtiter plates, it is recommended to use a sterilized magnetic stirrer bar at ~100 rpm to gently agitate the cell suspension stock and prevent cells from settling over lengthy periods of time.
3. Adjust dispense height to appropriate level. For Perkin Elmer View plate (6004460) a dispense height of 13.5 mm is required owing to the raised outer rim of the plate. A lower dispense height of 10 mm can be used for GreinerBio-One Screenstar and Perkin Elmer Cell Carrier.
4. Adjust dispense speed to the appropriate level; there are three options (high, medium, and slow). High speed can improve the accuracy of dispensing but must be optimized for each assay to ensure that the monolayer is not disrupted.

5. Seed cells at appropriate seeding density according to the optimized assay conditions (usually between 1000 and 4000 cells/well).

3.2.2 Using the Certus Flex™

1. The Certus Flex™ (Fritz Gyger AG) has the ability to use multiple microvalves for single-reagent dispensing, which is recommended for speed of dispensing cells to multiple plates. The 0.2/0.1 mm microvalve is recommended for cell dispensing with the Certus Flex™ into 1536-well plates.
2. As with the MD Combi, flush and prime the valve sequentially with ~5 mL of sterile H_2O, then ~5 mL of 70% (v/v) ethanol, and finally ~5 mL of cell culture medium.
3. Use of an angled Certus Flex™ head is recommended for cell dispensing. Liquid is dispensed down the side wall, rather than directly into the well, leading to a reduction in frothing with high-protein solutions, such as cell culture medium.
4. Use of an angled Certus Flex™ head is also recommended for adding additional reagents to plates containing cells, as the angled dispensing can reduce damage to the cell monolayer caused by direct dispensing, especially important for less adherent cells.

3.3 1536-Well Plate Washing of Live Cells and After Fixation, Permeabilization, and Immunofluoresence Staining

1. Using the BlueWasher™ the predefined wash program "Gentle Spin Wash: 50 × *g* RPM, 5s Clockwise, 5s CounterClockwise, 6 μL dispense volume, pressure level 5" is recommended for use with HEK293 cells in 1536-well microtiter plates coated with poly-D-lysine. This program leaves a residual volume of ~1 μL/well.
2. The number of wash cycles required should be optimized for each individual assay. The reduced residual volume with centrifugal plate washing can reduce liquid-handling steps in cell assays. We exemplify this with a reduction in washing cycles in a HEK293 reporter assay (2× centrifugal wash cycles) while maintaining assay quality ($Z' = 0.53$, %CV = 6.5%) and cell monolayer (Fig. 2).
3. A direct comparison of assay quality using either tip-based aspiration or centrifugation in a pancreatic β-cell proliferation assay after fixation, permeabilization, and fluorescent staining revealed that a reduction in the number of washing cycles (from three with tip-based aspiration to one with centrifugation) resulted in identification of similar cell numbers in channel 1 and EdU-positive cells in channel 2 (Fig. 3a).
4. Employing three washing cycles with tip-based aspiration yielded $Z' = 0.6$, whereas one wash cycle with centrifugation resulted in $Z' = 0.7$ (Fig. 3b).

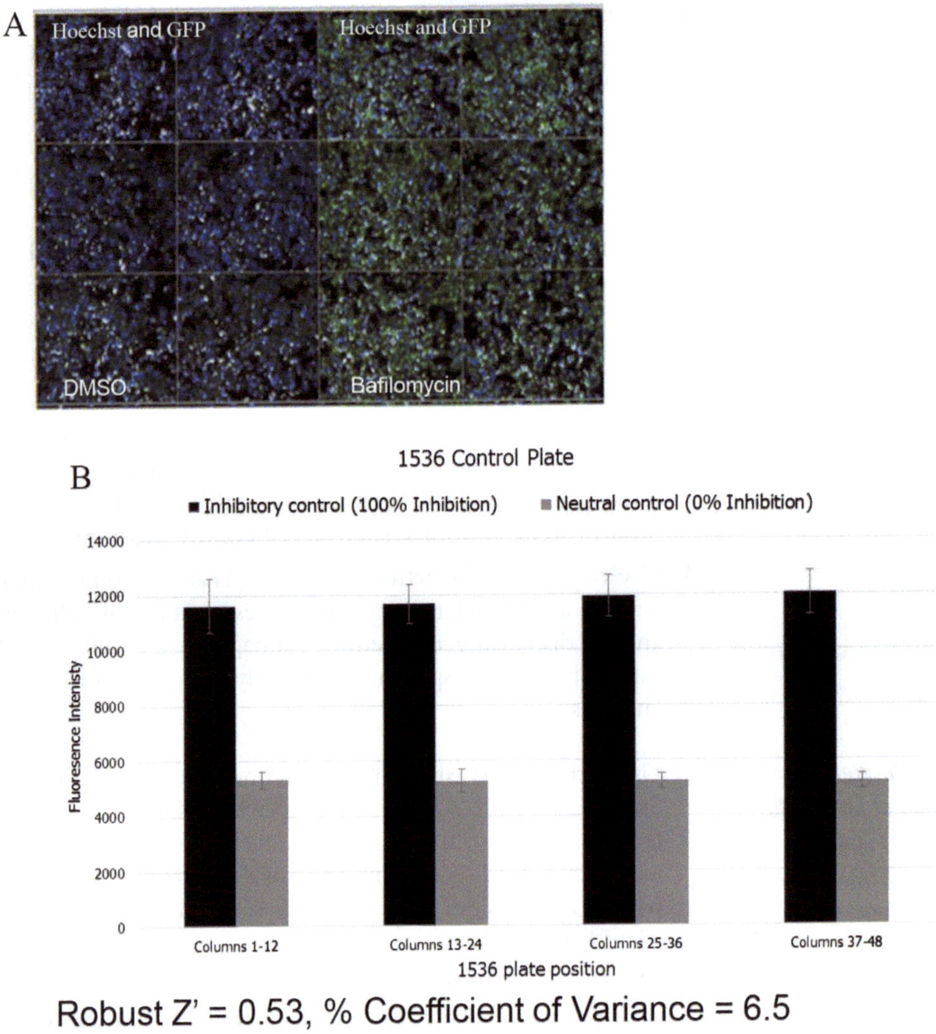

Fig. 2 1536-Well screening in HEK293 LDLR-GFP cells. (**a**) Cells were treated with neutral (DMSO) or inhibitor controls. Hoechst was used to identify cells in channel 1 and GFP expression was observed in channel 2. Images were captured at 10× using Operetta CLS™ High Content Analysis System. One field of view is shown. (**b**) HEK293 LDLR-GFP cells were treated with either neutral (DMSO) or inhibitor controls at different positions across a full 1536-well plate. Centrifugation was used for two wash cycles post-fixation. Fluorescence intensity was measured using EnVision® (Perkin Elmer) multilabel plate reader. Each histogram bar represents at least 128 data points. Assay *Z'* and % CV of inhibitor control is shown

5. Wash steps after medium change in live cells may also be eliminated but will need optimization for each individual assay.
6. The β-cell proliferation assay required a glucose starvation step, and when the centrifugal plate washer was used to remove medium, the low residual volume meant that no additional washing step was required prior to adding glucose starvation medium and showed improved results ($Z' = 0.6–0.7$) over tip-based aspiration ($Z' = 0.5$; Fig. 3b).

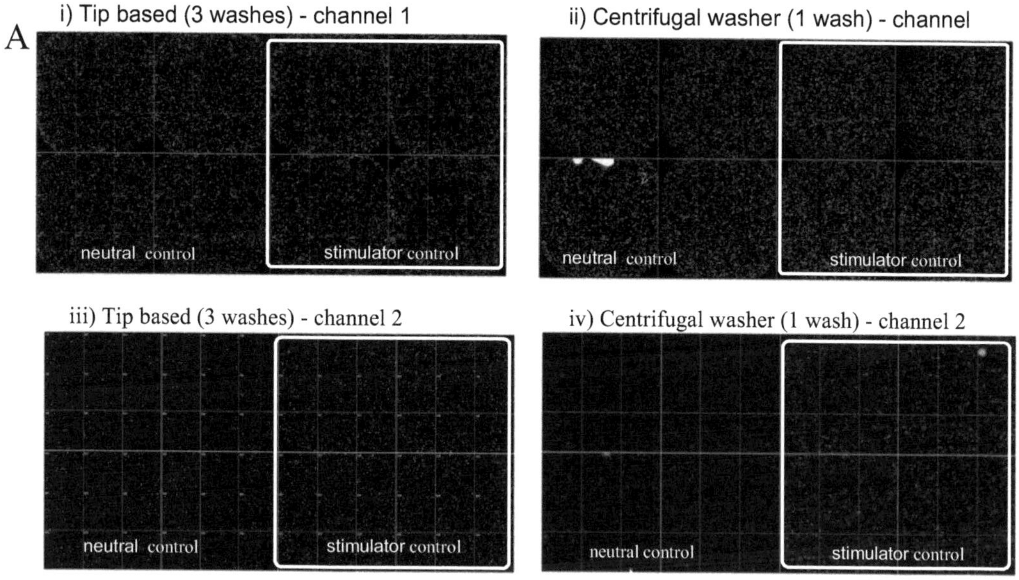

B

Wash Protocol	Media change	RZ'
Tip based (3 washes)	Centrifugation (no intermediate wash)	0.6
Centrifugation (1 wash)	Centrifugation (no intermediate wash)	0.7
Tip based (3 washes)	Tip based (1 wash)	0.5

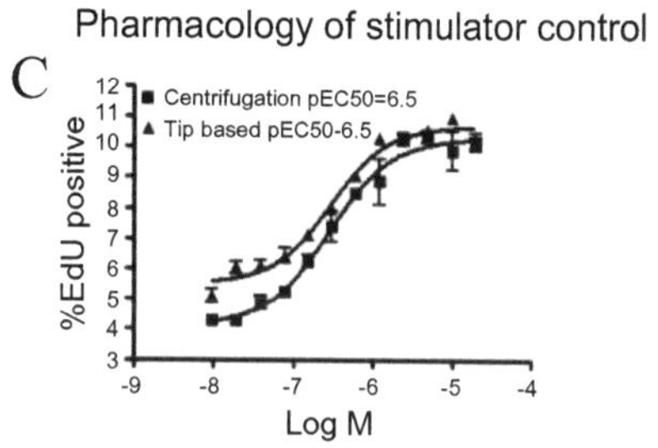

Fig. 3 Screening in pancreatic β cells. (**a**) Cells were treated with stimulator control to induce proliferation. Nuclear mask was used to identify cells in channel 1 (i and ii) and EdU-positive cells were identified in channel 2 (iii and iv). 3 wash cycles (iii and i) using tip-based method were used post-fixation, permeabilization, and fluorescence staining. 1 wash cycle (ii) and (iv) with centrifugation was used post-fixation; permeabilization and fluorescence staining images were captured at 10× magnification using CellInsight™. Whole-well images/9 fields of view are shown. Stimulator control wells are highlighted with a white outline. (**b**) Concentration response of stimulator control. Curves were generated using tip-based washing (1 wash cycle) or centrifugation (no intermediate wash cycle). Wells were treated with stimulator in 10-point concentration response curves ranging from 30 to 0.06 μM. Data was plotted and pEC_{50}s calculated using Prism (GraphPad, La Jolla CA). (**c**) Summary of wash protocols for pancreatic β-cell proliferation assay. Assay quality (Z') obtained when wash protocols post-fixation, permeabilization, and fluorescence staining were carried out using either tip-based method (3 wash cycles) or centrifugation (1 wash cycle). The Z' of the assay was also compared using centrifugation for medium change (no wash cycle) or tip-based method (1 wash cycle)

7. A pEC50 of 6.5 was measured in both protocols using a stimulator control (Fig. 3c), highlighting the suitability of deploying the centrifugal plate washer for the β-cell proliferation assay.

3.4 Addition of Compounds to 1536-Well Cell Assay Plates

Accurate addition of compounds or other reagents to 1536-well cell plates can be both challenging and time consuming (*see* **Note 1** for additional troubleshooting hints).

1. Automated liquid-handling platforms such as the Bravo™ or VPrep™ (both supplied by Agilent Technologies) are vertical tip-based pipetting stations suitable for addition of reagents to 1536-well plates.
2. 5, 10, or 15 μL disposable tips must be used, as standard 30 μL tips are too wide to enter the well of a 1536-well plate. For example, the 10 μL tips are suitable for accurate addition of between 0.5 and 10 μL volumes. Accurate teaching and quality control checking of tip-based pipetting devices are especially important for 1536-well format to avoid damaging tips through collisions with the microtiter plate.
3. All tip-based devices have the potential for compound carry-over, requiring a tip-wash step to be included between each addition step. Wash solutions such as 10% (v/v) ethanol or DMSO are generally suitable reagents for tip washing.
4. Wash solutions should be tested on an individual assay basis, and the number of wash cycles required also needs to be determined.
5. Generally we use up to three wash cycles of either 10% (v/v) ethanol or DMSO to eliminate compound carryover.
6. Most tip-based devices have a 384-well format head, requiring quadranting for 1536-well plate additions. This can result in the compound addition step being quite lengthy, and this needs to be accounted for when processing larger numbers of plates.
7. The CyBi®-well Vario (Analytik Jena AG) is currently the only commercially available device with a 1536-well format pipetting head, and is suitable for working volumes of between 100 nL and 10 μL.
8. It is important to routinely calibrate tip-based devices prior to their use for compound addition, as the pipetting performance of such dispensers is characterized by their accuracy. The optimization process has previously been described in detail (*see* technical notes for Agilent Bravo™ calibration [7], and A Quality Control Procedure for the Agilent Bravo™ Platform [8].
9. Tip-based devices allow the possibility of mix steps if these are required, ensuring thorough mixing of the well contents. Care must be taken to avoid introducing bubbles into the wells, especially when using larger volumes near the maximum tip capacity.

10. To mitigate the risk of the tip head being damaged by fluid being drawn into the pistons, aspirate air prior to taking up the required volume of liquid, and then ensure that the *total* aspirated volume is ejected from the tip. An example method for adding 2 μL compound to a cell plate is as follows:
    (a) *Step 1*: It is recommended to use V-bottomed polypropylene plates (e.g., Greiner Bio-One #781280) as the compound source plate, to minimize the amount of compound solution needed. Aspirate 4.5 μL air (at height above well contents), move tips down into the well, aspirate 3.5 μL compound solution 2 mm above the plate base, and then dispense 1 μL compound solution back into the source plate at 2 mm above the plate base. This step ensures that the pipetting head o-rings are in the correct direction for increased accuracy of pipetting, especially important when adding smaller volumes.
    (b) *Step 2*: Move the pipetting head to above the cell assay plate and dispense 2 μL compound solution into the well contents at a height just above the cell monolayer, e.g., 1.4 mm. It is important to ensure that the monolayer is not damaged by addition of the compound. At this stage a mix step can be included if required, by gently aspirating and dispensing liquid at a suitable height above the cell monolayer. Mixing is generally only required for shorter incubation steps (up to a few hours), whereas for longer incubation periods (24 h or longer), diffusion is sufficient.
    (c) *Step 3:* After this step move the tips to a wash station, and empty out the remaining contents of the tips to waste. Wash the tips up to 3 cycles with a suitable wash solution.
    (d) *Step 4:* After an addition or mix step it is useful to pulse centrifuge the cell assay plate for 1 s at 100 × *g* to ensure all contents are at the base of the well.
11. Acoustic reagent dispensers such as the Echo® 555 (Labcyte, Sunnyvale, CA) have the capacity to transfer compounds in aqueous solutions, thereby avoiding locally high concentrations of compound vehicle in the well, which could potentially damage the cells.
12. Acoustic dispensing of nL volumes reduces the amount of compound required, but a mix step is not an option using this method of non-contact addition. Again as with tip-based addition methods, the length of time to add compounds acoustically to a 1536-well plate can be a potential issue when performing an assay at scale.
13. The Echo® 525 liquid handler is able to acoustically dispense higher volumes than the Echo® 555 and may decrease the compound addition time for assays. However, we have not

found it to be suitable as a large-volume dispenser for addition of reagents to assay plates.

14. Equipment such as the VPrep™, Bravo™, and Echo® 555 can be used within a Biosafety Level II cabinet to maintain a sterile environment, essential for phenotypic assays with lengthy incubations at 37 °C. It is important to check when using a device in such a cabinet that airflow is not disrupted.

3.5 Reducing Edge-Effect Patterns in 1536-Well Phenotypic Assays

Edge effects occur when the conditions across the plate are different; small changes over time can have large and significant effects on the assay data. In our experience, the most frequent cause of edge effects is when liquid evaporates from the outer wells of a microtiter plate during extended incubations, and is particularly evident in 1536-well phenotypic assays where smaller assay volumes and longer incubation steps at 37 °C are required. Several methods can be employed to minimize these patterns. *See* **Note 2** for suggestions of quality-control experiments which help determine if edge effects or plate patterns are present.

1. After dispensing cells into microtiter plates, it is recommended to allow plates to equilibrate at ambient temperature for ~30 min prior to transferring plates to a 37 °C incubator to decrease the risk of microcirculation inside the wells, especially at the edge wells.
2. This circulation can lead to an uneven distribution of cells across the well bottom. Careful optimization is recommended for each assay to ensure that ambient temperature incubation has no effects on assay signal/background and pharmacology.
3. When incubating plates at 37 °C, use of a 95% humidified incubator with a rotating carousel, rather than fixed shelving, can reduce edge effects by maintaining a more uniform environment during extended incubation steps.
4. Increasing assay volumes can also help to reduce patterns. As an example, a 1536-well assay using HEK293 cells transfected with human LDLR-GFP showed an improved *Z*′ from <0 to 0.58 when assay volumes were increased from 6 to 8 μL while using an equivalent seeding density of 3000 cells/well (Fig. 4a).
5. Use of microtiter plate seals can improve edge effects, and a range of both impermeable and gas-permeable seals are commercially available. We have had success with both clear, gas-impermeable seals (TopSeal-A™ Microplate Press-On Adhesive Sealing Perkin Elmer) and gas-permeable seals (Gas Permeable Clear Seal, 4titude).
6. We have shown that variability in a 1536-well CellTiter-Blue™ viability assay using THP-1 cells was reduced from 14.1 to 7.5%

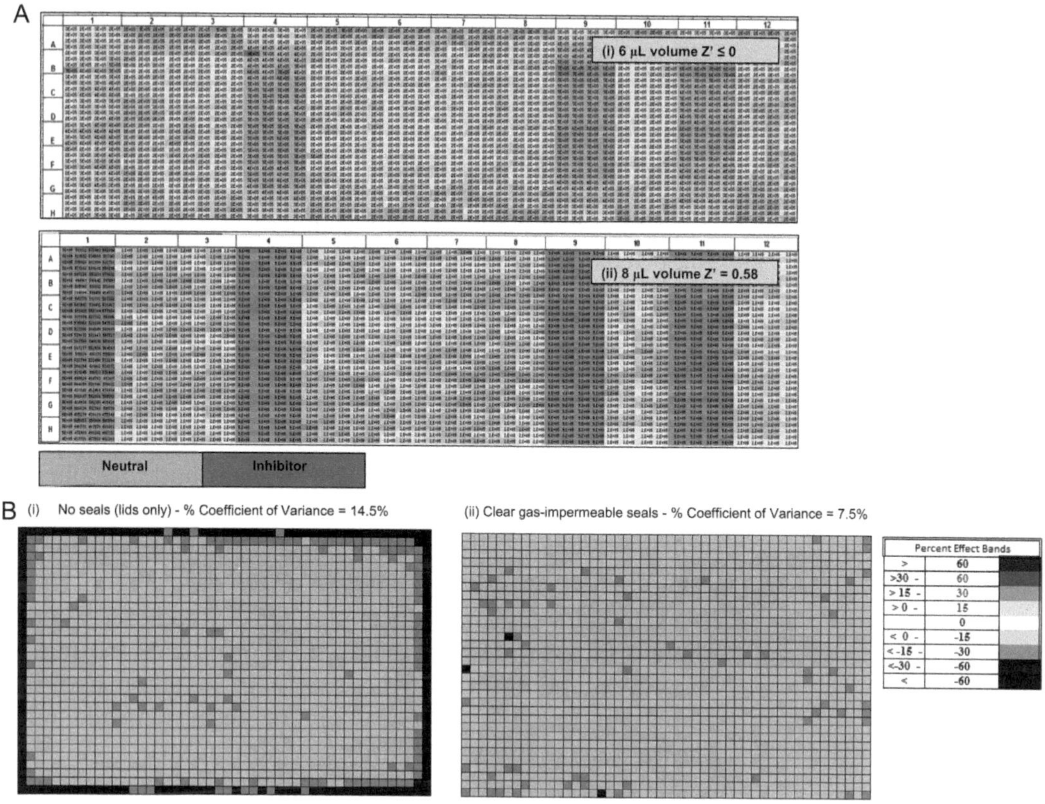

Fig. 4 Edge effects in 1536-well format. (**a**) Heat maps showing the effect of well volume on edge effects in a 1536-well format assay using HEK293 cells transfected with human a GFP reporter gene. Cells were treated with either neutral (DMSO) or inhibitor (Bafilomycin A1) controls. Fluorescence intensity measurements were made using a final volume of either 6 μL (i) or 8 μL (ii). An equivalent seeding density of 3000 cells/well was used for both assay volumes. (**b**) Heat maps illustrating the effect of plate seals on edge effect plate patterns and assay performance. THP-1 cells were seeded in 50 × 1536-well plate batches with either (i) lids only or (ii) clear gas-impermeable seals. All wells were treated with 0.5% (v/v) DMSO (neutral control). Fluorescence intensity measurements indicated level of cell viability. Each well was colored according to % difference from the plate median. % CV values are indicated

CV when using clear gas-impermeable seals during a 48-h incubation at 37 °C (Fig. 4b). Surprisingly, this result demonstrated that over fairly long assays the gas exchange rarely impacts cell viability and growth.

7. It is important to perform all assay development with the plate seals present, and to confirm that they have no impact on pharmacology.

3.6 Automating 1536-Well Phenotypic Assays

If there is a requirement for robotic handling of 1536-well plates, there are key points to consider in choice of plate type.

1. Ensure that the plate height is compatible with both the robot gripper, and that lidded plates are able to be picked successfully by the robot.
2. We recommend a gap of at least 2 mm between lid and plate to allow access by robot grippers; standard lids such as Greiner Bio-One (with a lid height of 6 mm) can be used successfully on automation platforms with standard SBS 1536-well imaging plates.
3. Use of non-SBS standard 1536-well imaging plates can cause problems, as the reduced plate height makes them difficult to handle in an automated environment, especially when a lid is present (*see* **Notes 3** and **4**).
4. We recommend using standard SBS 1536-well imaging plates such as Perkin Elmer View plate when robotic handling is required for handling 1536-well plates. However, as outlined in Subheading 3.7, the ability to image the whole plate may be compromised by the raised well elevation of 1.9 mm.
5. If using plate seals to reduce edge effects, consider whether there is a need to remove the seal to allow further steps in the assay to be performed. If the assay is to be automated, it is important that the plate seal can be easily removed using devices such as the XPeel® (Brooks Automation).
6. Some seals are difficult to remove and may leave a sticky residue, resulting in problems on automation platforms with lids adhering to plates.
7. Both the impermeable and gas-permeable seals recommended earlier do not leave this residue when they are removed from plates for downstream processing in assays.

3.7 High-Content Imaging for 1536-Well Phenotypic Assays

1536-Well high-content screening plates have to fulfill several (often conflicting) demands, and careful consideration of both the type of plate and image instrumentation is required for optimal performance of 1536-well phenotypic assays with imaging endpoints.

1. The 1536-well plate well elevation (distance from the bottom of the plate to the inner bottom surface of the well) of <0.5 mm is optimal and the objective lens must be considered in combination, since they are dependent on each other for optimal image acquisition.
2. In our evaluations, we have determined that the Perkin Elmer Cell Carrier plates (well elevation of 0.49 mm) provide optimal whole-plate imaging at 4×, 10×, 20×, and 60× on a Yokogawa CV7000 (Fig. 5, *see* **Notes 4** and **5**).

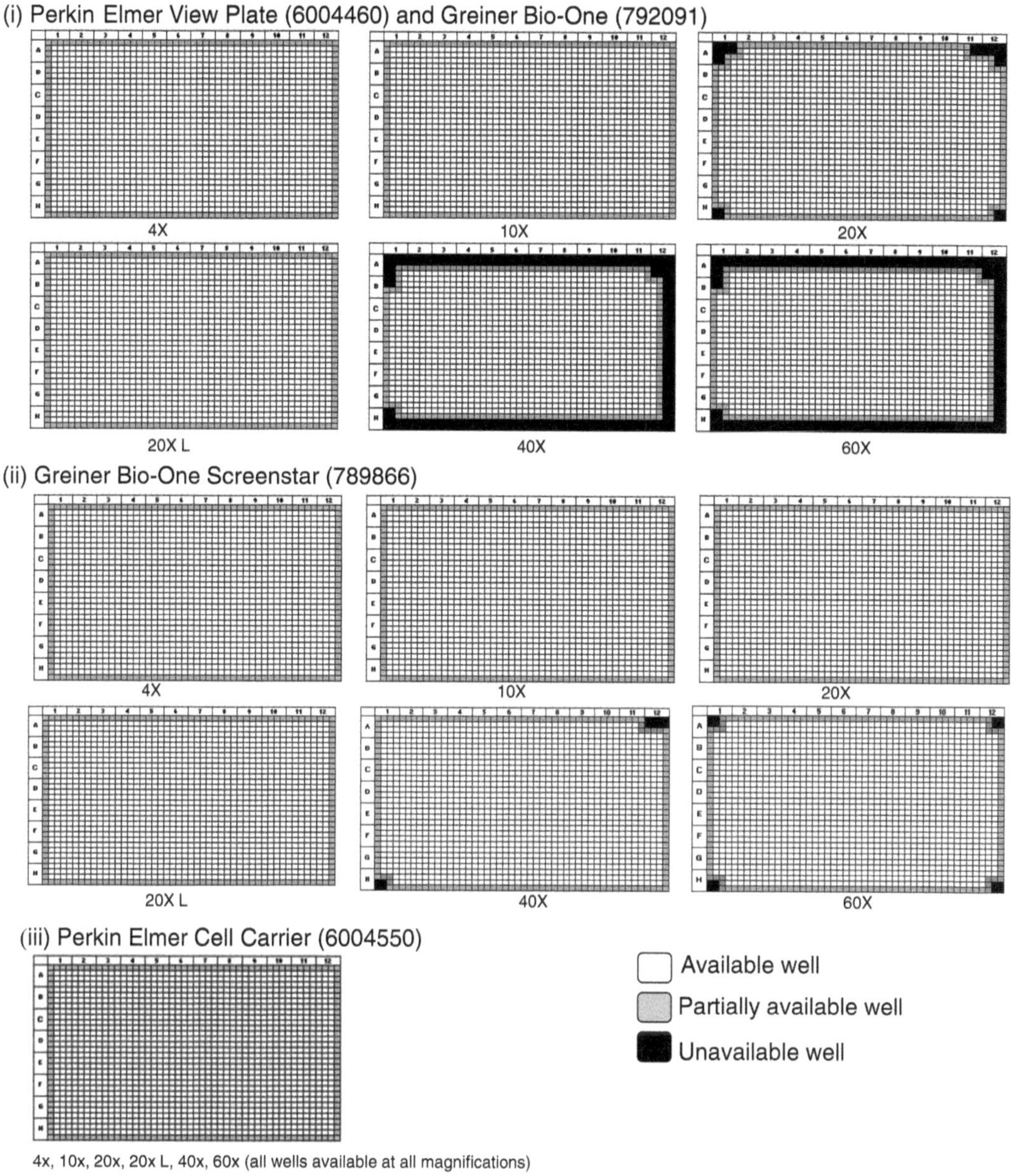

Fig. 5 Heat maps illustrating availability of wells of 1536-well plates to objectives of Yokogawa CV7000. Ability to image at 4×, 10×, 20×, and 60× was assessed for (i) Perkin-Elmer View plate, (ii) Greiner Bio-One Screenstar, and (iii) Perkin-Elmer Cell Carrier

3. The outer wells of the Greiner Bio-One Screenstar plate (well elevation of 0.7 mm) were partially available at all magnifications and some wells were unavailable at 40× and 60× (Fig. 5).
4. Perkin Elmer View Plate had the largest number of unavailable wells, most likely due to the raised well elevation of 1.9 mm, which partially prevents objective access to the well (Fig. 6).

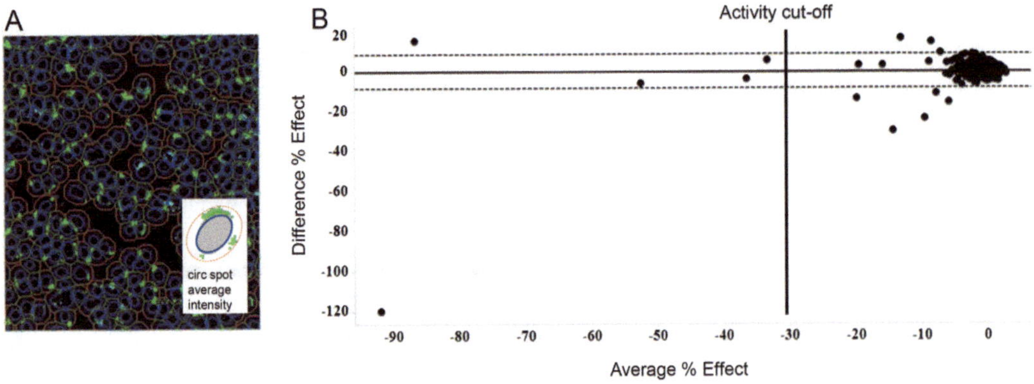

Fig. 6 Immunofluorescent HEK293 LDLR-GFP reporter assay in 1536-well format. (**a**) Compartmental Analysis BioApplication identified cells (primary objects) in channel 1 and quantified the level of GFP in channel 2 using circ spot intensity (average intensity of punctate objects) within the circ compartment (a cellular region derived from the area covered by the primary object). Images were acquired using CellInsight™ using 10× magnification. Hoechst was used as a nuclear marker. (**b**) Validation of HEK293 LDLR-GFP reporter assay in 1536-well format. A set of 1408 compounds were tested at 10 μM on two occasions and the level of GFP expression quantified using CellInsight™. Activity of test compounds was normalized to the median values of neutral and inhibitor controls. The Bland Altman plot shows the difference versus the average of each paired % effect data point. The solid line indicates the average value of the difference, dashed lines indicate 95% limits of agreement for each comparison (average difference ±2.0 standard deviation of the difference). One outlier is observed with difference >100% and represents a technical fault with dispensing for that well only

5. Specific experimental requirements will also dictate the choice of 1536-well plate. When we miniaturized a HEK293-GFP reporter assay to 1536-well format by acquiring images at 10× on CellInsight™ CX5 High Content Screening (HCS) Platform (ThermoFisher Scientific) and using Greiner Bio-One custom-made plates (well elevation 3 mm), sufficient image quality was obtained to achieve a robust HTS imaging assay using the whole plate.
6. To assess the compatibility of the 1536-well HEK293 reporter assay with the CellInsight™ imager, an algorithm was developed using the Compartmental Analysis BioApplication, which identified cells (primary objects) using Hoechst nuclear marker in channel 1 and quantified the level of GFP in channel 2 using the circ spot intensity (average intensity of punctate objects) within the circ compartment (a cellular region derived from the area covered by the primary object; Fig. 6a).
7. This assay was subsequently used to screen a set of 1408 compounds at 10 μM with compounds dispensed into two plates in different well positions on each plate.
8. Control plates containing neutral (DMSO) and inhibitor controls demonstrated the robustness of the assay ($Z' = 0.58$) (Fig. 4a).

9. Active compounds were reproducibly identified based on a cutoff of ≥30% inhibition (Fig. 6b).
10. Further recent technological advances which can facilitate complex 1536-well phenotypic screening are outlined in **Notes 6**, **7**, and **8**.

4 Notes

1. Centrifugation of 1536-well plates (100 × *g* for up to 30 s) after liquid dispensing can reduce variability, particularly if wicking is observed or bubbles prevent liquid from settling to the bottom of the well.
2. Uniform maximum and minimum signal control plates, where each signal is run uniformly across the entire plate, are strongly recommended to identify plate patterns generated by dispensing errors and/or edge effects. In addition, accessing the pharmacology of any known key positive control compounds is recommended to ensure assay consistency between 384-well and 1536-well formats.
3. The lower plate height of many imaging-compatible 1536-well plates, such as Greiner Bio-One Screenstar (8 mm) and Perkin Elmer Cell Carrier (7 mm), makes the choice of lid challenging. There is a requirement to prevent evaporation when working with smaller well volumes; hence standard lids are preferable but must not compromise robotic handling (see Subheading 3.6).
4. To further facilitate 1536-well image-based assays at HTS scale, it is necessary for manufacturers to consider generating plates which combine the optimal plate height of 10.4 mm with a well elevation of <0.5 mm. This will allow whole-plate imaging without compromising the ability to robotically handle the plates.
5. The Yokogawa CV7000™ (Wako Automation) offers the opportunity to run complex confocal imaging assays in 1536-well format. With a typical read time of 18 min at 10×, 1FOV, three channels (50 ms exposure), it offers the ability to screen up to 70 × 1536-well plates per day in an automated HTS setting.
6. The use of 3D assays such as tumor spheroids is increasingly required to obtain a desired phenotype in cellular assays. Recent advances such as the development of a prototype Ultra Low Attachment 1536-well plate (Corning) could prove beneficial for miniaturization of 3D assays. Use of magnetic 3D bioprinting technology such as the NanoShuttle™ from n3B Bioscience, Inc., will also facilitate the

miniaturization of spheroid assays to 1536-well plates by simplifying liquid handling and encouraging more uniform spheroid formation.

7. Advances in flow-cytometry such as the iQue Screener HD™ (Intellicyt), which has a 1536-well capability, facilitate miniaturization of suspension cell assays while obtaining rich multiplexed data including secreted protein information.
8. Application of emerging technologies such as bio-printing and acoustic dispensing offers improved control of cell density and precision on spatial placement of cells and may help to further deploy miniaturization of phenotypic assays for drug discovery to 1536-well format.

References

1. Vincent F, Loria P, Pregel M, Stanton R, Kitching L, Nocka K et al (2015) Developing predictive assays: the phenotypic screening "rule of 3". Sci Transl Med 7:293ps15
2. Fuller JA, Shaw GC, Bonnet-Wersinger D, Hansen BS, Berlincke CA, Inglese J et al (2014) A high content screening approach to identify molecules neuroprotective for photoreceptor cells. Adv Exp Med Biol 801:773–781
3. Li J, Crowley ST, Duskey J, Khargharia S, Wu M, Rice KG (2015) Miniaturization of gene transfection assays in 384- and 1536-well microplates. Anal Biochem 470:14–21
4. Luense S, Denner P, Fernández-Montalván A, Hartung I, Husemann M, Stresemann C et al (2015) Quantification of histone H3 Lys27 trimethylation (H3K27me3) by high-throughput microscopy enables cellular large-scale screening for small-molecule EZH2 inhibitors. J Biomol Screen 20:190–201
5. Knight S, Plant H, McWilliams L, Murray D, Dixon-Steele R, Varghese A et al (2016) Enabling 1536-well high-throughput cell-based screening through the application of novel centrifugal plate washing. SLAS Discov 22:732–742
6. Zhang JH, Chung TD, Oldenburg KR (1999) A simple statistical parameter for use in evaluation and validation of high throughput screening assays. J Biomol Screen 4:67–73
7. http://www.artel-usa.com/wp-content/uploads/2013/09/12A6480_Bravo-Optimization-Application-Note.pdf. Technical note for Agilent Bravo™ calibration
8. http://cn.agilent.com/cs/library/technicaloverviews/public/5991-8020EN.pdf. A quality control procedure for the Agilent Bravo™ platform

Index

A

Adipogenesis 118, 119, 126
Adipose tissue 115
Aging 78, 79, 85, 149, 156, 158
Antibacterials 1, 2, 19–21, 36
Arabidopsis thaliana 162–164
Asexual blood-stage 42, 44–47
ATP 24, 54, 203, 250, 259
Automated microscopy 97, 139

B

Bioactivity data integration 196, 198, 200

C

Caenorhabditis elegans .... v, 129–131, 141
Cell-based assays .... v, 67–74, 107, 196, 207
Cell-based profiling 171
Cell sorting 79, 81, 83, 84
Cell viability 55–58, 60, 62, 63, 65, 84, 105, 112, 126, 212, 220, 239, 254, 256, 273
Central nervous system (CNS) 101, 110, 147
Chemical probe 195, 196, 223
Chemical proteomics 117, 196
Chemogenomics 209, 210

D

3D culture 260
Diabetes therapeutics 88, 116
Differentiation 85, 90, 101, 102, 107, 110, 112, 117, 119, 120, 125, 126, 260
3D imaging 162, 167
Dimethylenastron (DME) 68–71, 73, 74
Drug development 49, 208, 223
Drug testing 261

E

Electrophysiology 235–251

F

Fat screening 129
Fluorescent bioprobe .... vi, 223–232
Freezing behavior 148, 155

G

Gametocyte-stage parasite 42
Genetics of obesity 129–145
Glucose bioprobe (GB2-Cy3) 225, 227–232
Growth inhibition 2, 3, 8, 37, 42, 49, 56, 60, 61

H

Heterogeneity 171, 197, 208, 213, 214, 216
High-content analysis 268
High-throughput drug discovery 148
Human islets 88–91, 93–95, 97–99
Human pancreatic beta cell 87–99
Huntington's disease (HD) .... v, 22, 28, 35, 209, 211–213, 220–222, 278

I

Image-based screening 131, 143, 171
Immunostaining 67, 70–72, 74, 88, 92, 94–97, 111
In planta 161
In situ 43, 161
Intrinsic root coordinate system (iRoCs) 161, 165
Ion channels 110, 235, 236, 239–243, 245, 249

K

KNIME 183–186, 189, 191, 192

L

Live-cell screening 223
Liver-stage parasite 42, 43

M

Malaria .... v, 41–46, 49, 50
Miniaturization 43, 49, 263, 264, 275, 276, 278

Bridget Wagner (ed.), *Phenotypic Screening: Methods and Protocols*, Methods in Molecular Biology, vol. 1787,
https://doi.org/10.1007/978-1-4939-7847-2, © Springer Science+Business Media, LLC, part of Springer Nature 2018

Mitosis 68, 74, 88
Monopolar spindle structure 68
Myeloproliferative neoplasms (MPNs) 53, 54

N

Neural progenitor cells (NPCs) 101
Neurogenesis .... v, 101–113
Neuronal differentiation 101, 102
Nutrient stress .... v, 1–17

O

Open PHACTS .... v, 183–192
Organoids 253–261

P

Patch clamp .... vi, 236, 239, 246, 248, 249
Pharmacotyping 253
Photoreactive small molecules 116
Pipeline Pilot 183–191
Pittsburgh heterogeneity index (PHI) 213, 214, 216, 221
Planar electrophysiology 236
Plasmodium falciparum 41–44, 46–50
Precision medicine 253, 254
Proliferation .... v, 43, 53, 77, 110, 112, 254, 267–269
Protein complex 161, 162, 189
Protein-protein interactions .... v, 161
Proximity ligation assay (PLA) 161–170

Q

Quantitative systems pharmacology (QSP) .... v, 207–221

R

Ribocil 20–23, 25, 26, 29, 31, 35, 37
Riboflavin 19–39
Riboswitch reporter assay 21
RNA aptamer-binding 21, 23–24
Root apical meristem 162, 165

S

Satellite cells .... v, 77–79, 84
Selectivity 62, 64, 126, 196–198, 200–202, 204
Seoul-fluor 44 (SF44) 225, 229
Skeletal muscle 77–86
Small-molecule chemical library 103
Spindle orientation .... v, 67, 69
Stem cells 53, 77, 260, 263
Strobe light response 148, 150, 155, 158

T

Target identification 109, 116, 117, 120, 121, 195, 223, 228
Thrombopoietin receptor (TpoR) .... v, 53–65
Tool compound 106, 107, 195–205, 266
Tumor models 254

V

Voltage gated calcium channels 236, 239, 240

W

1536-Well assays 266–267, 270–276
Whole-organism screening .... v, 147, 148
Workflow tool 183, 195–197

MIX
Papier aus verantwortungsvollen Quellen
Paper from responsible sources
FSC® C105338
FSC
www.fsc.org

If you have any concerns about our products,
you can contact us on
ProductSafety@springernature.com

In case Publisher is established outside the EU,
the EU authorized representative is:
Springer Nature Customer Service Center GmbH
Europaplatz 3, 69115 Heidelberg, Germany

Printed by Libri Plureos GmbH
in Hamburg, Germany